PETROLEUM REFINING
Technology and Economics
Second Edition, Revised & Expanded

James H. Gary

Professor of Chemical and
Petroleum Refining Engineering
Colorado School of Mines
Golden, Colorado

Glenn E. Handwerk

Consulting Chemical Engineer
Petroleum Refining and Gas Processing
Golden, Colorado

MARCEL DEKKER, INC.   New York and Basel

Library of Congress Cataloging in Publication Data

Gary, James H., [date]
   Petroleum Refining

   Includes bibliographical references and index.
   1. Petroleum—Refining. I. Handwerk, Glenn E.
II. Title.
TP690.G33 1984   665.5'3   84-1722
ISBN 0-8247-7150-8

910505

Marcel Dekker, Inc.
270 Madison Avenue, New York, New York 10016

Current printing (last digit):
10 9 8 7 6

Printed in the United States of America

PREFACE

The petroleum refining industry has undergone tremendous expansion and change since 1950. Enormous increases in the size of process units, new catalytic processes, shifting product demands, and new sources of petroleum from tar sands and oil shales have made present-day technology and economics of petroleum refining a very complex and sophisticated science.

American journal literature has recorded these changes in substantial detail but not, however, in a systematic or convenient manner for reference.

This book presents the basic aspects of current petroleum refining technology and economics in a systematic manner suitable for ready reference by technical managers, practicing engineers, university faculty members, and graduate or senior students in chemical engineering.

The physical and chemical properties of petroleum and petroleum products are described, along with major modern refining processes. Data for determination of product yields, investment, and operating costs are presented for all major refining processes. Similar data also are given for supporting processes, such as hydrogen generation and elemental sulfur recovery. Ecological problems of petroleum refining operations are described in general terms. Reaction chemistry is described in basic terms with reference to desirable thermodynamic conditions.

The investment, operating cost and utility data given herein are typical average 1982 data. As such, this information is suitable for approximating the economics of various refining configurations. The information is not sufficiently accurate for definitive comparisons of competing processes.

The yield data for reaction processes have been extended to allow complete material balances to be made from physical properties. Insofar as possible, data for catalytic reactions represent average yields from competing proprietary catalysts and processes.

The yield data combined with the cost data will serve the practicing engineer and refinery management as a convenient tool for developing prelimin-

iii

ary economic feasibility studies. Examples of such calculations are given as an aid to students.

The subject material is organized in such a way that the course can be taught by the case-study method. By furnishing crude analyses to the students, each student can proceed through the refinery operations calculating yields, utility requirements, operating costs, and product specifications. An example case-study problem begins in Chapter 4 (Crude Distillation) and concludes in Chapter 16 (Economic Evaluation).

Special processes such as lubrication oil and petrochemical production, and relatively obsolete processes such as thermal cracking, have been omitted.

The appendixes contain basic engineering data and a glossary of refining terms. Valuable literature references are noted throughout the book.

The authors have held responsible positions in refinery operation, design, and evaluation, as well as in practical approaches to teaching of many refining problems. The current work relies heavily on their direct knowledge of refining in addition to the direct knowledge of the authors' many associates. The authors express their appreciation to the many people who contributed data and suggestions incorporated into this book. Permission to use information from works of other authors also is gratefully acknowledged. Especially helpful were the efforts of D. R. Lohr, V. D. Kliewer, M. A. Prosche, J. R. McConaghy, P. M. Geren, and D. F. Tolen, who assisted in the petroleum refining courses at the Colorado School of Mines, and J.Z. Gary, who helped greatly in improving the clarity of presentation.

James H. Gary
Glenn E. Handwerk

CONTENTS

Contents

Contents

Chapter 1
INTRODUCTION

Modern refinery operations are very complex and, to a person unfamiliar with the industry, it seems to be an impossible task to reduce the complexity to a coordinated group of understandable processes. It is the purpose of this book to present the refinery processes, as far as possible, in the same order in which the crude flows through the refinery, to show the purposes and interrelationships of the units. The case-study method of presentation is best for quick understanding, and we recommend that a crude oil be selected and yield and cost calculations be made as you study the refining processes in order. An example problem is given in Chapter 15 for a refinery of low complexity.

The process flow and products for a complete refinery of high complixity are shown in Figure 1.1. (See also Photo 1, Appendix F.) The processing equipment indicated is for processing crude oils of average gravities and sulfur contents. Crude oils with low API gravities and high sulfur contents require additional hydrotreating equipment.

The quality of crude oils processed by U.S. refineries is expected to worsen slowly in the future while the demand for heavy fuel oil is expected to decrease. This will require the refineries to process the entire barrel of crude rather than just the material boiling below 1050°F (566°C). Sulfur restrictions on fuels (coke and heavy fuel oils) will affect bottom-of-the-barrel processing as well. These factors will require extensive refinery additions and modernization and the shift in market demand between gasoline and distillates for transportation fuels will challenge catalyst suppliers and refinery engineers to develop innovative solutions to these problems.

The environmental impacts of fuel preparation and consumption will probably require that a significant shift take place in product distribution (i.e., less gasoline and more distillate fuel in terms of percent on crude). This will have a major effect on refinery processing operations and will

1

place a burden on refinery construction along with producing the need to provide increased capacity for high sulfur and heavier crude oils.

The language of the refining industry is unfamiliar to those not in it and to ease the entry into an unfamiliar world, feedstock and product specifications are discussed before the refinery processing units.

Appendix A contains a glossary of refining terms and will assist in understanding the descriptions. In many cases, however, there is no standard definition, and a term will have different meanings in different companies, and even in different refineries of the same company. It is always important, therefore, to define terms with respect to the individual writing or talking.

1.1 OVERALL REFINERY FLOW

Figure 1.1 shows the processing sequence in a modern refinery, indicating major process flows between operations.

The crude oil is heated in a furnace and charged to an atmospheric distillation tower, where it is separated into butanes and lighter wet gas, unstabilized full-range gasoline, heavy naphtha, kerosine, heavy gas oil, and topped crude. The topped crude is sent to the vacuum tower and separated into a vacuum gas oil overhead stream and reduced crude bottoms.

The reduced crude bottoms from the vacuum tower is thermally cracked in a delayed coker to produce wet gas, coker gasoline, gas oil, and coke.

The atmospheric and vacuum crude unit gas oils and coker gas oil are used as feedstocks for the catalytic cracking or hydrocracking units. These units crack the heavy molecules into compounds boiling in the gasoline and distillate fuel ranges. The products from the hydrocracker are saturated. The unsaturated catalytic cracker products are saturated and improved in quality by catalytic reforming or hydrotreating.

The gasoline streams from the crude tower, coker, and cracking units are fed to the catalytic reformer to improve their octane numbers. The products from the catalytic reformer are blended into regular and premium gasolines for sale.

The wet gas streams from the crude unit, coker, and cracking units are fractionated in the vapor recovery section into fuel gas, liquefied petroleum gas (LPG), unsaturated hydrocarbons (propylene, butylenes, and pentenes), normal butane, and isobutane. The fuel gas is burned in refinery furnaces and the normal butane is blended into gasoline or LPG. The unsaturated hydrocarbons and isobutane are sent to the alkylation unit for processing.

The alkylation unit uses either sulfuric or hydrofluoric acid as catalyst to react olefins with isobutane to form isoparaffins boiling in the gasoline range. The product is called alkylate, and is a high-octane product blended into premium motor gasoline and aviation gasoline.

The middle distillates from the crude unit, coker, and cracking units are blended into diesel and jet fuels and furnace oils.

In some refineries, the heavy vacuum gas oil and reduced crude from paraffinic or naphthenic base crude oils are processed into lubricating oils. After removing the asphaltenes in a propane deasphalting unit, the reduced crude bottoms are processed in a blocked operation with the heavy vacuum gas oils to produce lube-oil base stocks.

The heavy vacuum gas oils and deasphalted stocks are first solvent-extracted to remove the heavy aromatic compounds and then dewaxed to im-

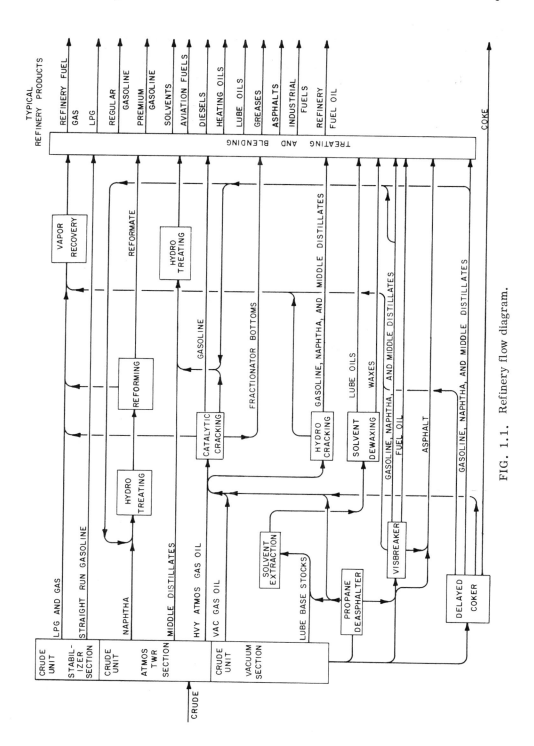

FIG. 1.1. Refinery flow diagram.

prove the pour point. They are then treated with special clays to improve their color and stability before being blended into lubricating oils.

Each refinery has its own unique processing scheme which is determined by the equipment available, operating costs, and product demand. The optimum flow pattern for any refinery is dictated by economic considerations and no two refineries are identical in their operations.

Chapter 2
REFINERY PRODUCTS

While the average consumer tends to think of petroleum products as consisting of a few items such as motor gasoline, jet fuel, kerosine, etc., a survey conducted by the American Petroleum Institute (API) of the petroleum refineries and petrochemical plants revealed over 2,000 products made to individual specifications [1]. Table 2.1 shows the number of individual products in 17 classes [2].

In general, the products which dictate refinery design are relatively few in number, and the basic refinery processes are based on the large-quantity products such as gasoline, jet fuel, and diesel fuel. Storage and waste disposal are expensive, and it is necessary to sell or use all of the items produced from crude oil even if some of the materials, such as heavy fuel oil, must be sold at prices less than the cost of crude oil. Economic balances are required to determine whether certain crude fractions should be sold as is (i.e., straight-run) or further processed to produce products having greater value. In general, the lowest value of a hydrocarbon product is its heating value or fuel oil equivalent (FOE). This value is always established by location, demand, availability, combustion characteristics, sulfur content, and prices of competing fuels.

Knowledge of the physical and chemical properties of the petroleum products is necessary for an understanding of the need for the various refinery processes. To provide an orderly portrayal of refinery products, they are described in the following paragraphs in order of increasing specific gravity and decreasing volatility.

The petroleum industry uses a shorthand method of listing hydrocarbon compounds which characterized the materials by number of carbon atoms and unsaturated bonds in the molecule. For example, propane is shown as C_3 and propylene as $C_3^=$. The corresponding hydrogen atoms are assumed to be present unless otherwise indicated. This notation will be used throughout this book.

5

TABLE 2.1

Products Made by the U.S. Petroleum Industry

Class	Number
Fuel gas	1
Liquefied gases	13
Gasolines	40
Motor	19
Aviation	9
Other (tractor, marine, etc.)	12
Gas turbine (jet) fuels	5
Kerosines	10
Distillates (diesel fuels and light fuel oils)	27
Residual fuel oils	16
Lubricating oils	1,156
White oils	100
Rust preventives	65
Transformer and cable oils	12
Greases	271
Waxes	113
Asphalts	209
Cokes	4
Carbon blacks	5
Chemicals, solvents, misc.	300
	2,347

2.1 LOW-BOILING PRODUCTS

The classification low-boiling products encompasses the compounds which are in the gas phase at ambient temperatures and pressures: methane, ethane, propane, butane, and the corresponding olefins.

Methane (C_1) is usually used as a refinery fuel, but can be used as a feedstock for hydrogen production by pyrolytic cracking. Its quantity is generally expressed in terms of pounds, standard cubic feet (scf) at 60°F and 14.7 psia, or in barrels FOE (equivalent barrels of fuel oil having the same heating value). The physical properties of methane are given in Table 2.2.

Ethane (C_2) can be used as refinery fuel or as a feedstock to produce hydrogen or ethylene, which are used in petrochemical processes. Ethylene and hydrogen are sometimes recovered in the refinery and sold to petrochemical plants.

Propane (C_3) is frequently used as a refinery fuel but is also sold as a liquefied petroleum gas (LPG), whose properties are specified by the Natural Gas Processors Association (NGPA) [3]. Typical specifications include a maximum vapor pressure of 210 psig at 100°F and 95% boiling point of -37°F or lower at 760 mmHg atmospheric pressure. In some locations, propylene is separated for sale to polypropylene manufacturers.

The butanes present in crude oils and produced by refinery processes are used as components of gasoline and in refinery processing as well as in LPG. Because it has a lower vapor pressure than isobutane (i-C_4), normal butane (n-C_4) is usually preferred for blending into gasoline to regulate its vapor pressure and promote better starting in cold weather. Normal butane

TABLE 2.2

Physical Properties of Paraffins

	C_n	Boiling point (°F)	Melting point (°F)	Gravity (60/60°F)	Gravity °API
Methane	1	-258.7	-296.5	0.30	340
Ethane	2	-128.5	-297.9	0.356	265.5
Propane	3	-43.7	-305.8	0.508	147.2
Butane					
normal	4	31.1	-217.1	0.584	110.6
iso	4	10.9	-225.3	0.563	119.8
Octane					
normal	8	258.2	-70.2	0.707	68.7
2,2,4	8	210.6	-161.3	0.696	71.8
2,2,3,3	8	223.7	219.0	0.720	65.0
Decane, normal	10	345.5	-21.4	0.734	61.2
Cetane, normal	16	555.0	64.0	0.775	51.0
Eicosane, normal	20	650.0	98.0	0.782	49.4
Triacontane					
normal	30	850.0	147.0	0.783	49.2
2,6,10,14,18,22	30	815.0	-31.0	0.823	40.4

Generalizations:
1. Boiling point rises with increase in molecular weight.
2. Boiling point of a branched chain is lower than for a straight chain hydrocarbon of the same molecular weight.
3. Melting point increases with molecular weight.
4. Melting point of a branched chain is lower than for a straight chain hydrocarbon of the same molecular weight unless branching leads to symmetry.
5. Gravity increases with increase of molecular weight.
6. For more complete properties of paraffins, see Table B.2.

has a Reid vapor pressure (RVP) of 52 psi as compared with the 71 psi RVP of isobutane, and more n-C_4 can be added to the gasoline without exceeding the RVP specification of the gasoline product. On a volume basis, gasoline has a sales value higher than that of LPG, thus, it is desirable to blend as much normal butane as possible into the gasoline. Normal butane is also used as a feedstock to isomerization units to form isobutane.

Isobutane has its greatest value when used as a feedstock to alkylation units, where it is reacted with unsaturated materials (propenes, butenes, and pentenes) to form high-octane compounds in the gasoline boiling range. Although isobutane is present in crude oils, its principal sources of supply are from hydrocracking units and natural gas processing plants. Isobutane not used for alkylation unit feed can be sold as LPG or used as a feedstock for propylene manufacture. When butanes are sold as LPG, they conform to the NGPA specifications for commercial butane. These include a vapor pressure of 70 psig or less at 100°F and a 95% boiling point of 36°F or lower at 760 mmHg atmospheric pressure. Butane as LPG has the disadvantage of a fairly high boiling point (32°F at 760 mmHg) and during the winter is not

TABLE 2.3

Properties of Commercial Propane and Butane

Property	Coml Propane	Coml Butane
Vapor pressure, psig:		
70° F	124	31
100° F	192	59
130° F	286	97
Specific gravity of liquid, 60/60° F	0.509	0.582
Initial boiling point at 14.7 psia, ° F	-51	15
Dew point at 14.7 psia, ° F	-46	24
Specific heat of liquid at 60° F, Btu/(lb)(° F)	0.588	0.549
Specific heat of gas at 60° F, Btu/(lb)(° F)	0.404	0.382
Limits of flammability, % gas in air		
Lower limit	2.4	1.9
Upper limit	9.6	8.6
Latent heat of vaporization at boiling pt, Btu/lb	185	165
Gross heating values:		
Btu/lb of liquid	21,550	21,170
Btu/ft^3 of gas (60° F, 14.7 psia)	2,560	3,350

satisfactory for heating when stored outdoors in areas which frequently have temperatures below freezing.

Butane-propane mixtures are also sold as LPG, and their properties and standard test procedures are also specified by the NGPA.

Average properties of commercial propane and butane are given in Table 2.3.

2.2 GASOLINE

Although an API survey [1] reports that 40 types of gasolines are made by refineries, about 90% of the total gasoline produced in the United States is used as fuel in automobiles. Most refiners produce gasoline in two grades, regular and premium and in addition supply a low-lead or non-leaded gasoline to comply with antipollution requirements. The principal difference between the regular and premium fuels is the antiknock performance. In 1982 the posted method octane number (PON) of regular gasolines (see Section 2.3) was about 88 and that of premium gasolines about 91. The non-leaded gasolines averaged about 87 RON. Posted octane numbers are arithmetic averages of the motor and research octane numbers and average about four numbers below the research octane number (RON).

Gasolines are a complex mixture of hydrocarbons having a boiling range from 100 to 400°F as determined by the ASTM method. Components are blended to promote high antiknock quality, ease of starting, quick warm-up, low tendency to vapor lock, and low engine deposits. Gruse and Stevens [4] give a very comprehensive account of properties of gasoline and the manner in which they are affected by the blending components. For the purposes of preliminary plant design, however, the components used in blending motor gasoline can be limited to light straightrun gasoline, catalytic reformate, catalytically cracked gasoline, hydrocracked gasoline, polymer, alkylate, and n-butane. Additives such as antioxidants, metal deactivators, and anti-stall agents are not considered individually at this time, but are included with the cost of the antiknock chemicals added. The quantity and cost of antiknock agents, if used, must be determined by making octane blending calculations.

Light straight-run (LSR) gasoline consists of the C_5 $-190°F$ fraction of the naphtha cuts from the crude still. Some refiners cut at 180 or 200°F instead of 190°F but, in any case, this is the fraction that cannot be materially upgraded in octane by catalytic reforming. As a result, it is processed separately from the heavier straight-run gasoline fractions and generally requires only caustic washing or light hydrotreating to produce a gasoline blending stock. In some cases, it is necessary to sweeten the LSR gasoline to convert residual mercaptan to disulfides to give a doctor sweet product. For maximum octane with no lead addition, some refiners have installed isomerization units for processing the LSR gasoline fraction.

Catalytic reformate is the C_5^+ gasoline product of the catalytic reformer. Heavy straight-run (HSR) and coker gasolines are used as feed to the catalytic reformer and, when the octane needs require, catalytic-cracked and hydrocracked gasolines may also be processed by this unit to increase octane levels. The processing conditions of the catalytic reformer are controlled to give the desired product antiknock properties in the range of 90 to 100 RON clear (lead-free).

The catalytic-cracker and hydrocracker gasolines are generally used directly as gasoline blending stocks, but in some cases are upgraded by catalytic reforming before being blended into motor gasoline.

Polymer gasoline is manufactured by polymerizing olefinic hydrocarbons to produce paraffins in the gasoline boiling range. Refinery technology favors alkylation processes rather than polymerization because larger quantities of higher octane product can be made from the light olefins available.

Alkylate gasoline is the product of the reaction of isobutane with propylene, butylene, or pentylene to produce a branched-chain hydrocarbon in the gasoline boiling range. Alkylation of a given quantity of olefin produces twice as much higher octane motor fuel as can be produced by polymerization. In addition, the blending octane number (PON) of alkylate is higher and the lead susceptibility greater than that of polymer gasoline.

Normal butane is blended into gasoline to give the desired vapor pressure. The vapor pressure (expressed as RVP) of gasoline is a compromise between a high RVP to improve starting characteristics and a low RVP to prevent vapor lock and reduce evaporation losses. As such, it changes with the season of the year and varies at sea level between 10 psi in the summer and 13.5 psi in the winter. As butane has a high blending octane number, it is a very desirable component of gasoline and refiners put as much in their gasolines as vapor pressure limitations permit. Isobutane can be used for this purpose but is not as desirable because its higher vapor pressure permits a lesser amount to be incorporated than n-butane.

Since the 1940s, motor gasoline has been the principal product of refineries and, in 1962, gasoline production was the largest of any of the basic industries in the United States. The 204 million tons of gasoline produced exceeded the output of steel and lumber and other high-volume products [5]. Of this, over 90% was used in passenger cars and trucks.

The aviation gasoline market is relatively small and accounts for less than 3% of the gasoline market. For this reason, it is generally not considered in the preliminary refinery design.

2.3 GASOLINE SPECIFICATIONS

There are several important properties of gasoline, but the two that have the greatest effects are the boiling range and antiknock characteristics.

The boiling range of gasoline governs ease of starting, rate of acceleration, loss by crankcase dilution, and tendency toward vapor lock. Engine warm-up time is affected by the percent distilled at 158°F and the 90% ASTM distillation temperature. Warm-up is expressed in terms of the miles of operation required to develop full power without excessive use of the choke. A two- to four-mile warm-up is considered satisfactory and the relationship between outside temperature and percent distilled to give acceptable warm-up properties is:

% distilled @ 158°F	3	11	19	28	38	53
Min ambient temp, °F	80	60	40	20	0	−20

Crankcase dilution is controlled by the 90% ASTM distillation temperature and is also a function of outside temperature. To keep crankcase dilution within acceptable limits, the volatility should be:

Min ambient temp, °F	80	60	40	20	0	−20
90% ASTM distillation, °F	370	350	340	325	310	300

Tendency to vapor lock is directly related to the RVP of the gasoline. In order to control vapor lock the vapor pressure of the gasoline should not exceed the following limits:

Ambient temp (°F)	Max allowable RVP (psia)
60	12.7
70	11.0
80	9.4
90	8.0

The Reid vapor pressure is approximately the vapor pressure of the gasoline at 100°F in pounds per square inch absolute (ASTM designation: D-323).

Altitude affects several properties of gasoline, the most important of which are losses by evaporation and octane requirement. Octane number requirement is greatly affected by altitude and, for a constant spark advance, is about three units lower for each 1,000 feet of elevation. In practice, however, the spark is generally advanced at higher elevations to improve engine performance and the net effect is to reduce the RON of the gasoline marketed by about 3 numbers for a 5,000 ft increase in elevation. Octane requirements for the same model of engine will vary by 7 to 12 RON

TABLE 2.4

Effects of Variables on Octane Requirements

Variable	Effect on Octane Requirement
Altitude	-3 RON per 1,000 ft increase in altitude
Humidity	-0.5 RON per 10% increase in rel humidity @ 70° F
Engine speed	-1 RON per 300 rpm increase
Air temperature	+1 RON per 20° F rise
Spark advance	+1.5 RON per 1° advance
Coolant temperature	+1 RON per 10° F increase
Combustion chamber deposits	+1 to 2 RON per 1,000 miles up to 6,000 miles

because of differences in tuneup, engine deposits, and clearances. Table 2.4 lists some typical effects of variables on engine octane requirements.

There are two types of octane numbers for gasoline engines: those determined by the "motor method" (MON) and those determined by the "research method" (RON). Both methods use the same basic test engine but operate under different conditions. The RON (ASTM D-908) represents the performance during low-speed driving when acceleration is relatively frequent, and the MON (ASTM D-357) is a guide to engine performance at high speeds or under heavy load conditions. The difference between the research and motor octane numbers of a gasoline is an indicator of the changes in performance under both city and highway driving and is known as the "sensitivity" of the fuel.

2.4 DISTILLATE FUELS

Distillate fuels can be divided into three types: jet fuels, diesel fuels, and heating oils. These products are blended from a variety of refinery streams to meet the desired specifications.

The consumption of heating oils ranks second only to gasoline in refinery product volume, but the kerosine-type jet fuel is expected to have the maximum growth potential for the future. This will reflect the expansion of air travel and its change to larger and faster aircraft.

2.5 JET FUELS

Commercial jet fuel is a material in the kerosine boiling range and must be clean burning. The ASTM specifications for commercial jet fuels are given in Table 2.5. One of the critical specifications of jet fuel is its smoke point, and this limits the percentage of cracked products high in aromatics that can be incorporated. Specifications limit the aromatic concentrations to 20%. Hydrocracking saturates the aromatics in the cracked products and raises the smoke point. The freeze point specification is very low (−40°F to −58°F max) and hydrocracking is also used to isomerize the paraffins and lower the freeze

TABLE 2.5

Characteristics of Aircraft Turbine Fuels
(ASTM D-1655-81)

Property	Type A	Type B	Type A-1
Gravity, °API			
Max	51	57	51
Min	37	45	37
Distillation temp, °F			
10% evap, max	400	—	400
20% evap, max	—	290	—
50% evap, max	—	370	—
90% evap, max	—	470	—
FBP, max	572	—	572
Flash point, °F			
Min	100	—	100
Freezing point, °F, max	−40	−58	−53
Sulfur wt. %, max	0.3	0.3	0.3
Aromatics, vol %, max	20	20	20

point. Hydrocracking normally produces a very low (14 to 16) smoke point jet fuel when the cracking is done in the presence of a small amount of hydrogen sulfide.

Jet fuel is blended from low sulfur or desulfurized kerosine, cracked, and hydrocracked blending stocks.

The two basic types of jet fuels are naphtha and kerosine. Naphtha jet fuel is produced primarily for the military and is a wide-boiling-range stock which extends through the gasoline and kerosine boiling ranges. In case of a national emergency, there would be a tremendous demand for jet fuels and to meet the requirements, both gasoline and kerosine production would be needed. The military JP-4 jet fuel is very similar to the ASTM Type B fuel, and the limiting specifications are generally freezing point (−60 to −76°F max), gravity (45 to 57° API), RVP (2.0 to 3.0 psi), and aromatics (20 to 25% max). The wide boiling range (150 to 550°F) permits a great flexibility in blending and kerosine, naphtha, and low-octane gasolines can be used. The amount of each component is typically:

Component	Percent in blend	Limiting specifications
Naphtha	90-100	API, distillation
Gasoline	25-40	RVP, API
Kerosine	35-45	Freeze point

The jet fuels are blended from the various components to arrive at the lowest-cost blend that meets specifications.

Safety considerations limit commercial jet fuels to a narrower boiling range product (350 to 550°F) which is sold as JP-5, ASTM Types A and A-1, or Jet 50. The principal differences among these are freezing points, which range from −40 to −58°F max. In addition to freezing point, the limiting specifications are flash point (110 to 150°F), distillation, and aromatics content (20% max).

2.6 AUTOMOTIVE DIESEL FUELS

Volatility, ignition quality, and viscosity are the important properties of automotive diesel fuels. No. 1 diesel fuel (sometimes called super diesel) is generally made from virgin stocks having cetane numbers above 50. It has a boiling range from 360 to 600°F and is used in high-speed engines in trucks and buses.

No. 2 diesel fuel is very similar to No. 2 fuel oil, and has a wider boiling range (350 to 650°F) than No. 1. It usually contains some cracked stocks and may be blended from naphtha, kerosine, and light cracked cycle oils. Limiting specifications are flash point (120 to 125°F min), sulfur content (0.5% max), distillation range, cetane number (52 min), and pour point (−10 to +10 max).

The ignition properties of diesel fuels are expressed in terms of cetane number. This is very similar to the octane number in gasoline and expresses the volume percentage of cetane ($C_{16}H_{34}$, high-ignition quality) in a mixture with alpha-methyl-naphthalene ($C_{11}H_{10}$, low-ignition quality). The fuel is used to operate a standard diesel test engine according to ASTM test method D-613-62.

2.7 RAILROAD DIESEL FUELS

The largest single market for diesel fuels is for railroad diesel engines [6] and, in 1964, this market accounted for more than one-third of all diesel-fuel sales. Railroad diesel fuels are similar to the heavier automotive diesel fuels but have higher boiling ranges (up to 700°F end point) and lower cetane numbers (40 to 45).

2.8 HEATING OILS

Although the consumption of petroleum products for space heating ranked second to gasoline in 1965, the consumption varied widely according to locality and climate. The ASTM specifications for heating oils are given in Table 2.6. The principal distillate fuel oil consist of No. 1 and No. 2 fuel oils. No. 1 fuel oil is very similar to kerosine, but generally has a higher pour point and end point. Limiting specifications are distillation, pour point, flash point, and sulfur content.

No. 2 fuel oil is very similar to No. 2 diesel fuel, contains cracked stock, and is blended from naphtha, kerosine, diesel, and cracked-cycle oils. Limiting specifications are sulfur content (0.25 to 0.5% max), pour point (−10 to +5 max), distillation, and flash point (120°F min).

TABLE 2.6

Heating Oil Specifications (ASTM D-396-80)

	No. 1	No. 2	No. 4	No. 6
Flashpoint, °F, min	100	100	130	140
Pour point, °F, max	0	20	20	—
Distillation temp, °F				
10% evap, max	420	—	—	—
90% evap, max	550	640	—	—
min	—	540	—	—
Viscosity, cSt, At 100°F				
Max	2.2	3.6	26.4	638*
Min	1.4	2.0	5.8	92*
Gravity, °API, min	35	30	—	—
Carbon residue on				
10% bottoms, % max	0.15	0.35	—	—
Ash, % by wt, max	—	—	0.10	—
Sulfur, % by wt, max	0.5	0.5	—	—
Water and sediment,				
vol % max	0.05	0.05	0.50	2.0

*At 122°F.

2.9 RESIDUAL FUEL OILS

Over 60% of the residual fuel oil used in the United States is imported. It is composed of the heaviest parts of the crude and is generally the tower bottoms from vacuum distillation. It sells for a very low price (historically about 70% of the price of the crude from which it is produced) and is considered a byproduct. Critical specifications are viscosity (110 SSF at 122°F, max) and sulfur content (1.0% max). This maximum sulfur specification will undoubtedly be lowered in the next few years because of atmospheric pollution regulations.

Currently only low-sulfur fuel oils can be burned in some areas and this trend is expanding rapidly. On the U.S. West Coast, 0.5 wt % sulfur is the maximum allowable.

NOTES

1. Amer. Petrol. Inst. Inform. Bull. No. 11 (Philadelphia, Pa., 1958).
2. W. F. Bland and R. L. Davidson, Eds., *Petroleum Processing Handbook* (McGraw-Hill Book Company, New York, 1967), p. 1-11.

3. *Publication 2140–72, Liquefied Petroleum Gas Specifications and Test Methods* (Natural Gas Processors Assoc., Tulsa, Okla., 1972).
4. W. A. Gruse and D. R. Stevens, *Chemical Technology of Petroleum*, 3rd Ed. (McGraw-Hill Book Company, New York, 1960), pp. 424-472.
5. W. F. Bland and R. L. Davidson, Eds., *Petroleum Processing Handbook* (McGraw-Hill Book Company, New York, 1967), p. 11-12.
6. Ibid., p. 11-39.

Chapter 3
REFINERY FEEDSTOCKS

The basic raw material for refineries is petroleum or crude oil, even though in some areas synthetic crude oils from other sources (Gilsonite, tar sands, etc.) are included in the refinery feedstocks. The chemical compositions of crude oils are suprisingly uniform even though their physical characteristics vary widely. The elementary composition of crude oil usually falls within the following ranges.

Element	% by wt
Carbon	84-87
Hydrogen	11-14
Sulfur	0-3
Nitrogen	0.2

In the United States, crude oils are classified as paraffin base, naphthene base, asphalt base, or mixed base. There are some crude oils in the Far East which have up to 80% aromatic content, and these are known as aromatic-base oils. The U.S. Bureau of Mines [1, 2] has developed a system which classifies the crude according to two key fractions obtained in distillation: No. 1 from 250 to 275°C at atmospheric pressure and No. 2 from 275 to 300°C at 40 mmHg pressure. The gravity of these two fractions is used to classify crude oils into types as shown on the following page. The paraffinic and asphaltic classifications in common use are based on the properties of the residuum left from nondestructive distillation and are more descriptive to the refiner because they convey the nature of the products to be expected and the processing necessary.

	Key fractions, °API	
	1	2
Paraffin	≥ 40	≥ 30
Paraffin, intermediate	≥ 40	20-30
Intermediate, paraffin	33-40	≥ 30
Intermediate	33-40	20-30
Intermediate, naphthene	33-40	≤ 20
Naphthene, intermediate	≤ 33	20-30
Naphthene	≤ 33	≤ 20

3.1 CRUDE OIL PROPERTIES

Crude petroleum is very complex and, except for the low-boiling components, no attempt is made by the refiner to analyze for the pure components contained in the crude oil. Relatively simple analytical tests are run on the crude and the results of these are used with empirical correlations to evaluate the crude oils as feedstocks for the particular refinery. Each crude is compared with the other feedstocks available and, based upon the product realization, is assigned a value. The more useful properties are:

Gravity, °API

The density of petroleum oils is expressed in terms of API gravity rather than specific gravity; it is related to specific gravity in such a fashion that an increase in API gravity corresponds to a decrease in specific gravity. The units of API gravity are °API and can be calculated from specific gravity by the following:

$$°API = \frac{(141.5)}{sp\ gr} - 131.5$$

In the above equation, specific gravity and API gravity refer to the weight per unit volume at 60°F. Crude oil gravity may range from less than 10°API to over 50°API but most crudes fall in the 20 to 45°API range. API gravity always refers to the liquid sample at 60°F (15.6°C).

Sulfur Content

Sulfur content and API gravity are two properties which have had the greatest influence on the value of crude oil, although nitrogen and metals contents are increasing in importance. The sulfur content is expressed as percent sulfur by weight and varies from less than 0.1% to greater than 5%. Crudes with greater than 0.5% sulfur generally require more extensive processing than those with lower sulfur content. Although the term "sour" crude initially had reference to those crudes containing dissolved hydrogen sulfide independent of total sulfur content, it has come to mean any crude oil with a sulfur content high enough to require special processing. There is no sharp dividing line between sour and sweet crudes, but 0.5% sulfur content is frequently used as the criterion.

Pour Point

The pour point of the crude oil, in °F, is a rough indicator of the relative paraffinicity and aromaticity of the crude. The lower the pour point the lower the paraffin content and the greater the content of aromatics.

Carbon Residue

Carbon residue is determined by distillation to a coke residue in the absence of air. The carbon residue is roughly related to the asphalt content of the crude and to the quantity of the lubricating oil fraction that can be recovered. In most cases the lower the carbon residue the more valuable the crude. This is expressed in terms of the weight percent carbon residue by either the Ramsbottom (RCR) or Conradson (CCR) ASTM test procedures (D-524 and D-189).

Salt Content

If the salt content of the crude, when expressed as NaCl, is greater than 10 lb/1,000 bbl, it is generally necessary to desalt the crude before processing. If the salt is not removed, severe corrosion problems may be encountered. If residua are processed catalytically, desalting is desirable at even lower salt contents of the crude.

Characterization Factors

There are several correlations between yield and the aromaticity and paraffinicity of crude oils, but the two most widely used are the UOP or "Watson characterization factor" (K_w) and the U.S. Bureau of Mines "correlation index" (CI).

$$K_w = (T_B)^{1/3}/G$$

$$CI = (87,552/T_B) + 473.7G - 456.8$$

where

T_B = mean average boiling point, °R

G = specific gravity at 60°F.

The Watson characterization factor ranges from less than 10 for highly aromatic materials to almost 15 for highly paraffinic compounds. Crude oils show a narrower range of K_w and vary from 10.5 for a highly naphthenic crude to 12.9 for a paraffinic base crude.

The correlation index is useful in evaluating individual fractions from crude oils. The CI scale is based upon straight-chain paraffins have a CI value of 0 and benzene having a CI value of 100. The CI values are not quantitative, but the lower the CI value the greater the concentrations of paraffin hydrocarbons in the fraction, and the higher the CI value the greater the concentrations of naphthenes and aromatics [3].

Nitrogen Content

A high nitrogen content is undesirable in crude oils because organic nitrogen compounds cause severe poisoning of catalysts used in processing. Crudes containing nitrogen in amounts above 0.25% by weight require special processing to remove the nitrogen.

Distillation Range

The boiling range of the crude gives an indication of the quantities of the various products present. The most useful type of distillation is known as a true boiling point (TBP) distillation and generally refers to a distillation performed in equipment that accomplishes a reasonable degree of fractionation. There is no specific test procedure called a TBP distillation, but the U.S. Bureau of Mines Hempel and ASTM D-285 distillations are the tests most commonly used. Neither of these specify either the number of theoretical plates or the reflux ratio used and, as a result, there is a trend toward using the results of a 15:5 distillation (D-2892) rather than the TBP. The 15:5 distillation is carried out using 15 theoretical plates at a reflux ratio of 5:1

The crude distillation range also has to be correlated with ASTM distillations because product specifications are generally based on the simple ASTM distillation tests D-86 and D-1160. The TBP cut point for various fractions can be approximated by use of Figure 3.1. A more detailed procedure for correlation of ASTM and TBP distillations is given in the "API Technical Data Book—Petroleum Refining."

Metals Content

The metals content of crude oils can vary from a few parts per million to more than 1,000 ppm and, in spite of their relatively low concentrations, are of considerable importance [4]. Minute quantities of some of these metals (nickel, vanadium, and copper) can severely affect the activities of catalysts and result in a lower-value product distribution. Vanadium concentrations above 2 ppm in fuel oils can lead to severe corrosion to turbine blades and deterioration of refractory furnace linings and stacks [2].

Distillation concentrates the metallic constituents of crude in the residues, but some of the organometallic compounds are actually volatilized at refinery distillation temperatures and appear in the higher-boiling distillates [5].

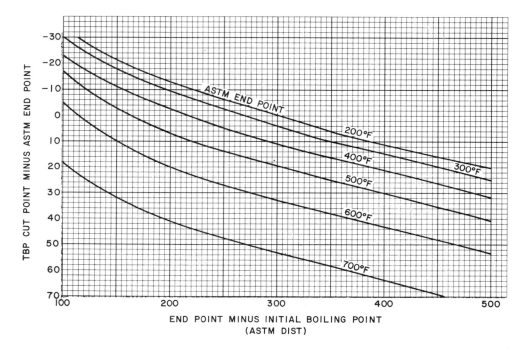

FIG. 3.1.* TBP cut point vs ASTM end point.

The metallic content may be reduced by solvent extraction with propane or similar solvents as the organometallic compounds are precipitated with the asphaltenes and resins.

3.2 COMPOSITION OF PETROLEUM

Crude oils and high-boiling crude oil fractions are composed of many members of a relatively few homologous series of hydrocarbons [6]. The composition of the total mixture, in terms of elementary composition, does not vary a great deal, but small differences in composition can greatly affect the physical properties and the processing required to produce salable products. Petroleum is essentially a mixture of hydrocarbons, and even the non-hydrocarbon elements are generally present as components of complex molecules predominantly hydrocarbon in character, but containing small quantities of oxygen, sulfur, nitrogen, vanadium, nickel, and chromium [4]. The hydrocarbons present in crude petroleum are classified into three general types: paraffins, naphthenes, and aromatics. In addition, there is a fourth type, olefins, that is formed during processing by the dehydrogenation of paraffins and naphthenes.

---

*See Appendix G for full size reproductions of graphical figures.

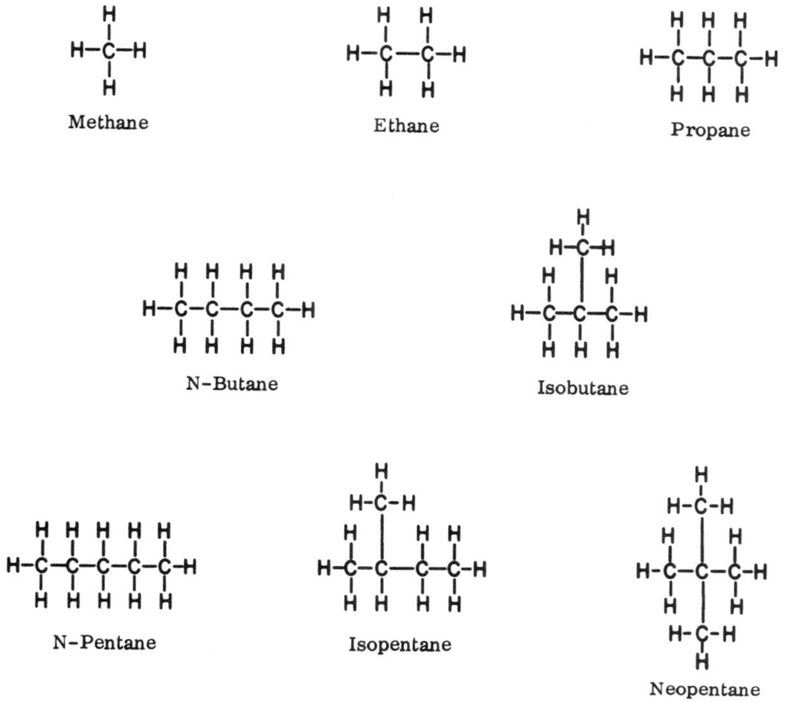

FIG. 3.2. Paraffins in crude oil.

Paraffins

The paraffin series of hydrocarbons is characterized by the rule that the carbon atoms are connected by a single bond and the other bonds are saturated with hydrogen atoms. The general formula for paraffins is C_nH_{2n+2}.

The simplest paraffin is methane, CH_4, followed by the homologous series of ethane, propane, normal and isobutane, normal, iso-, and neopentane, etc. (Fig. 3.2.). When the number of carbon atoms in the molecule is greater than three, several hydrocarbons may exist which contain the same number of carbon and hydrogen atoms but have different structures. This is because carbon is capable not only of chain formation, but also of forming single- or double-branched chains which give rise to isomers that have significantly different properties. For example, the motor octane number of n-octane is −17 and that of isooctane (2,2,4-trimethyl pentane) is 100.

The number of possible isomers increases in geometric progression as the number of carbon atoms increases. There are two paraffin isomers of butane, three of pentane and, by the time the number of carbon atoms has increased to eight, 17 structural isomers of octane. Crude oil contains molecules with up to 70 carbon atoms, and the number of possible paraffinic hydrocarbons is very high.

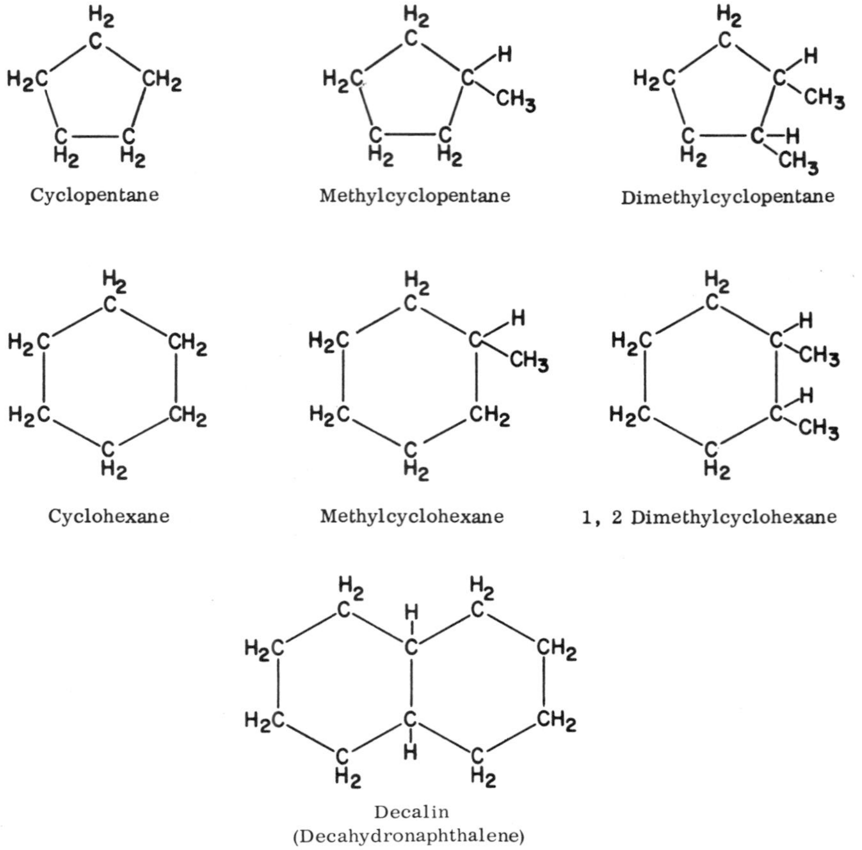

Cyclopentane   Methylcyclopentane   Dimethylcyclopentane

Cyclohexane   Methylcyclohexane   1, 2 Dimethylcyclohexane

Decalin
(Decahydronaphthalene)

FIG. 3.3. Naphthenes in crude oil.

Olefins

Olefins do not naturally occur in crude oils but are formed during the processing. They are very similar in structure to paraffins but at least two of the carbon atoms are joined by double bonds. The general formula is C_nH_{2n}. Olefins are generally undesirable in finished products because the double bonds are reactive and the compounds are more easily oxidized and polymerized.

Some diolefins (containing two double bonds) are also formed during processing, but they react very rapidly with olefins to form high-molecular-weight polymers consisting of many simple unsaturated molecules joined together.

Naphthenes (Cycloparaffins)

Cycloparaffin hydrocarbons in which all of the available bonds of the carbon atoms are saturated with hydrogen are called naphthenes. There are many types of naphthenes present in crude oil but, except for the lower-

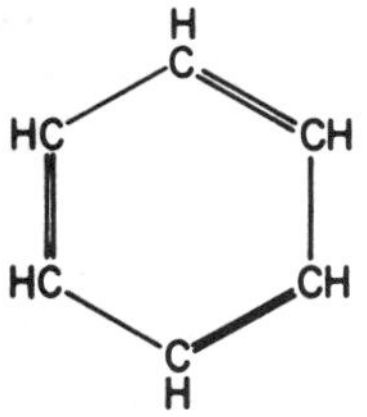

Benzene

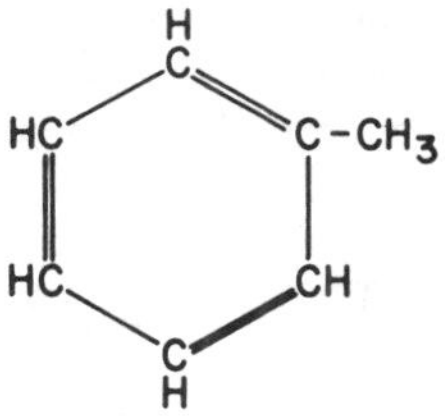

Toluene

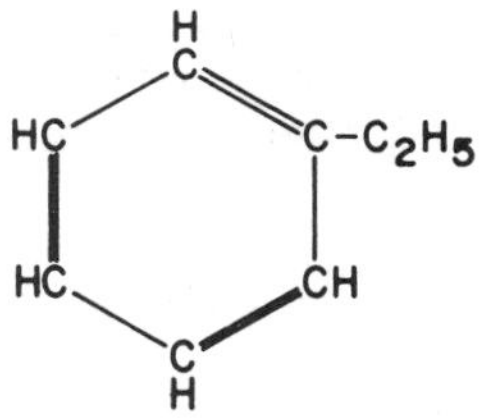

Ethylbenzene

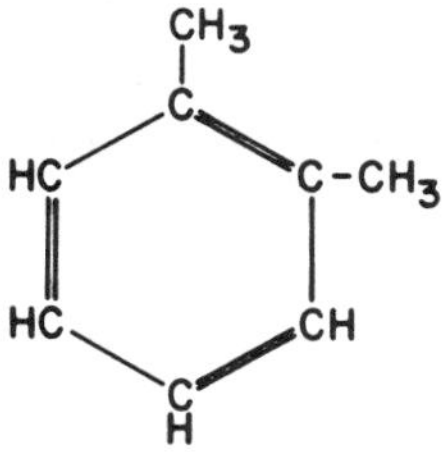

Ortho-xylene

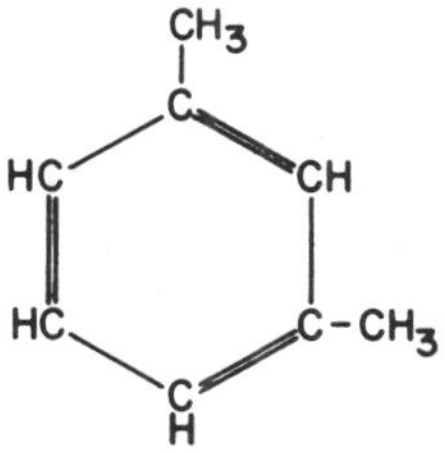

Meta-xylene

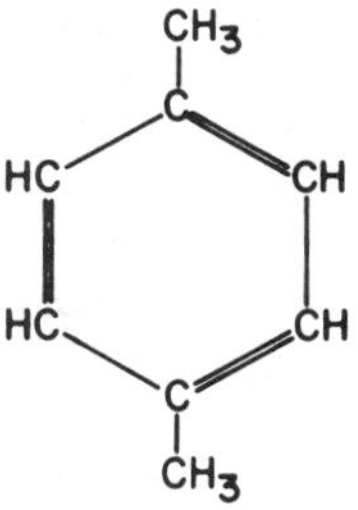

Para-xylene

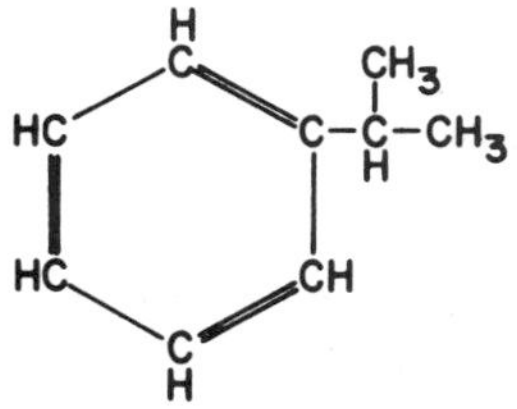

Cumene

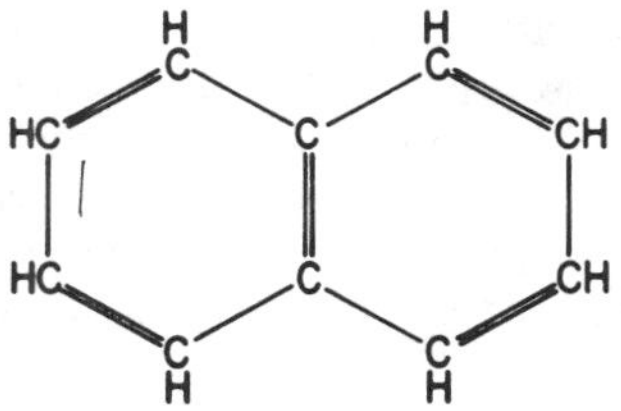

Naphthalene

FIG. 3.4. Aromatic hydrocarbons in crude oil.

molecular-weight compounds such as cyclopentane and cyclohexane, are generally not handled as individual compounds. They are classified according to boiling range and their properties determined with the help of correlation factors such as the K_W factor, or CI. Some typical naphthenic compounds are shown in Fig. 3.3.

Aromatics

The aromatic series of hydrocarbons is chemically and physically very different from the paraffins and naphthenes. Aromatic hydrocarbons contain a benzene ring which is unsaturated but very stable and frequently behaves as a saturated compound. Some typical aromatic compounds are shown in Figure 3.4.

The cyclic hydrocarbons, both naphthenic and aromatic, can add paraffin side chains in place of some of the hydrogen attached to the ring carbons

and form a mixed structure. These mixed types have many of the chemical
and physical characteristics of both of the parent compounds, but generally
are classified according to the parent cyclic compound.

3.3 CRUDES SUITABLE FOR ASPHALT MANUFACTURE

It is not possible to predict with 100% accuracy whether or not a particular
crude will produce specification asphalts without actually separating the
asphalts from the crude and running the tests. There are, however, certain characteristics of crude oils that indicate if they are possible sources of
asphalt. If the crude oil contains a residue (750°F mean average boiling
point) having a Watson characterization factor of less than 11.8 and the
gravity is below 35° API, it is usually suitable for asphalt manufacture [7].
If, however, the difference between the characterization factors for the
750°F and 550°F fraction is greater than 0.15, the residue may contain too
much wax to meet most asphalt specifications.

3.4 CRUDE DISTILLATION CURVES

When a refining company evaluates its own crude oils to determine the most
desirable processing sequence to obtain the required products, its own labo-
ratories will provide data concerning the distillation and processing of the
oil and its fractions [8]. In many cases, information not readily available is
desired concerning the processing qualities of crude oils. In such instances,
true boiling point (TBP) and gravity mid-percent curves can be developed
from U.S. Bureau of Mines crude petroleum analysis data sheets (Fig. 3.5).

The U.S. Bureau of Mines has carried out Hempel distillations on
thousands of crude oil samples from wells in all major producing fields. Although the degree of fractionation in a Hempel assay is less than that in a
15:5 distillation, the results are sufficiently similar that they can be used
without correction. If desired, correction factors developed by Nelson [9]
can be applied.

The major deficiency in a Bureau of Mines assay is the lack of information concerning the low-boiling components. The materials not condensed by
water-cooled condensers are reported as "distillation loss." An estimate of
the composition of the butane and lighter components is frequently added to
the low-boiling end of the TBP curve to compensate for the loss during distillation.

The Bureau of Mines analysis is reported in two parts: the first is the
portion of the distillation performed at atmospheric pressure and up to 527°F
end point, the second, at 40 mmHg total pressure, to 572°F end point. The
portion of the distillation at reduced pressure is necessary to prevent excessive pot temperatures, which cause cracking of the crude oil.

The distillation temperatures reported in the analysis must be corrected
to 760 mmHg pressure. Generally, those reported in the atmospheric distillation section need not be corrected, but if carried out at high elevations
it may also be necessary to correct these. The distillation temperatures at
40 mmHg pressure can be converted to 760 mmHg by use of Procedure
5A1.13 in the API Technical Data Book. Two of the curves from this procedure are included here as Figures 3.6 and 3.7.

The 572°F end point at 40 mmHg pressure corresponds to 790°F at
760 mmHg. Refinery crude oil distillation practices take overhead streams

CRUDE PETROLEUM ANALYSIS

Bureau of Mines ..Bartlesville............ Laboratory
Sample ..53016.........................

IDENTIFICATION

Hastings Field

Texas
Brazoria County

GENERAL CHARACTERISTICS

Gravity, specific, ...0.867......   Gravity, ° API, ...31.7...........   Pour point, ° F., ..below 5.....
Sulfur, percent, ...0.15........   Color, ..brownish green.......
Viscosity, Saybolt Universal at ....100°.....................   Nitrogen, percent, ...

DISTILLATION, BUREAU OF MINES ROUTINE METHOD

STAGE 1—Distillation at atmospheric pressure, ..751...... mm. Hg
First drop, ...84.... ° F.

Fraction No.	Cut temp. ° F.	Percent	Sum, percent	Sp. gr. 60/60° F.	° API. 60° F.	C. I.	Refractive index, n_D at 20° C.	Specific dispersion	S. U. visc., 100° F.	Cloud test, ° F.
1......	122	0.8	0.8	0.673	78.8					
2.....	167	1.0	1.8	.685	75.1	15				
3.....	212	3.0	4.8	.725	63.7	24	1.39574	127.7		
4.....	257	3.4	8.2	.755	55.9	29	1.41756	128.6		
5.....	302	3.1	11.3	.777	50.6	32	1.42985	135.4		
6.....	347	3.9	15.2	.798	45.8	35	1.44192	137.8		
7.....	392	4.9	20.1	.817	41.7	38	1.45217	139.9		
8.....	437	6.8	26.9	.833	38.4	40	1.46057	140.3		
9.....	482	8.0	34.9	.848	35.4	41	1.46875	148.0		
10.....	527	10.9	45.8	.864	32.3	44	1.47679	149.8		

STAGE 2—Distillation continued at 40 mm. Hg

Fraction No.	Cut temp. ° F.	Percent	Sum, percent	Sp. gr. 60/60° F.	° API. 60° F.	C. I.	Refractive index, n_D at 20° C.	Specific dispersion	S. U. visc., 100° F.	Cloud test, ° F.
11.....	392	7.3	53.1	0.873	30.6	45	1.48274	155.2	42	Below 5
12.....	437	7.8	60.9	.879	29.5	44	1.48474	156.2	50	do
13.....	482	6.2	67.1	.889	27.7	45	1.49058	152.7	71	do
14.....	527	5.7	72.8	.901	25.6	48			125	10
15.....	572	6.9	79.7	.916	28.0	52			280	20
Residuum.	......	20.3	100.0	.945	18.2					

Carbon residue, Conradson: Residuum, .4.7. percent; crude, .1.0. percent.

APPROXIMATE SUMMARY

	Percent	Sp. gr.	° API	Viscosity
Light gasoline..	4.8	0.708	68.4	
Total gasoline and naphtha	20.1	0.771	52.0	
Kerosine distillate	--	--	--	
Gas oil..	36.9	0.858	33.4	
Nonviscous lubricating distillate	10.2	.879-.895	29.5-26.6	50-100
Medium lubricating distillate	5.8	.895-.908	26.6-24.3	100-200
Viscous lubricating distillate	6.7	.908-.924	24.3-21.6	Above 200
Residuum..	20.3	0.945	18.2	
Distillation loss	0			

U. S. GOVERNMENT PRINTING OFFICE   16—57835-3

FIG. 3.5. U.S. Bureau of Mines crude petroleum analysis.

with end points from 950 to 1,050°F at 760 mmHg. Estimates of the shape of the TBP curve above 790°F can be obtained by plotting the distillation temperature versus percentage distilled on probability graph paper and extrapolating to 1,100°F [10]. (See Fig. 3.8.) The data points above 790°F can be transferred to the TBP curve.

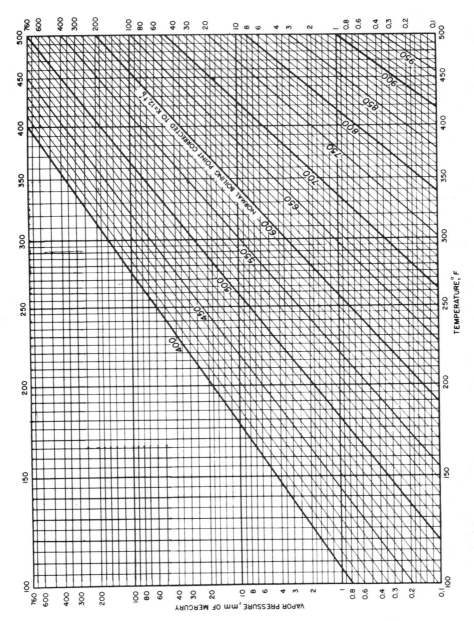

FIG. 3.6. Vapor pressure of pure hydrocarbons and narrow-boiling petroleum fractions (low-temperature range). From "API Technical Data Book" (American Petroleum Institute, Division of Refining, 1970).

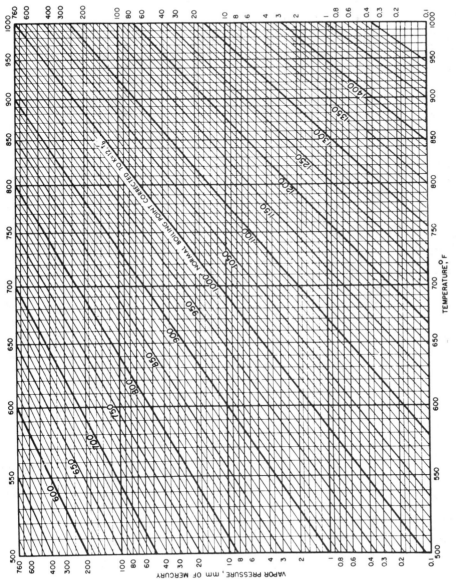

FIG. 3.7. Vapor pressure of pure hydrocarbons and narrow-boiling petroleum fractions (high-temperature range) From "API Technical Data Book" (American Petroleum Institute, Division of Refining, 1970).

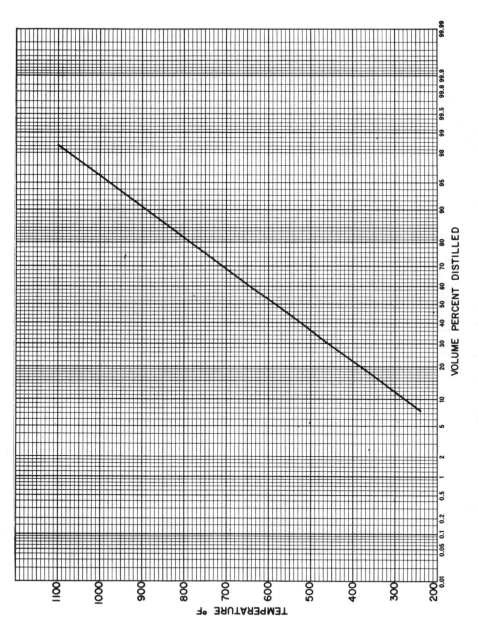

FIG. 3.8. Crude distillation curve.

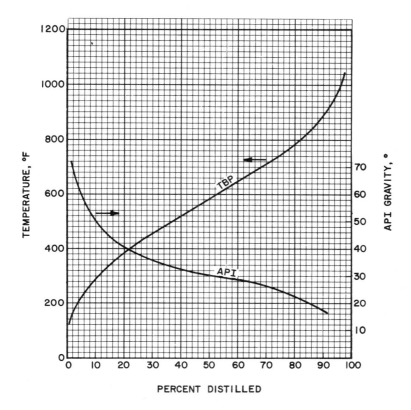

FIG. 3.9. TBP and gravity curves. Crude: Hastings Field, Texas; gravity:
31.7° API; sulfur: 0.15 wt %.

   The gravity mid-percent curve is plotted on the same chart with the
TBP curve. The gravity should be plotted on the average volume percent
of the fraction, as the gravity is the average of the gravities from the first
to the last drops in the fraction. For narrow cuts, a straight-line relationship
can be assumed and the gravity used as that of the mid-percent of the
fraction.
   Smooth curves are drawn for both the TBP and gravity mid-percent
curves. Figure 3.9 illustrates these curves for the crude oil reported in
Figure 3.5.

PROBLEMS

1. Develop a TBP and gravity mid-percent curve for one of the crude oils
    given in Appendix D.

2. Using the TBP and gravity curves from problem 1, calculate the Watson
    characterization factors for the fractions having mean average boiling
    points of 550°F and 750°F. Is it probable this crude oil will produce a
    satisfactory quality asphalt?

3. Using the U.S. Bureau of Mines method for classifying crude oils from
    the gravities of the 250 to 275°C and 275 to 300°C fractions, classify the
    crude oil used in problem 1 according to type.

NOTES

1. W. A. Gruse and D. R. Stevens, *Chemical Technology of Petroleum*,
   3rd Ed. (McGraw-Hill Book Company, New York, 1960), p. 16.
2. E. C. Lane and E. L. Garton, U. S. Bureau of Mines Rept. Inves.
   3279 (1935).
3. H. M. Smith, U. S. Bureau of Mines Tech. Paper 610 (1940).
4. M. C. K. Jones and R. L. Hardy, Ind. Eng. Chem. *44*, 2615 (1952).
5. R. A. Woodle and W. B. Chandlev, Ind. Eng. Chem. *44*, 2591 (1952).
6. W. A. Gruse and D. R. Stevens, *Chemical Technology of Petroleum*,
   3rd. Ed. (McGraw-Hill Book Company, New York, 1960), p. 13.
7. W. L. Nelson, Oil Gas J. *68*(44), 92-96 (1970).
8. J. R. Dosher, Chem. Eng. *77*(8), 96-112 (1970).
9. W. L. Nelson, *Petroleum Refinery Engineering*, 4th Ed. (McGraw-Hill
   Book ompany, New York, 1958), p. 114.
10. W. L. Nelson, Oil Gas J. *66*(13), 125-126 (1968).

Chapter 4
CRUDE DISTILLATION

The crude stills are the first major processing units in the refinery. They are used to separate the crude oils by distillation into fractions according to boiling point, so that each of the following processing units will have feedstocks that meet their particular specifications. Higher efficiencies and lower costs are achieved if the crude oil separation is accomplished in two steps: first, by fractionating the total crude oil at essentially atmospheric pressure; then, by feeding the high-boiling bottoms fraction (topped crude) from the atmospheric still to a second fractionator operated at a high vacuum. (See Photo 2, Appendix F.)

The vacuum still is employed to separate the heavier portion of the crude oil into fractions because the high temperatures necessary to vaporize the topped crude at atmospheric pressure cause thermal cracking to occur, with the resulting loss to dry gas, discoloration of the product, and equipment fouling due to coke formation.

Typical fraction cut points and boiling ranges for atmospheric and vacuum still fractions are given in Tables 4.1 and 4.2.

Relationships among the volume average, molal average, and mean average boiling points of the crude oil fractions are shown in Fig. 4.1.

Nitrogen and sulfur contents of petroleum fractions as functions of original crude oil contents are given in figs. 4.2, 4.3, 4.4, and 4.5.

4.1 DESALTING CRUDE OILS

If the salt content of the crude oil is greater than 10 lb/1,000 bbl (expressed as NaCl), the crude requires desalting to minimize fouling and corrosion caused by salt deposition on heat transfer surfaces and acids formed by decomposition of the chloride salts. In addition, some metals which can cause

31

TABLE 4.1

Boiling Ranges of Typical Crude Oil Fractions

Fraction	Boiling Ranges	
	ASTM (° F)	TBP (° F)
Butanes and lighter		
Light straight-run gasoline (LSR)	90-220	90-190
Naphtha (heavy straight-run gasoline)	180-400	190-380
Kerosine	330-540	380-520
Light gas oil (LGO)	420-640	520-610
Atmospheric gas oil	550-830	610-800
Vacuum gas oil (VGO)	750-1,050	800-1,050
Vacuum reduced crude	$1,000^+$	$1,050^+$

TABLE 4.2

TBP Cut Points for Various Crude Oil Fractions

Cut	IBP (° F)	EP (° F)	Processing use
LSR gasoline	90	180	Min light gasoline cut
	90	190	Normal LSR cut
	80	220	Max LSR cut
HSR gasoline	180	380	Max reforming cut
(naphtha)	190	330	Max jet fuel opr
	220	330	Min reforming cut
Kerosine	330	520	Max kerosine cut
	330	480	Max jet-50 cut
	380	520	Max gasoline operation
Light gas oil	420	610*	Max diesel fuel
	480	610*	Max jet fuel
	520	610*	Max kerosine
Heavy gas oil (HGO)	610	800	Catalytic cracker or hydrocracker feed
Vacuum gas oil	800	1,050	Deasphalter or catalytic cracker feed
	800	950	Catalytic cracker or hydrocracker feed

Note: In some specific locations, economics can dictate that all material between 330° F IBP and 800° F EP be utilized as feed to a hydrocracker.

*For maximum No. 2 diesel-fuel production, end points as high as 650° F can be used.

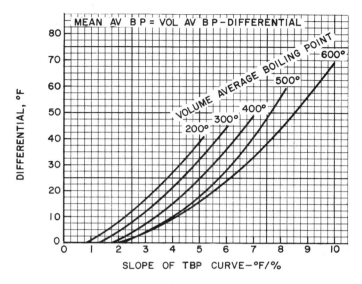

FIG. 4.1a. Mean average boiling point of petroleum fractions [1].

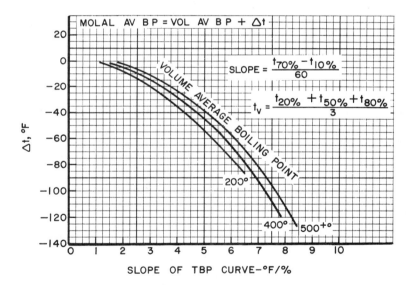

FIG. 4.1b. Molal average boiling point of petroleum fractions [1].

catalyst deactivation in catalytic processing units are partially rejected in the desalting process.

The trend toward running heavier crude oils has increased the importance of efficient desalting of crudes. Until recently, the criteria for desalting crude oils was 10 lb salt/1000 bbl (expressed as NaCl) or more but now many companies desalt all crude oils. Reduced equipment fouling and corrosion and longer catalyst life provide justification for this additional treatment. Two-stage desalting is used if the crude oil salt content is more than 20 lb/1000 bbl.

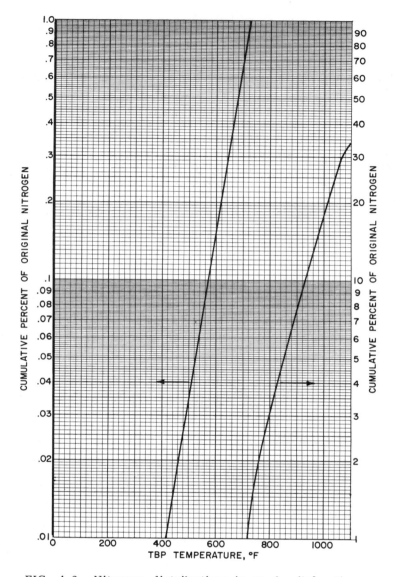

FIG. 4.2. Nitrogen distributions in crude oil fractions.

The salt in the crude is in the form of dissolved or suspended salt crystals in water emulsified with the crude oil. The basic principle is to wash the
salt from the crude oil with water. Problems occur in obtaining efficient and
economical water/oil mixing, water-wetting of suspended solids, and separation of the wash water from the oil. The pH, gravity, and viscosity of the
crude oil, as well as the volume of wash water used per volume of crude oil,
affect the separation ease and efficiency.

A secondary but important function of the desalting process is the removal of suspended solids from the crude oil. These are usually very fine
sand, clay, and soil particles, iron oxide and iron sulfide particles from

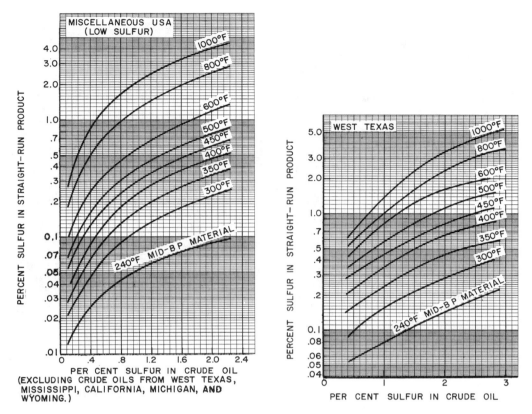

FIG. 4.3. Left: Sulfur content on products from miscellaneous U.S. crude
oils [6]. Right: Sulfur content of products from West Texas
crude oils [2].

pipelines, tanks, or tankers, and other contaminants picked up in transit or
production. Total suspended solids removal should be 60% or better [3] with
80% removal of particles greater than 0.8 micron size.

Desalting is carried out by mixing the crude oil with from 3 to 10 vol
per cent water at temperatures from 90 to 150°C (200 to 300°F). Both the
ratio of the water to oil and the temperature of operation are functions of
the density of the oil. Typical operating conditions are [3]:

°API	Water Wash, vol %	Temp °C (°F)
API > 40	3-4	115-125 (240-260)
30 > API < 40	4-7	125-140 (260-280)
API < 30	7-10	140-150 (280-300)

The salts are dissolved in the wash water and the oil and water
phases separated in a settling vessel either by adding chemicals to assist in
breaking the emulsion or by developing a high-potential electrical field

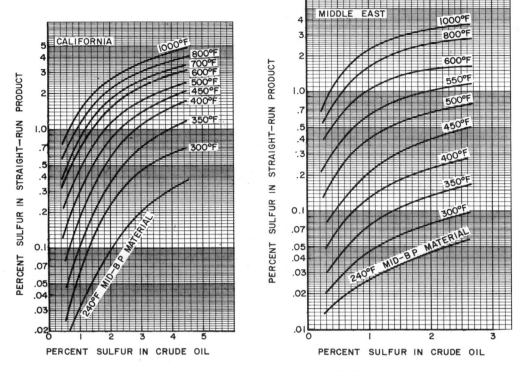

FIG. 4.4. Left: Sulfur content of products from California crude oils [2].
           Right: Sulfur content of products from Middle East crude oils [2].

across the settling vessel to coalesce the droplets of salty water more rapidly.
Either AC or DC fields may be used and potentials from 16,000 to 35,000
volts are used to promote coalescence. For single-stage desalting units 90
to 95% efficiencies are obtained and two-stage processes achieve 99% efficiency.

One process uses both AC and DC fields to provide high dewatering
efficiency. An AC field is applied near the oil-water interface and a DC
field in the oil phase above the interface. Efficiencies of up to 99% water removal in a single stage are claimed for the dual field process.

The dual field electrostatic process provides efficient water separation
at temperatures lower than the other processes and, as a result, higher
energy efficiencies are obtained.

Heavy naphthenic crudes form more stable emulsions than most other
crude oils and desalters usually operate at lower efficiencies when handling
them. The crude oil densities are close to the density of water and temperature above 135°C (280°F) are needed. It is sometimes necessary to adjust
the pH of the brine to obtain pH values of 7 or less in the water. If the pH
of the brine exceeds 7.0, emulsions can be formed because of the sodium
naphthenate and sodium sulfide present. For most crude oils it is desirable
to keep the pH below 8.0. Better dehydration is obtained in electrical desalters when they are operated in the pH range of 6 to 8 with the best

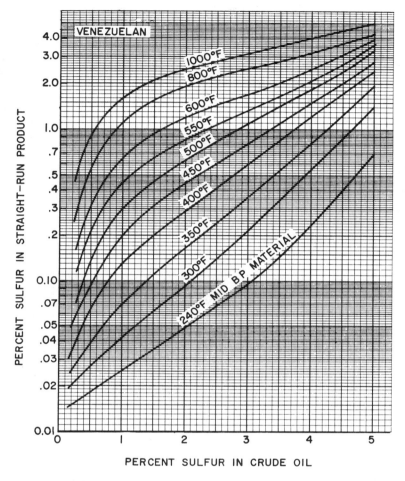

FIG. 4.5. Sulfur content of products from Venezuelan crude oils [2].

dehydration obtained at a pH near 6. The pH value is controlled by the addition of acid to the inlet or recycled water.

Make-up water averages 4 to 5% on crude oil charge and is added to the second stage of a two-stage desalter. For very heavy crude oils (<15 °API), addition of gas oil as a diluent to the second stage is recommended to provide better separation efficiencies.

Frequently, the wash water used is obtained from the vacuum crude unit barometric condensers or other refinery sources containing phenols. The phenols are preferentially soluble in the crude oil thus reducing the phenol content of the water sent to the refinery waste water handling system.

Suspended solids are one of the major causes of water-in-oil emulsions. Wetting agents are frequently added to improve the water wetting of solids and reduce oil carry-under in desalters. Oxyalkylated phenols and sulfates are the most frequently used wetting agents.

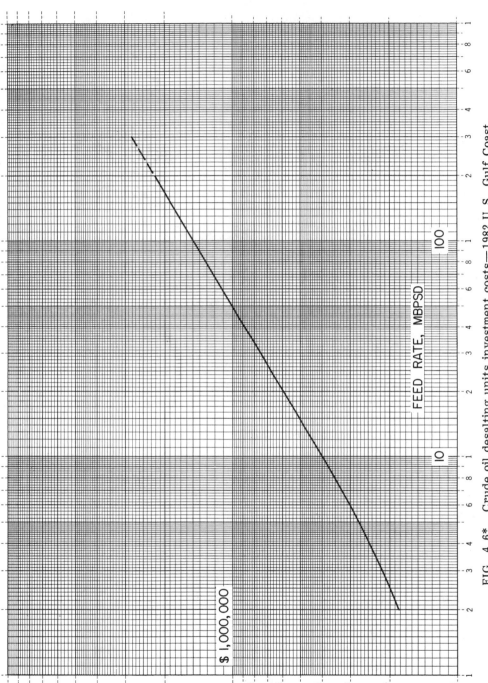

$1,000,000

FEED RATE, MBPSD

10

100

FIG. 4.6*. Crude oil desalting units investment costs—1982 U.S. Gulf Coast.
*See Table 4.3, page 39.

TABLE 4.3*

Desalter

Costs Included:
1. Conventional electrostatic desalting unit.
2. Water injection.
3. Caustic injection.
4. Water preheating and cooling.

Costs Not Included:
1. Waste water treating and disposal.
2. Cooling water and power supply.

Utility Data (per bbl feed):

Power, kWh	0.052 to 0.15
Cooling water circulation†, gal.	4.0
Water injection, gallons	1 to 3
Demulsifier chemical, lb	0.001
Caustic, lb	0.001 to 0.015

Note: Treating rate: 2 to 4 lb/1,000 bbl; 1982 price approx. $1.30/lb.
*See Fig. 4.6, page 38.
†30°F rise

The following analytical methods are used to determine the salt content of crude oils:

1. HACH titration with mercuric nitrate after water extraction of the salt.
2. Potentiometric titration after water extraction.
3. Mohr titration with silver nitrate after water extraction.
4. Potentiometric titration in a mixed solvent.
5. Conductivity.

Although the conductivity method is the one most widely used for process control, it is probably the least accurate of the above methods. Whenever used it should be standardized for each type of crude oil processed.

Installed costs of desalting units are shown in Figure 4.6*, and utility and chemical requirements are given by Table 4.3.

4.2 ATMOSPHERIC TOPPING UNIT

After desalting, the crude oil is pumped through a series of heat exchangers and its temperature raised to about 550°F by heat exchange with product and

---

*Throughout this text the symbol M is used to represent 1,000, in accordance with standard U. S. engineering practice. For example, MBPSD means 1,000 BPSD and MMBtu means 1,000,000 Btu.

reflux streams [3, 4]. It is then further heated to about 750°F in a furnace (i.e., direct-fired heater or "pipe-still") and charged to the flash zone of the atmospheric fractionators. The furnace discharge temperature is sufficiently high (650 to 750°F) to cause vaporization of all products withdrawn above the flash zone plus about 20% of the bottoms product. This 20% "over-flash" allows some fractionation to occur on the trays just above the flash zone by providing internal reflux in excess of the sidestream withdrawals.

Reflux is provided by condensing the tower overhead vapors and returning a portion of the liquid to the top of the tower, and by pump-around and pumpback streams lower in the tower. Each of the sidestream products removed from the tower decreases the amount of reflux below the point of drawoff. Maximum reflux and fractionation is obtained by removing all heat at the top of the tower, but this results in an inverted cone-type liquid loading which requires a very large diameter at the top of the tower. To reduce the top diameter of the tower and even the liquid loading over the length of the tower, intermediate heat-removal streams are used to generate reflux below the sidestream removal points. To accomplish this, liquid is removed from the tower, cooled by a heat exchanger, and returned to the tower or, alternatively, a portion of the cooled sidestream may be returned to the tower. This cold stream condenses more of the vapors coming up the lower and thereby increases the reflux below that point.

The energy efficiency of the distillation operation is also improved by using pump-around reflux. If sufficient reflux was produced in the overhead condenser to provide for all side-stream drawoffs as well as the required reflux, all of the heat energy would be exchanged at the bubble-point temperature of the overhead stream. By using pump-around reflux at lower points in the column, the heat transfer temperatures are higher and a higher fraction of the heat energy can be recovered by preheating the feed.

Although crude towers do not normally use reboilers, several trays are generally incorporated below the flash zone and steam is introduced below the bottom tray to strip any remaining gas oil from the liquid in the flash zone and to produce a high-flash-point bottoms. The steam reduces the partial pressure of the hydrocarbons and thus lowers the required vaporization temperature.

The atmospheric fractionator normally contains 30 to 50 fractionation trays. Separation of the complex mixtures in crude oils is relatively easy and generally five to eight trays are needed for each sidestream product plus the same number above and below the feed plate. Thus, a crude oil atmospheric fractionation tower with four liquid sidestream drawoffs will require from 30 to 42 trays.

The liquid sidestream withdrawn from the tower will contain low-boiling components which lower the flashpoint, because the lighter products pass through the heavier products and are in equilibrium with them on every tray. These "light ends" are stripped from each sidestream in a separate small stripping tower containing four to ten trays with steam introduced under the bottom tray. The steam and stripped light ends are vented back into the vapor zone of the atmospheric fractionator above the corresponding side-draw tray (Fig. 4.7).

The overhead condenser on the atmospheric tower condenses the pentane-and-heavier fraction of the vapors that pass out of the top of the tower. This is the light gasoline portion of the overhead, containing some propane and butanes and essentially all of the higher-boiling components in the tower overhead vapor. Some of this condensate is returned to the top of the

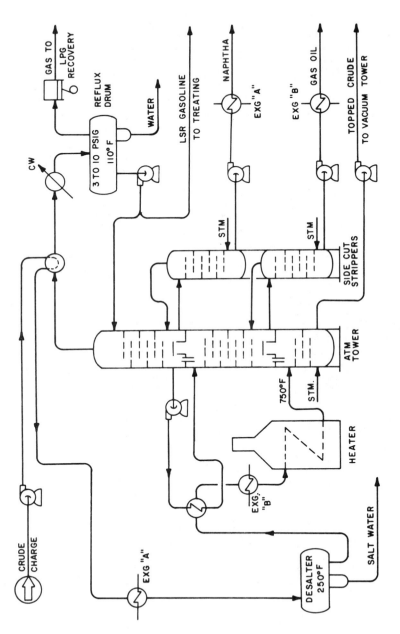

FIG. 4.7. Crude distillation. For simplicity, only two side strippers are shown. Usually at least four are provided to produce extra cuts such as kerosine and diesel.

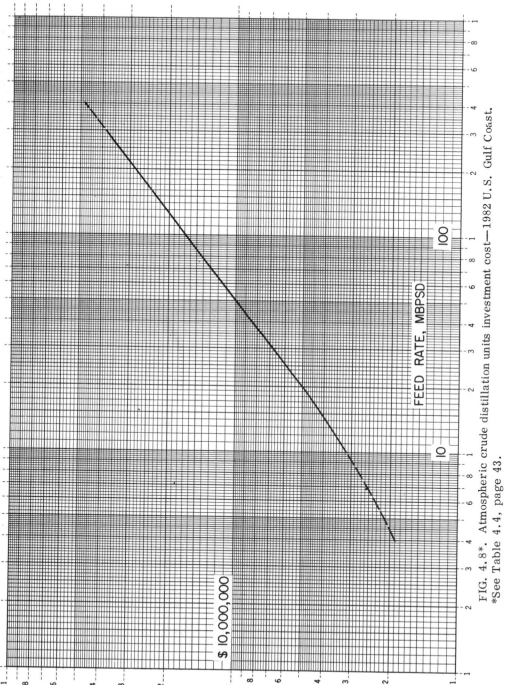

FIG. 4.8*. Atmospheric crude distillation units investment cost—1982 U.S. Gulf Coast.
*See Table 4.4, page 43.

TABLE 4.4 *

Atmospheric Crude Distillation

Costs Included:
1. Two side cuts with strippers (add 10% for each additional side cut).
2. All BL process facilities.
3. Sufficient heat exchange to cool top products and side cuts to ambient temperature.
4. Central control system.

Costs Not Included:
1. Cooling water, steam, and power supply.
2. Desalting.
3. Cooling on reduced crude (bottoms).
4. Sour water treating and disposal.
5. Feed and product storage.
6. Naphtha stabilization.
7. Light ends recovery.

Utility Data (per bbl feed):

Steam (300 psig), lb	6.0
Power, kWh	0.5
Cooling water circulation, gal[a]	6.0
Fuel, MMbtu[b]	0.07

*See Fig. 4.8, page 42.

[a]30°F rise

[b]LHV basis, heater efficiency taken into account. (All fuel data in this text are on this same basis.)

tower as reflux, and the remainder is sent to the stabilization section of the refinery gas plant where the butanes and propane are separated from the C_5-180° LSR gasoline.

Installed costs of atmospheric crude distillation units are shown in Figure 4.8, and utility requirements are given by Table 4.4.

4.3 VACUUM DISTILLATION

The furnace outlet temperatures required for atmospheric pressure distillation of the heavier fractions of crude oil are so high that thermal cracking would occur, with the resultant loss of product and equipment fouling. These materials are therefore distilled under vacuum because the boiling temperature decreases with a lowering of the pressure. Distillation is carried out with absolute pressures in the tower flash zone area of 25 to 40 mmHg (Fig. 4.9). To improve vaporization, the effective pressure is lowered even further (to 10 mmHg or less) by the addition of steam to the furnace inlet and at the bottom of the vacuum tower. Addition of steam to the furnace inlet increases the furnace tube velocity and minimizes coke formation in the furnace as well as decreasing the total hydrocarbon partial pressure in the vacuum tower. The amount of stripping steam used is a function of the boiling range of the feed and the fraction vaporized, but generally ranges from 10 to 50 lb/bbl feed [5].

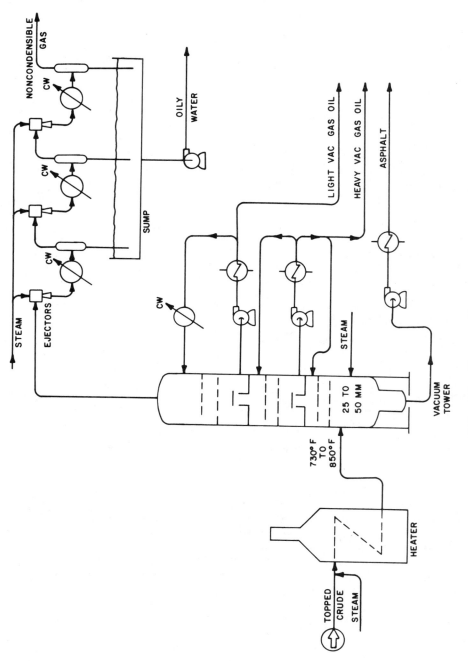

FIG. 4.9. Vacuum distillation.

Furnace outlet temperatures are also a function of the boiling range of the feed and the fraction vaporized as well as of the feed coking characteristics. High tube velocities and steam addition minimize coke formation, and furnace outlet temperatures in the range of 730 to 850°F are generally used [6].

The effective pressure (total absolute pressure-partial pressure of the steam) at the flash zone determines the fraction of the feed vaporized for a given furnace outlet temperature, so it is essential to design the fractionation tower, overhead lines, and condenser to minimize the pressure drop between the vacuum-inducing device and the flash zone. A few millimeters decrease in pressure drop will save many dollars in operating costs.

The lower operating pressures cause significant increases in the volume of vapor per barrel vaporized and, as a result, the vacuum distillation columns are much larger in diameter than atmospheric towers. It is not unusual to have vacuum towers up to 40 feet in diameter.

The desired operating pressure is maintained by the use of steam ejectors and barometric or surface condensers. The size and number of ejectors and condensers used is determined by the vacuum needed and the quality of vapors handled. For a flash zone pressure of 25 mmHg, three ejector stages are usually required. The first stage condenses the steam and compresses the noncondensable gases, while the second and third stages remove the noncondensable gases from the condensers. The vacuum produced is limited to the vapor pressure of the water used in the condensers. If colder water is supplied to the condensers, a lower absolute pressure can be obtained in the vacuum tower.

Although more costly than barometric condensers, a recent trend is to the use of surface condensers in order to reduce the contamination of water with oil.

Installed costs of crude oil vacuum distillation units are shown in Figure 4.10, and utility requirements are given by Table 4.5.

4.4 AUXILIARY EQUIPMENT

In many cases, a flash drum is installed between the feed preheat heat exchangers and the atmospheric pipe-still furnace. The lower boiling fractions which are vaporized by heat supplied in the preheat exchangers are separated in the flash drum and flow directly to the flash zone of the fractionator. The liquid is pumped through the furnace to the tower flash zone. This results in a smaller and lower-cost furnace and lower furnace outlet temperatures for the same quantity of overhead streams produced.

A stabilizer is incorporated in the crude distillation section of some refineries instead of being placed with the refinery gas plant. The liquid condensed from the overhead vapor stream of the atmospheric pipe-still contains propane and butanes which make the vapor pressure much higher than is acceptable for gasoline blending. To remove these, the condensed liquid in excess of reflux requirements is charged to a stabilizing tower where the vapor pressure is adjusted by removing the propane and butanes from the LSR gasoline stream. Later, in the product-blending section of the refinery, n-butane is added to the gasoline stream to provide the desired Reid vapor pressure.

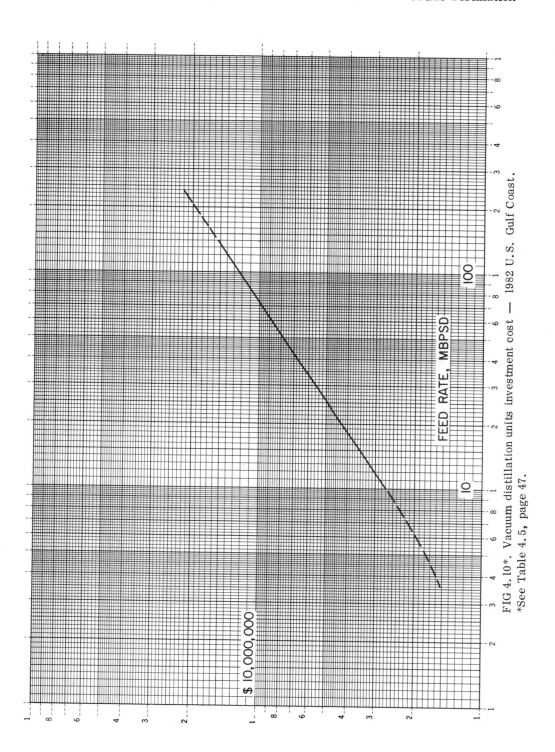

FIG 4.10*. Vacuum distillation units investment cost — 1982 U.S. Gulf Coast.
*See Table 4.5, page 47.

TABLE 4.5*

Vacuum Distillation

Costs Included:
1. All facilities required for producing a clean vacuum gas oil (single cut).
2. Three-stage jet system for operation of flash zone at 30 to 40 torr.
3. Coolers and exchangers to reduce VGO to ambient temperature.

Costs Not Included:
1. Cooling water, steam, power supply.
2. Bottoms cooling below 400°F.
3. Feed preheat up to 670°F, ±10 (assumes feed is direct from atm crude unit).
4. Sour water treating and disposal.
5. Feed and product storage.
6. Multiple "cuts" or lube oil production.

Utility Data (per bbl feed):

Steam (300 psig), lb	9.0
Power, kWh	0.2
Cooling water circulation, gal[a]	60
Fuel, MMbtu[b]	0.04

*See Fig. 4.10, page 46.
[a]30°F rise
[b]LHV basis, heater efficiency taken into account.

4.5 CRUDE DISTILLATION UNIT PRODUCTS

In the order of increasing boiling points, the main products from a typical crude distillation unit are:

Fuel gas. The fuel gas consists mainly of methane and ethane. In some refineries, propane in excess of LPG requirements is also included in the fuel gas stream. This stream is also referred to as "dry gas."

Wet gas. The wet gas stream contains propane and butanes as well as methane and ethane. The propane and butanes are separated to be used for LPG and, in the case of butanes, for gasoline blending and alkylation unit feed.

LSR gasoline. The stabilized LSR gasoline stream is desulfurized and used in gasoline blending or processed in an isomerization unit to improve octane before blending into gasoline.

Naphtha or HSR gasoline. The naphtha cuts are generally used as catalytic reformer feed to produce high-octane reformate for gasoline blending and aromatics.

Gas oils. The light, atmospheric, and vacuum gas oils are processed in a hydrocracker or catalytic cracker to produce gasoline, jet, and diesel fuels. The heavier vacuum gas oils can also be used as feedstocks for lubricating oil processing units.

Residuum. The vacuum still bottoms can be processed in a visbreaker, coker, or deasphalting unit to produce heavy fuel oil or cracking and/or lube base stocks. For asphalt crudes, the residuum can be processed further to produce road and/or roofing asphalts.

4.6 CASE-STUDY PROBLEM: CRUDE UNITS

To illustrate the operation of a fuels refinery and the procedures for making a preliminary economic evaluation, a material balance will be made on each processing unit in a refinery and operating and construction costs will be estimated. The material balance for each unit is located at the end of the chapter that discusses that particular unit, and the evaluation of operating and construction costs is located at the back of Chapter 15 (Cost Estimating). An overall economic evaluation is given at the end of Chapter 16 (Economic Evaluation).

In order to determine the economics of processing a given crude oil or of constructing a complete refinery or individual processing units, it is necessary to make case studies for each practical processing plan and to select the one giving the best economic return. This is a very time-consuming process and can be very costly. The use of computer programs to optimize processing schemes is wide-spread throughout the refining industry. For the purpose of illustrating calculation methods, the process flow will not be optimized in this problem, but all units will be utilized (i.e., delayed coker, fluid catalytic cracker, and hydrocracker) even though certain ones cannot be economically justified under the assumed conditions.

In addition to the basic data contained in this book, it is necessary to use industry publications such as the Oil and Gas Journal, Chemical Week, Oil Daily, and Chemical Marketing (formerly Oil, Paint, and Drug Reporter) for current prices of raw materials and products. Chemical Week, for example, has an annual issue on plant sites which lists utility costs, wages, taxes, etc., by location within the United States.

In many processing units, there is a change in volume between the feed and the products, so it is not possible to make a volume balance around the unit. For this reason, it is essential that a weight balance be made. Even though in practice, the weights of individual streams are valid to only three significant figures, for purposes of calculation it is necessary to carry them out to at least the closest 100 pounds. For this example, they will be carried to the nearest 10 pounds.

In order to make the balance close on each unit, it is necessary to determine one product stream by difference. Generally, the stream having the least effect is determined by difference and, in most cases, this works out to be the heaviest product.

Conversion factors (e.g. lb/hr. per BPD) and properties of pure compounds are tabulated in Appendix B. Properties of products and intermediate streams are given in the chapters on products and the individual process units.

4.7 STATEMENT OF THE PROBLEM

Develop preliminary estimates of product yields, capital investment, operating costs, and economics of building a grassroots refinery on the West Coast to process 100,000 BPCD of North Slope, Alaska crude oil.

The major income products will be motor gasoline, jet fuel, and diesel fuel and the refinery will be operated to maximize gasoline yields within economic limits. The gasoline split will be 50/50 94 RON regular and 100 RON premium. An analysis of the crude oil is given in Table 4.6. It is estimated

TABLE 4.6

Crude Petroleum Analysis [7]

North Slope, Alaska, Crude Oil
Gravity, °API, 25.7   Gravity, specific, 0.900 (60/60° F)
Sulfur, wt %, 1.12   Pour point, °F, 20
Viscosity, SUS at 70° F, 182.5 sec; at 100° F, 94.1 sec
Conradson carbon: Crude, 5.99%: $1,000^+$, 19%

TBP Distillation

TBP cut (° F)	Vol % on crude Frac	Vol % on crude Sum	Sp gr (60/60° F)	°API (60° F)	Sulfur (wt %)	Nitrogen (ppm)	Metals (ppm) Ni	Metals (ppm) V
C_2	0.1	0.1	0.374	246.8				
C_3	0.4	0.5	0.509	146.5				
i-C_4	0.2	0.7	0.564	119.4				
n-C_4	0.7	1.4	0.584	110.8				
i-C_5	0.5	1.9	0.625	94.9				
n-C_5	0.7	2.6	0.631	92.7				
97-178	1.5	4.1	0.697	71.6				
178-214	2.1	6.2	0.740	59.7				
214-242	2.0	8.2	0.759	55.0				
242-270	2.0	10.2	0.764	53.8				
270-296	2.0	12.2	0.781	49.6				
296-313	1.0	13.2	0.781	49.6				
313-342	2.0	15.2	0.791	47.3				
342-366	1.9	17.1	0.797	46.0				
366-395	2.0	19.1	0.806	44.0				
395-415	2.0	21.1	0.831	38.8				
415-438	2.0	23.1	0.832	38.6				
438-461	2.0	25.1	0.839	37.2				
461-479	2.0	27.1	0.848	35.4				
479-501	2.0	29.1	0.856	33.9				
501-518	2.0	31.1	0.860	33.1				
518-538	2.0	33.1	0.864	32.2				
538-557	2.0	35.1	0.867	31.8				
557-578	2.0	37.1	0.868	31.6				
578-594	2.1	39.2	0.872	30.7				
594-610	2.0	41.2	0.878	29.6				
610-632	2.0	43.2	0.887	28.0				
632-650	1.8	45.0	0.893	26.9				
650-1,000	31.8	76.8	0.933	20.3	1.0	1,390		
$1,000^+$	23.2	100.0	0.995	10.7	2.4	4,920	56	97

that this crude oil will be available FOB the refinery at $30.00/bbl. Product
prices will be based on the average posted prices for 1982 less the following:

1. 0.5¢/gal on all liquid products except fuel oil
2. 5¢/bbl on fuel oil
3. 50¢/ton on coke

Utility prices will be those reported in the 1982 Plant Sites issue of Chemical
Week (October 1982) for the area in which the refinery will be built. Federal
and state income taxes will be 55% (48% U.S. and 7% Calif.) and land cost will
be 2% of the cost of the process units, storage, steam systems, cooling water
systems, and offsites.

4.8 GENERAL PROCEDURE

1. From the crude distillation data given in Table 4.6, plot TBP and
    gravity curves. These are shown in Figure 4.11.

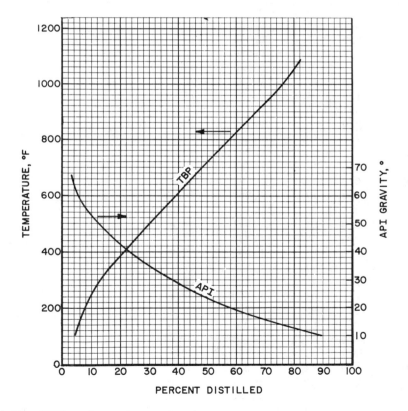

FIG. 4.11. TBP and gravity curves. Crude: North Slope, Alaska; gravity:
25.7°API; sulfur: 1.12 wt %.

2. From Table 4.2, select TBP cut-points of products to be made from atmospheric and vacuum pipe stills.
3. From TBP and gravity curves, determine percentages and gravities of fractions.
4. Using Tables B.1 and B.2, convert volumes to weights.
5. Determine weight of $1,000^+$ bottoms stream by difference. If volume of $1,000^+$ stream is taken from TBP curve, then gravity is calculated from weight and volume. In this case, since the gravity of $1,000^+$ stream was determined in the laboratory and the gravity is very sensitive to small changes in weight, it was decided to use the laboratory gravity and calculate the volume. This gives a total volume recovery of 100.6%. This is reasonable because there is a negative volume change on mixing of petroleum fractions, and it is possible on some crudes to obtain liquid products (more properly liquid-equivalent products because C_4 and lighter products are not liquids at 60°F and 1 atm pressure) having a total volume of up to 103% on crude.

TABLE 4.7

Crude Units Material Balance

Basis: 100,000 BPCD   North Slope, Alaska, Crude Oil

Component	Vol %	BPCD	°API	(lb/hr)/BPD	lb/hr	wt % S	lb/hr S
Feed:							
N.S. Crude	100.0	100,000	25.7	13.13	1,313,000	1.12	14,700
Products:							
C_2	0.1	100	246.8	5.18	550		
C_3	0.4	400	146.5	7.42	2,970		
$i-C_4$	0.2	200	119.4	8.22	1,640		
$n-C_4$	0.7	700	110.8	8.51	5,960		
C_5-190°	5.6	5,600	63.0	10.61	59,420		
190-380°	12.5	12,500	48.5	11.47	143,380	0.1	140
380-520°	13.0	13,000	38.0	12.18	158,340	0.3	480
520-650°	12.0	12,000	29.8	12.80	153,600	0.5	770
650-800°	13.5	13,500	23.0	13.36	180,360	0.8	1,440
800-1,000°	18.8	18,800	16.8	13.92	261,700	1.5	3,930
$1,000^+$	23.8	*23,760	10.7	14.52	*345,080	2.3	*7,940
Total	100.6	100,560			1,313,000		14,700

Utility Requirements (per day):

	Atm PS	Vac PS	Total
Steam (300 psig), Mlb	600	505	1,105
Power, MkWh	50	11	61
Cooling water circulation, Mgal	600	3,364	3,964
Fuel, MMBtu	7,000	2,200	9,200

Note: Vacuum still feed = $56,060$ BPCD (650-$1,000^+$ material)

*Obtained by difference.

The material balance and utility requirements for the crude units are given in Table 4.7.

6. For crude oils containing significant amounts of sulfur or nitrogen, it is also necessary to make sulfur and/or nitrogen balances around each unit. The North Slope crude has a sufficiently high quality of sulfur to require a sulfur balance, but since there is no nitrogen shown in the analysis, a nitrogen balance is not made. It is assumed that the North Slope crude is similar to those represented in Figure 4.3.

PROBLEMS

1. Using one of the crude oils in Appendix D, make TBP and gravity mid-percent curves. From these curves and the crude oil fraction specifications in Tables 4.1 or 4.2, make a complete material balance around an atmospheric crude still. Assume a 10,000 BPCD crude oil feed rate to the atmospheric crude still.

2. For the crude oil used in problem 1, make sulfur and nitrogen weight balances for the feed and products to the nearest pound.

3. Using the high-boiling fraction from problem 1 as feed to a vacuum pipe still, make an overall weight balance and nitrogen and sulfur balances. Assume the reduced crude bottoms stream from the vacuum pipe still has an initial TBP boiling point of 1,000°F.

4. Estimate the calendar day utility and chemical requirements for the crude oil desalter, atmospheric pipe still, and vacuum pipe still of problems 1 and 3.

5. Calculate the Watson characterization factors for the crude oil and product streams from problems 1 and 3.

6. Estimate the carbon content of the reduced crude bottoms product of problem 3. Express as weight percent Conradson carbon.

7. Using information in Table 4.2, tell where each product stream from the crude stills will be utilized (unit feed or product).

8a. Calculate the Btu required per hour to heat 100,000 BPSD of the assigned crude oil from 15.6°C (60°F) to 430°C (806°F) in the atmospheric distillation unit furnace.

8b. Assuming a furnace efficiency of 73%, how many barrels of 13.0°API residual fuel oil will be burned per barrel of crude oil charged?

9. Calculate the atmospheric distillation unit furnace exit temperature needed to vaporize 30% of the assigned crude oil in the tower flash zone.

10a. Determine the API gravity, characterization factor and temperature of the topped crude leaving the bottom of the tower in example 4.9.

10b. If the topped crude was used as the hot fluid in a train of feed preheater heat exchangers are exited at 38°C (100°F), what was the temperature to which the crude oil feed to the furnace was heated? (Assume crude inlet temperature is 60°F.)

11. If the crude oil leaving the preheater heat exchanger train in Prob. 4.10 was partially vaporized in a flash drum and the liquid from the flash drum heated to 400°C (752°F) in the furnace, calculate the percent on original crude oil feed of the topped crude exiting from the botton of the atmospheric crude tower.

12. If the assigned crude oil has an atmospheric crude unit furnace outlet temperature of 400°C (752°F) and a flash zone pressure of 1 atmosphere (101 kPa) calculate the flash zone temperature and the volume percent vaporized.

13. The crude oil in Prob. 4.11 can be heated to 275°C (527°F) in a preheater train and flashed in a flash drum before the liquid fraction is heated to 400°C (752°F) in the crude unit furnace. If the fractionating tower flash zone pressure is 1 atmosphere (101 kPa), what is the total volume percent of the crude oil vaporized in the flashdrum and the fractionating tower?

NOTES

1. J. B. Maxwell, *Data Book on Hydrocarbons* (D. Van Nostrand, New York, 1950), p. 14.
2. W. L. Nelson, G. T. Fombona, and L. J. Cordero, Proc. Fourth World Pet. Congr., Sec. V/A, pp. 13-23 (1955).
3. A. A. Kutler, Petro/Chem. Eng. *41*(5), 9-11 (1969). Also, David A. Cheshire, personal communication, November, 1982.
4. A. A. Kutler, Heat Eng. *44*(8), 135-141 (1970).
5. W. L. Nelson, *Petroleum Refinery Engineering*, 4th Ed. (McGraw-Hill Book Company, New York, 1958), p. 232.
6. Ibid, p. 228.
7. Oil Gas J. *67*(40), 91 (1969); *69*(20), 124-131 (1971).

Chapter 5
COKING AND THERMAL PROCESSES

Delayed coking is described in Sections 5.1 to 5.6. This is the most widely used coking process. Fluid coking and Flexicoking are described in Sections 5.7 to 5.10. These fluid bed processes have been under development by Exxon over the past 20 years and are now commercially operated in several refineries around the world [1, 2, 3].

The delayed coking process was devloped to minimize refinery yields of residual fuel oil by severe thermal cracking of stocks such as vacuum residuals and thermal tars. In early refineries severe thermal cracking of such stocks resulted in unwanted deposition of coke in the heaters. By gradual evolution of the art it was found that heaters could be designed to raise residual stock temperatures above the coking point without significant coke formation in the heaters. This required high velocities (minimum retention time) in the heaters. Providing an insulated surge drum on the heater effluent allowed sufficient time for the coking to take place before subsequent processing—hence the term "delayed coking."

From chemical reaction viewpoint coking can be considered as a severe thermal cracking process in which one of the end products is carbon (i.e., coke). Actually the coke formed contains some volatile matter or high boiling hydrocarbons. To eliminate essentially all volatile matter from petroleum coke it must be calcined at approximately 2,000°F to 2,300°F. Minor amounts of hydrogen remain in the coke even after calcining which gives rise to the theory held by some authors that the coke is actually a polymer.

During the 1940 to 1960 period, delayed coking was used primarily to pretreat vacuum residuals to prepare coker gas oil streams suitable for feed to a catalytic cracker. This reduced coke formation on the cracker catalyst and thereby allowed increased cracker throughputs. This also reduced the net refinery yield of low-priced residual fuel. Added benefit was obtained by reducing the metals content of the catalytic cracker feed stocks.

In recent years coking has also been used to prepare hydrocracker feedstocks, and to produce a high quality "needle coke" from stocks such as catalytic cracker slurry oil [4, 5]. Coal tar pitch is also processed in delayed coking units [6].

5.1 PROCESS DESCRIPTION—DELAYED COKING

This discussion relates to conventional delayed coking as shown in the following flow diagram (Fig. 5.1). See also Photo 3, Appendix F.

Hot fresh liquid feed is charged to the fractionator two or four trays above the bottom vapor zone. This accomplishes the following:

1. The hot vapors from the coke drum are quenched by the cooler feed liquid thus preventing any significant amount of coke formation in the fractionator and simultaneously condensing a portion of the heavy ends which are recycled.
2. Any remaining material lighter than the desired coke drum feed is stripped (vaporized) from the fresh feed liquid.
3. The fresh feed liquid is further preheated.

The remaining fresh feed liquid combined with the condensed recycle is pumped from the bottom of the fractionator through the coker heater where it is partially vaporized and then into one of the two coke drums. Steam is usually introduced in the heater tubes to control velocities and thus minimize coke deposition. The unvaporized portion of the heater effluent settles out in the coke drum where the combined effect of retention time and temperature causes the formation of coke.

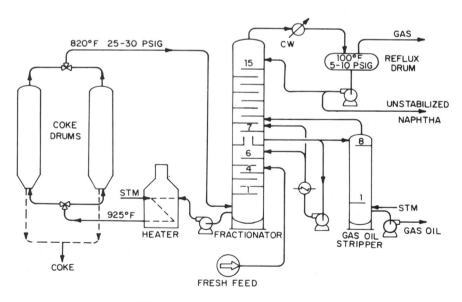

FIG. 5.1. Delayed coking unit.

Vapors from the top of the coke drum return to the base of the fractionator. These vapors consist of steam and the products of the thermal
cracking reaction: gas, naphtha, and gas oils. The vapors flow up through
the quench trays previously described. Above the fresh feed entry in the
fractionator there are usually two or three additional trays below the gas oil
drawoff tray. These trays are refluxed with partially cooled gas oil in order
to provide fine trim control of the gas oil end point and to minimize entrainment of any fresh feed liquid or recycle liquid into the gas oil product.

The gas oil side draw is a conventional configuration employing a six-
to eight-tray stripper with steam introduced under the bottom tray for
vaporization of light ends to control the initial boiling point (IBP) of the
gas oil.

Steam and vaporized light ends are returned from the top of the gas
oil stripper to the fractionator one or two trays above the draw tray. A
pump-around reflux system is provided at the draw tray to recover heat at
a high temperature level and minimize the low-temperature-level heat removed by the overhead condenser. This low-temperature-level heat cannot
normally be recovered by heat exchange and is rejected to the atmosphere
through a water cooling tower or aerial coolers.

Eight to ten trays are generally used between the gas-oil draw and
the naphtha draw or column top. If a naphtha side draw is employed, additional trays are required above the naphtha draw tray.

Major design criteria for coking units are described in the literature
[7, 8].

5.2 COKE REMOVAL—DELAYED COKING

When the coke drum in service is filled to a safe margin from the top, the
heater effluent is switched to the empty coke drum and the full drum is isolated, steamed to remove hydrocarbon vapors, cooled by filling with water,
opened, drained, and the coke removed.

The decoking operation is accomplished in some plants by a mechanical
drill or reamer [9], however most plants use a hydraulic system. The hydraulic system is simply a number of high pressure (2,000 to 4,500 psig)
water jets which are lowered into the coke bed on a rotating drill stem. A
small diameter hole, called a "rat hole" is first cut all the way through the
bed from top to bottom using a special jet. This is done to allow movement
of coke and water through the bed and the main drill stem.

The main bulk of coke is then cut from the drum, usually beginning at
the bottom. Some operators prefer to begin at the top of drum to avoid
the chance of dropping large pieces of coke which can cause problems in
subsequent coke handling facilities.

The coke which falls from the drum is often collected directly in railroad cars. Alternately, it is sluiced or pumped as a water slurry to a
stock-pile.

5.3 PROPERTIES AND USES OF PETROLEUM COKE

Most petroleum coke is produced as hard, porous, irregular-shaped lumps
ranging in size from 20 inches down to fine dust. This type of coke is called
sponge coke because of its appearance.

TABLE 5.1

Typical Coke Specifications

Type use	Sponge anodes	Needle electrodes
Calcined coke		
Moisture, wt %	<0.5	<0.5
Volatile matter, wt %	0.5	—
Ash, wt %	<0.5	0.5
Sulfur, wt %	<3.0	<1.5
Metals, ppm		
V	<350	—
Ni	<300	—
Si	<150	—
Fe	<270	—
Density, g/cc		
−200 Mesh RD	2.04-2.08	>2.12
VBD	>0.80	—
CTE, 1/°C X 10^{-7}	<40	<4.0

	wt % (as produced)	wt % (after calcining)
Water	2-4	Nd
Volatile matter	7-10	2-3
Fixed carbon	85-91	95+
Ash	0.5-1.0	1-2

The main uses of petroleum sponge coke are as follows:

1. Manufacture of electrodes for use in electric furnace production of elemental phosphorus, titanium dioxide, calcium carbide and silicon carbide.
2. Manufacture of anodes for electrolytic cell reduction of alumina.
3. Direct use as chemical carbon source for manufacture of elemental phosphorus, calcium carbide, and silicon carbide.
4. Manufacture of graphite.
5. Fuel.

It is important to note that petroleum coke does not have sufficient strength to be used in blast furnaces for the production of pig iron nor is it generally acceptable for use as foundry coke. Coal-derived coke is used for these purposes. Typical analyses of petroleum cokes and specifications for anode and electrode grades are summarized in Table 5.1.

The sulfur content of petroleum coke varies with the sulfur content of the coker feedstock. It is usually in the range of 0.3 to 1.5 wt %. It can sometimes, however, be as high as 6%. The sulfur content is not significantly reduced by calcining.

A second form of petroleum coke being produced in increasing quantities is needle coke. Needle coke derives its name from its microscopic elongated crystalline structure. Needle coke is produced from highly aromatic feedstocks when a coking unit is operated at high pressures (100 psig) and high recycle ratios (1:1). Needle coke is preferred over sponge coke for use in electrode manufacture because of its lower electrical resistivity and lower coefficient of thermal expansion.

Occasionally a third type of coke is produced unintentionally. This coke is called shot coke because of the clusters of shot-sized pellets which characterize it. Its production usually occurs during operational upsets, probably as a result of low coke drum temperatures or pressures or low feed API gravities. Strangely, its production apparently occurs only when processing California residuals.

5.4 OPERATION—DELAYED COKING

As indicated in the paragraph describing coke removal, the coke drums are filled and emptied on a time cycle. The fractionation facilities are operated continuously. Usually just two coke drums are provided but units having four drums are not uncommon. The following time schedule is typical

Operation	Hours
Fill drum with coke	24
Switch and steam out	3
Cool	3
Drain	2
Unhead and decoke	5
Head up and test	2
Heat up	7
Spare time	2
Total:	48

The main independent operating variables in delayed coking (Table 5.2) are the heater outlet temperature, the fractionator pressure, the temperature of the vapors rising to the gas oil drawoff tray, and the "free" carbon content of the feed as determined by the Conradson carbon test. As would be expected, high heater outlet temperatures increase the cracking and coking reactions, thus increasing yields of gas, naphtha, and coke and decreasing the yield of gas oil. An increase in fractionator pressure has the same effect as an increase in the heater outlet temperature. This is due to the fact that more recycle is condensed in the fractionator and returned to the heater and coke drums. The temperature of the vapors rising to the gas oil drawoff tray is controlled to produce the desired gas oil end point. If this temperature is increased more heavies will be drawn off in the gas oil leaving less material to

TABLE 5.2

Relation of Operating Variables in Delayed Coking

Dependent variable	Independent Variables							
	Heater outlet temp		Fractionator pressure		Hat temp[a]		Feed carbon content[b]	
	+	−	+	−	+	−	+	−
Gas yield	+	−	+	−	−	+	+	−
Naphtha yield	+	−	+	−	−	+	+	−
Coke yield	+	−	+	−	−	+	+	−
Gas oil yield	−	+	−	+	+	−	−	+
Gas oil end point	c	c	−	+	+	−	c	c
Gas oil metals content	c	c	−	+	+	−	c	c
Coke metals content	c	c	+	−	−	+	c	c
Recycle quantity	c	c	+	−	−	+	c	c

[a]Hat temperature is the temperature of the vapors rising to the gas oil drawoff tray in the fractionator.

[b]Carbon content is that determined by Conradson carbon test procedure.

[c]For these items the heater outlet temperature and the carbon content, per se, do not have a significant independent effect.

be recycled. Thus, the yield of gas oil increases and the yields of gas, naphtha, and coke decrease. An increase in the Conradson carbon content of the feed results in increased yields of coke, gas, and naphtha and a decreased yield of gas oil. The high-boiling material in the coke-drum vapors condensed in the base of the fractionator is termed recycle. It is frequently stated that an increase in recycle increases the cracking reaction, thereby resulting in the production of more gas, naphtha, and coke and less gas oil. This is a true statement; however, it is somewhat misleading because the recycle quantity is not an independent variable. For a given feed the recycle quantity is determined by the fractionator pressure and the temperature of the vapors rising to the gas oil drawoff tray. The relationship of the independent operating variables with product yields and properties is shown in Table 5.2. In this tabulation, a (+) indicates an increase in the variable and a (−) indicates a decrease.

5.5 YIELDS FOR DELAYED COKING

Yields for conventional delayed coking operations may be calculated from the following equations. Three sets of equations are given:

1. A generalized set in which the coke and gas yields are calculated on the basis of the actual Conradson carbon content of the coker feed (Table 5.3).
2. Yields for coking a typical East Texas crude residual (Table 5.4).
3. Yields for coking Wilmington, California crude residuals (Table 5.5).

TABLE 5.3

Coke Yields when Conradson Carbon is Known

Coke wt %	= 1.6 ×(wt % Conradson carbon[a])
Gas (C_4-) wt %	= 7.8 + 0.144(wt % Conradson carbon[a])
Gaso. wt %	= 11.29 + 0.343(wt % Conradson carbon[a])
Gas oil wt %	= 100 − wt % coke − wt % gas − wt % gaso.
Gaso. vol %	= $(\dfrac{186.5}{131.5 + °API})$ (gaso. wt %)[b]
Gas oil vol %	= $(\dfrac{155.5}{131.5 + °API})$ (gas oil wt %)[b]

Note: These yield correlations are based on the following conditions:
1. Coke drum pressure 35 to 45 psig.
2. Feed is "straight-run" residual.
3. Gas oil end point 875 to 925° F.
4. Gasoline end point 400° F.

[a]Use actual Conradson carbon when available.
[b]All °API are those for net fresh feed to coker.

These yield data have been developed from correlations of actual plant operating data and pilot plant data. Values calculated from these equations are sufficiently accurate for primary economic evaluation studies; however, for the actual design of a specific coking unit the yields should be determined by the pilot plant operation. In all cases, the weight and volume percents given are based on the net fresh feed to the coking unit and are limited to feedstocks having gravities of less than 18° API. The yields shown will vary significantly if the coker feed is derived from material other than straight-run crude residuals. The numerical values in the equations do not represent a high degree of accuracy but are included for the purpose of establishing a complete weight balance.

TABLE 5.4

Coke Yields, East Texas Crude Residuals

Coke wt %	= 45.76 − 1.78 × °API
Gas (C_4-) wt %	= 11.92 − 0.16 × °API
Gaso. wt %	= 20.5 − 0.36 × °API
Gas oil wt %	= 21.82 + 2.30 × °API
Gaso. vol %	= $(\dfrac{186.5}{131.5 + °API})$ (gaso. wt %)
Gas oil vol %	= $(\dfrac{155.5}{131.5 + °API})$ (gas oil wt %)

TABLE 5.5

Coke Yields, Wilmington Crude Residuals

Coke wt%	$= 39.68 - 1.60 \times °API$
Gas (C_4-) wt %	$= 11.27 - 0.14 \times °API$
Gaso. wt %	$= 20.5 - 0.36 \times °API$
Gas oil wt %	$= 28.55 + 2.10 \times °API$
Gaso. vol %	$= (\dfrac{186.5}{131.5 + °API})$ (gaso. wt %)
Gas oil vol %	$= (\dfrac{155.5}{131.5 + °API})$ (gas oil wt %)

TABLE 5.6

Typical Gas Composition for Delayed Coking
(Sulfur-Free Basis)

Component	Mol %
Methane	51.4
Ethene	1.5
Ethane	15.9
Propene	3.1
Propane	8.2
Butene	2.4
i-Butane	1.0
n-Butane	2.6
H_2	13.7
CO_2	0.2
	100.0

Note: MW = 22.12

TABLE 5.7

Sulfur and Nitrogen Distribution for Delayed Coking
(Basis: sulfur and nitrogen in feed to coker)

	Sulfur (%)	Nitrogen (%)
Gas	30	—
Naphtha	5	1
Kerosine	} 35	2 } 24
Gas oil		22
Coke	30	75
	100	100

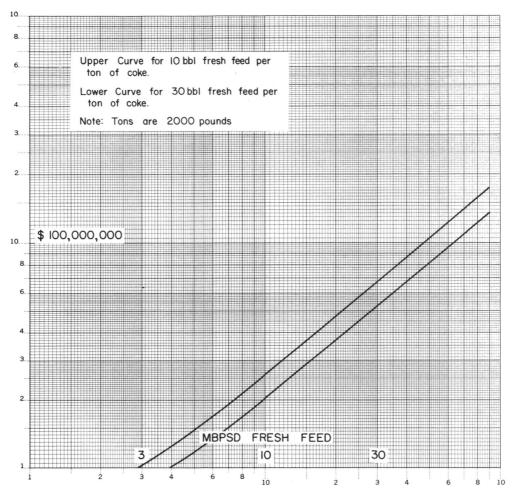

FIG. 5.2*. Delayed coking units investment cost — U.S. Gulf Coast.
*See Table 5.8, page 63.

Typical gas compositions and sulfur and nitrogen distributions in products produced by delayed coking of reduced crudes are given in Tables 5.6
and 5.7. The 1982 installed costs for delayed cokers in the Gulf Coast section
of the United States are given by Figure 5.2. Table 5.8 gives utility requirements for delayed coker operation.

5.6 CASE-STUDY PROBLEM: DELAYED COKER

See Section 4.6 for statement of problem and Table 4.7 for feed to delayed
coker.
The delayed coker material balance is calculated from the equations
given in Table 5.3. The results are tabulated in Table 5.9.

TABLE 5.8*

Delayed Coking Units

Costs Included:
1. Coker fractionator to produce naphtha, light gas oil and heavy gas oil.
2. Hydraulic decoking equipment.
3. Coke dewatering, crushing to -2" and separation of -1/4" from +1/4".
4. Three days covered storage for coke.
5. Coke drums designed for 50 to 60 psig.
6. Blowdown condensation and purification of waste water.
7. Sufficient heat exchange to cool products to ambient temperatures.

Costs Not Included:
1. Light ends recovery facilities.
2. Light ends sulfur removal.
3. Product sweetening.
4. Cooling water, steam, and power supply.
5. Off gas compression.

Utility Data:

Steam, lb/ton coke	500
Power, kWh/ton coke[a]	40
Cooling water, gal/bbl feed (30° FΔT)	100
Fuel, MMBtu/bbl feed[b]	0.16

*See Fig. 5.2, page 62.
[a]Includes electric motor drive for hydraulic decoking pump.
[b]Based on 600°F fresh feed. LHV basis, heater efficiency taken into account.

Calculations of yields:

Coke $= 1.6 (19) = 30.4$ wt %

Gas $(C_4^-) = 7.8 + 0.144 (19) = 10.5$ wt %

Gasoline $= 11.29 + 0.343 (19) = 17.8$ wt %

Gasoline $= \dfrac{186.5}{131.5 + 10.7} (17.8) = 23.3$ vol %

Gas oil $= 100.0 - (30.4 + 10.5 + 17.8) = 41.3$ wt %

$= \dfrac{155.5}{142.2} (41.3) = 45.2$ vol %

Sulfur distribution is obtained from Table 5.7 and gas (C_4^-) composition from table 5.6.

Coker Gas Balance:

Total gas	=	36,230 lb/hr
Sulfur in gas	=	2,380 lb/hr
Sulfur-free gas		33,850 lb/hr

TABLE 5.9

Delayed Coker Material Balance

Basis: 100,000 BPCD   North Slope, Alaska, Crude Oil
Conradson Carbon, $1,000^+RC$ = 19% (from Table 4.6)

Component	vol %	BPD	°API	(lb/hr)/BPD	lb/hr	wt % S	lb/hr S
Feed:							
$1,000^+RC$	100.0	23,760	10.7	14.52	345,080	2.3	7,940
Products:							
Gas (C_4-), wt %	(10.5)				36,230		2,380
Gaso.	23.3	5,540	54.6	11.09	61,420	.65	400
Gas oil	45.2	10,740	24.0	13.27	142,530	1.95	2,780
Coke, wt %	(30.4)				104,900	2.27	2,380
		16,280			345,080		7,940

Using composition of sulfur-free gas from Table 5.6, total lb mol/hr is
$33,850 \div 22.12 = 1,530.3$ and sulfur-free composition is calculated as follows:

Component	mol %	mol/hr
C_1	51.4	786.5
$C_2^=$	1.5	23.0
C_2	15.9	243.3
$C_3^=$	3.1	47.4
C_3	8.2	125.5
$C_4^=$	2.4	36.7
i-C_4	1.0	15.3
n-C_4	2.6	39.8
H_2	13.7	209.7
CO_2	0.2	3.1
	100.0	1,530.3

It is now necessary to adjust this gas composition to allow for the sulfur content. In actual operations some of the sulfur will be combined as mercaptan molecules (R-S-H) but for preliminary calculations it is sufficiently accurate to assume that all the sulfur in the gas fraction is combined as H_2S. Since there are 2,380 lb/hr of sulfur (equivalent to 74.4 mol/hr) the free hydrogen must be reduced by 74.4 mol/hr so the final coker gas balance is as follows:

Component	mol/hr	MW	lb/hr*	(lb/hr)/BPD	BPD
C_1	786.5	16	12,580		
$C_2^=$	23.0	28	640		
C_2	243.3	30	7,300		
$C_3^=$	47.4	42	1,990	7.61	261
C_3	125.5	44	5,520	7.42	744
$C_4^=$	36.7	56	2,060	8.76	235
$i\text{-}C_4$	15.3	58	890	8.22	108
$n\text{-}C_4$	39.8	58	2,310	8.51	271
H_2	135.3	2	270		
CO_2	3.1	44	140		
H_2S	74.4	34	2,530		
	1,530.3		36,230		

*Rounded to nearest 10

Coker utility requirements are given in Table 5.10.

TABLE 5.10

Coker Utility Requirements
(Basis: 52 tons coke per hour)

Steam, lb/hr	26,000
Power, MkWh/day	50
Cooling water, gpm	1,650
Fuel, MMBtu/day	3,802

5.7 PROCESS DESCRIPTION—FLEXICOKING

The Flexicoking process is shown on Figure 5.3 [3].

Feed can be any heavy oil such as vacuum resid, coal tar, shale oil, or tar sand bitumen. The feed is preheated to about 600 to 700°F and sprayed into the reactor where it contacts a hot fluidized bed of coke. This hot coke is recycled to the reactor from the coke heater at a rate which is sufficient to maintain the reactor fluid bed temperature between 950 and 1000°F. The coke recycle from the coke heater thus provides sensible heat and heat of vaporization for the feed and the endothermic heat for the cracking reactions.

The cracked vapor products pass through cyclones in the top of the reactor to separate most of the entrained coke particles and are then quenched in the scrubber vessel located on top of the reactor. Some of the high boiling (+925°F) cracked vapors are condensed in the scrubber and recycled to the reactor. The balance of the cracked vapors flow to the coker fractionator where the various cuts are separated. Wash oil circulated over baffles in the scrubber provides the quench cooling and also serves to further reduce the amount of entrained coke particles.

The coke produced by cracking is deposited as a thin film on the existing coke particles in the reactor fluidized bed. The coke is stripped with steam in a baffled section at the bottom of the reactor to prevent reaction products, other than coke, from being entrained with the coke leaving the reactor. Coke flows from the reactor to the heater where it is reheated to about 1100°F. The coke heater is also a fluidized bed and its primary function is to transfer heat from the gasifier to the reactor.

Coke flows from the coke heater to the gasifier where it is gasified with air and steam in a third fluidized bed. The coke gas product consists of N_2, H_2O, H_2, CO, CO_2, and some H_2S, COS, and NH_3. This gas flows from the top of the gasifier to the bottom of the heater where it serves to fluidize the heater bed and provide a part of the heat needed in the reactor. The balance of the reactor heat requirement is supplied by recirculating hot coke from the gasifier to the heater.

The system can be designed and operated to gasify about 60 to 97% of the coke produced in the reactor. The overall coke inventory in the system is maintained by withdrawing a stream of purge coke from the heater.

The coke gas leaving the heater is cooled in a waste heat steam generator before passing through external cyclones and a venturi type wet scrubber. The coke fines collected in the cyclones and venturi scrubber plus the purge coke from the heater represent the net coke yield and contain essentially all of the metal and ash components from the reactor feed stock.

After removal of entrained coke fines the coke gas is treated for removal of H_2S in a Stretford unit and then used for refinery fuel. The treated coke gas has a lower heating value in the range of 100 to 130 BTU/SCF and therefore modifications of boilers and furnaces may be required for efficient combustion of this gas.

5.8 PROCESS DESCRIPTION—FLUID COKING

Fluid Coking is a simplified modification of Flexicoking. In the Fluid Coking process only enough of the coke produced in the reactor is burned to satisfy the heat requirements of the reactor. Typically, this is about 20 to 25% of the

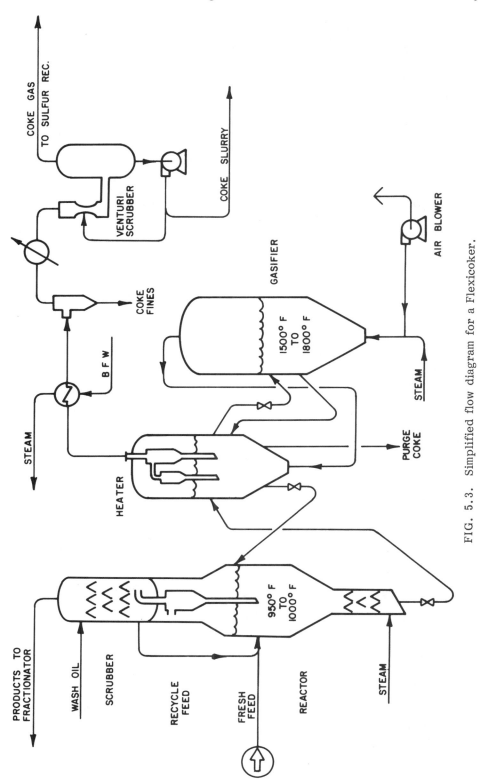

FIG. 5.3. Simplified flow diagram for a Flexicoker.

coke produced in the reactor. The balance of the coke is withdrawn from the burner vessel and is not gasified as it is in a Flexicoker. Therefore only two fluid beds are used in a Fluid Coker—a reactor and a burner.

The primary advantage of the Flexicoker over the more simple Fluid Coker is that most of the heating value of the coke product is made available as low sulfur gas which can be burned without an SO_2 removal system on the resulting stack gas, whereas such a system would be necessary if coke which contains 3 to 8 weight % sulfur is burned directly in a boiler. In addition, the coke gas can be used to displace liquid and gaseous hydrocarbon fuels in the refinery process heaters and does not have to be used exclusively in boilers as is the case with fluid coke.

5.9 YIELDS FROM FLEXICOKING AND FLUID COKING

As with delayed coking, the yields from the fluidized bed coking processes can be accurately predicted only from pilot plant data for specific feeds. Typical yields for many various feeds are available from the Licensor. One set of yields is shown in Table 5.11.

If specific yield data are not available, it can be assumed for preliminary estimates that the products from Flexicoking and Fluid Coking are the same as those from delayed coking except for the amount of reactor coke product which is burned or gasified. Thus the coke yield from a Fluid Coking unit would be about 75 to 80 percent of the coke yield from a delayed coker and the yield of coke from a Flexicoker would be in the range of 2 to 40% of the delayed coker yield.

TABLE 5.11

Comparison of Flexicoking vs. Fluid Coking Yields

	Fluid Coking	Flexicoking
Feed Properties	Arabian Heavy 1050°F+	
Conradson Carbon, wt. %	24.4	24.4
Gravity, °API	4.4	4.4
Sulfur, wt. %	5.34	5.34
Feed Rate, MBPSD	25.0	25.0
Yields, % on Fresh Feed		
C_4^-, wt. %	13.1	13.1
C_5/950°F, LV%	65.1	65.1
Gross Coke, wt. %	30.7	30.7
Net Coke, wt. %	24.9	0.61
Tons/SD*	1130	28
Coke Gas, LV% (FOEB)	—	18.8

*1 Ton = 2000 pounds

The coke which is gasified in a Flexicoker produces coke gas of the following approximate composition after H_2S removal:

Component	Mol %
H_2	15
CO	20
CH_4	2
CO_2	10
N_2	53
	100

The above composition is on a dry basis. The coke gas as actually produced from the H_2S removal step is water saturated which means it contains typically 5 to 6 mol % water vapor.

Assuming that the coke is about 98 weight % carbon on a sulfur free and ash free basis, the calculated amount of coke gas produced having the above composition is 194 MSCF per ton of coke gasified. Combustion of this gas results in production of about 80% of the equivalent coke heating value. Recovery of sensible heat from the coke gas leaving the coke heater increases the recoverable heat to about 85% of the equivalent coke heating value.

5.10 CAPITAL COSTS AND UTILITIES FOR FLEXICOKING AND FLUID COKING

As a rough approximation it can be assumed that the investment for a Fluid Coking unit is about the same as that for a delayed coking unit for a given feedstock and that a Flexicoker costs about 30% more.

The utility requirements for Fluid Coking are significantly higher than those for delayed coking primarily because of the energy required to circulate the solids between fluid beds. The air blower in a Flexicoker requires more power than that for a Fluid Coker. The process Licensor should be consulted to determine reasonably accurate utility requirements.

5.11 VISBREAKING

Visbreaking is a relatively mild thermal cracking operation mainly used to reduce the viscosities and pour points of vacuum tower bottoms to meet No. 6 fuel oil specifications or to reduce the amount of cutting stock required to dilute the resid to meet these specifications. Refinery production of heavy fuel oils can be reduced from 20-35% and cutter stock requirements from 20-30% by visbreaking. Visbreaking is also used to increase cat cracker feed stocks and gasoline yields.

Long paraffinic side chains attached to aromatic rings are the primary cause of high pour points and viscosities for paraffinic base residua. Vis-

breaking is carried out at conditions to optimize the breaking off of these
long side-chains and their subsequent cracking to shorter molecules with
lower viscosities and pour points. The amount of cracking is limited, however, because if the operation is too severe the resulting product becomes
unstable and forms polymerization products during storage which cause filter
plugging and sludge formation. The objective is to reduce the viscosity as
much as possible without significantly affecting the fuel stability. For most
feedstocks, this reduces the severity to the production of less than 10%
gasoline and lighter materials.

The degree of viscosity and pour point reduction is a function of the
composition of the residua feed to the visbreaker. Waxy feed stocks achieve
pour point reductions from 15-35°F and final viscosities from 25-75% of the
feed. High asphaltene content in the feed reduces the conversion ratio at
which a stable fuel can be made [10] which results in smaller changes in the
properties. The properties of the cutter stocks used to blend with the visbreaker tars also have an affect on the severity of the visbreaker operation.
Aromatic cutter stocks, such as catalytic gas oils, have a favorable effect on
fuel stability and permit higher visbreaker conversion levels before reaching
fuel stability limitations [11].

The molecular structures of the compounds in petroleum which have
boiling points above 1000°F are highly complex and historically have been
classified arbitrarily as oils, resins, and asphaltenes according to solubility
in light paraffinic hydrocarbons. The oil fraction is soluble in propane while
the resin fraction is soluble (and the asphaltene fraction insoluble) in either
pentane, hexane, n-heptane, or octane, depending upon the investigator.
Usually either pentane or n-heptane is used. The solvent selected does have
an effect on the amounts and properties of the fractions obtained but normally
little distinction is made in terminology. Chapter 9, *Catalytic Hydrocracking
and Hydroprocessing*, contains a more detailed discussion of the properties
of these fractions.

Many investigators believe the asphaltenes are not in solution in the oil
and resins but are very small, perhaps molecular size, solids held in suspen-
sion by the resins and there is a definite critical ratio of resins to asphaltenes
below which the asphaltenes will start to precipitate. During the cracking
phase some of the resins are cracked to lighter hydrocarbons and others are
converted to asphaltenes. Both reactions affect the resin-asphaltene ratio,
the resultant stability of the visbreaker tar product, and serve to limit the
severity of the operation.

The principal reactions [11] which occur during the visbreaking operation are:

1. Cracking of the side-chains attached to cycloparaffin and aromatic
   rings at or close to the ring so the chains are either removed or
   shortened to methyl or ethyl groups.
2. Cracking of resins to light hydrocarbons (primarily olefins) and
   compounds which convert to asphaltenes.
3. At temperatures above 900°F, some cracking of naphthene rings.
   There is little cracking of naphthenic rings below 900°F.

The severity of the visbreaking operation can be expressed in several
ways: the yield of material boiling below 330°F, the reduction in product
viscosity, and the amount of standard cutter stock needed to blend the vis-

TABLE 5.12

Visbreaking
Time-Temperature Relationship [14]
(Equal conversion conditions)

Time, min	Temperature	
	°C	°F
1	485	905
2	470	878
4	455	850
8	440	825

breaker tar to No. 6 fuel oil viscosity specifications as compared to the amount needed for the feedstock [12]. Usually, the severity is expressed in the U.S. as the vol % product gasoline in a specified boiling range and in Europe as the wt % yield of gas plus gasoline (product boiling below 165°C or 330°F).

There are two types of visbreaker operations, coil and furnace cracking and soaker cracking. As in all cracking processes, the reactions are time-temperature dependent (see Table 5.12), and there is a trade-off between temperature and reaction time. Coil cracking uses high furnace outlet temperatures (885-930°F) and reaction times from one to three minutes while

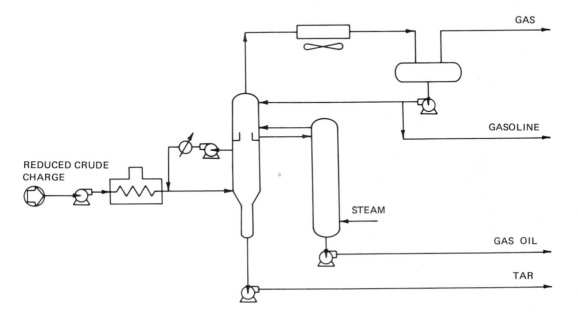

FIG. 5.4. Visbreaker.

TABLE 5.13

Visbreaking Results
Kuwait Long Resid [11]

	Feed	Product
Yields, wt %:		
Butane & lighter		2.5
C_5-330°F naphtha		5.9
Gas oil, 660°F EP		13.5
Tar		78.1
Product properties:		
Naphtha		
°API		65.0
Sulfur, wt %		1.0
RONC		—
Gas Oil		
°API		32.0
Sulfur, wt %		2.5
Tar or feed:		
% on crude	—	
°API	14.4	11.0
Sulfur, wt %	4.1	4.3
Viscosity, cS, 50°C	720	250

soaker cracking uses lower furnace outlet temperatures (800-830°F) and longer reaction times. The product yields and properties are similar but the soaker operation with its lower furnace outlet temperatures has the advantages of lower energy consumption and longer run times before having to shut down to remove coke from the furnace tubes. Run times of 3-6 months are common for furnace visbreakers and 6-18 months for soaker visbreakers. This apparent advantage for soaker visbreakers is at least partially balanced by the greater difficulty in cleaning the soaking drum [13].

A process flow diagram is shown in Fig. 5.4. The feed is introduced into the furnace and heated to the desired temperature. In the furnace or coil cracking process the feed is heated to cracking temperature (885-930°F) and quenched as it exits the furnace with gas oil or tower bottoms to stop the cracking reaction. In the soaker cracking operation, the feed leaves the furnace between 800 and 820°F and passes through a soaking drum, which provides the additional reaction time, before it is quenched. Pressure is an important design and operating parameter with units being designed for pressures as high as 750 psig for liquid-phase visbreaking and as low as 100-300 psig for 20-40 % vaporization at the furnace outlet [15]. Typical yields and product properties from visbreaking operations are shown in Tables 5.13 and 5.14.

For furnace cracking, fuel consumption accounts for about 80% of the operating cost with a net fuel consumption equivalent of 1-1.5 wt % on feed.

TABLE 5.14

Visbreaking Results
Agha-Jari Short Resid [11]

	Feed	Product
Yields, wt %:		
Butane & lighter		2.4
C_5-330°F naphtha		4.6
Gas oil, 660°F EP		14.5
Tar		78.5
Product properties:		
Naphtha		
°API		—
Sulfur, wt %		—
RONC		—
Gas Oil		
°API		32.2
Sulfur, wt %		—
Tar or feed:		
% on crude	—	
°API	8.2	5.5
Sulfur, wt %	—	—
Viscosity, cS, 122°F (50°C)	100,000	45,000

Fuel requirements for soaker visbreaking are about 30-35% lower [14]. See Table 5.15.

Many of the properties of the products of visbreaking vary with conversion and the characteristics of the feedstocks. However, some properties, such as diesel index and octane number, are more closely related to feed qualities and others, such as density and viscosity of the gas oil, are relatively independent of both conversion and feedstock characteristics [11].

TABLE 5.15

Coil and Soaker Visbreaking

	Coil	Soaker
Furnace outlet temperature, °F (°C)	900 (480)	805 (430)
Fuel consumption, relative	1.0	0.85
Capital cost, relative	1.0	0.90

PROBLEMS

1. For the crude oil in Figures 3.5 and 3.9 estimate the Conradson carbon of the $1,000^{+}°F$ residual crude oil fraction.

2. Estimate the coke yields for the crude oil fraction of problem 1 and make a material balance around the delayed coking unit. The $1,000^{+}°F$ fraction contains 0.38% sulfur by weight.

3. Using the information from problem 2, estimate the capital cost of a 7,500 BPSD delayed coking unit and its utility requirements.

4. Using Bureau of Mines distillation data from Appendix D, calculate the coke yield and make a material balance from a Wilmington, California crude oil residuum having an API gravity of 10.4° and a sulfur content of 3.2% by weight.

5. Estimate the capital and operating costs for a 10,000 BPSD delayed coker processing the reduced crude of problem 4. Assume four men per shift at average of $20.50/hr per man.

6. Calculate the long tons per day of coke produced by charging 30,000 barrels of $1050^{+}°F$ residuum from the assigned crude oil to a delayed coking unit.

7. If 95% of the hydrogen sulfide present in the coker gas product stream can be converted to elemental sulfur, how many long tons of sulfur will be produced per day when charging 30,000 barrels of $1050^{+}°F$ residuum from the assigned crude oil to the delayed coker?

8. Make a material balance around the delayed coker for the charge rate and the crude oil of problems 6 and 7. Also estimate the utility requirements.

9. Estimate the capital and operating costs for a 30,000 BPSD delayed coker processing the assigned reduced crude. Assume four men per shift at an average of 20.50/hr/man.

NOTES

1. Anon, Hydrocarbon Process, 47(9), 152 (1968).
2. J. McDonald, Refining Engineer, p. C-15 (Sept. 1957).
3. D. E. Allan et al., Chem. Engr. Prog., 77(12), 40 (1981).
4. D. H. Stormont, Oil Gas J., p. 75 (17 March 1969).
5. F. L. Shea, U.S. Patent No. 2, 775, 549.
6. R. Remirez, Chem. Eng., p. 74 (24 Feb. 1969).
7. V. Mekler, Refining Engineer, p. C-7 (Sept. 1957).
8. K. E. Rose, Hydrocarbon Process. 50(7), 85 (1971).
9. J. H. Eppard, Petrol. Refiner, p. 98 (July 1953).
10. M. Notarbartolo, C. Menegazzo, and J. Kuhn, Hydro Proc., p. 114-118, Sept. 1979.

11. A. Rhoe and C. de Blignieres, Hydro. Proc., p. 131-136, Jan. 1979.
12. Martin Hus, OGJ, p. 109-120, April 13, 1981.
13. D. E. Allan, C. H. Martinez, C. C. Eng, and W. J. Barton, Chem.
     Engr. Prog. 79(1), 85-89 (1983).
14. M. Akbar and H. Geelen, Hydro. Proc., p. 81-85, May, 1981.
15. R. Hournac, J. Kuhn, and M. Notarbartolo, Hydro. Proc., p. 97-102,
     Dec. 1979.

ADDITIONAL READING

Frank Stolfa, Hydro. Proc., p. 101-109, May, 1980.
T. A. Cooper and W. P. Ballard, *Advances in Petroleum Chemistry
and Refining*, Vol. 6, p. 171-238 (1962).
A. N. Sachanen, *Conversion of Petroleum: Production of Motor
Fuels by Thermal and Catalytic Processes*, 2nd ed., Reinhold
Publishing Corp., New York, 1948.

Chapter 6
CATALYTIC REFORMING AND ISOMERIZATION

The demand of today's automobiles for high-octane gasolines has stimulated the use of catalytic reforming. Catalytic reformate furnishes approximately 40% of the United States gasoline requirements and with the advent of low-lead and lead-free gasolines, this can be expected to increase.

In catalytic reforming, the change in the boiling point of the stock passed through the unit is relatively small as the hydrocarbon molecules are not cracked but their structures are rearranged to form higher octane aromatics. Thus catalytic reforming primarily increases the octane of motor gasoline rather than increasing its yield.

Feedstocks: The typical feedstocks to catalytic reformers are heavy straight-run (HSR) gasolines and naphthas (180 to 375°F). These are composed of the four major hydrocarbon groups: paraffins, olefins, naphthenes, and aromatics (PONA). Typical feedstocks and reformer products have the following PONA analyses:

	Vol %	
Component	Feed	Product
Paraffins	45-55	30-50
Olefins	0-2	0
Naphthenes	30-40	5-10
Aromatics	5-10	45-60

The paraffins and naphthenes undergo two types of reactions in being converted to higher octane components: cyclization and isomerization. The ease and probability of either of these occurring increases with the number of carbon atoms in the molecules and it is for this reason that only the HSR gasoline is used for reformer feed. The LSR gasoline (C_5-180°F) is largely composed of lower-molecular-weight paraffins that tend to crack to butane and lighter fractions and it is not economical to process this stream in a catalytic reformer. Hydrocarbons boiling above 400°F are easily hydrocracked and cause an excessive carbon laydown on the catalyst.

Reactions: As in any series of complex chemical reactions, reactions occur which produce undesirable products in addition to those desired. Reaction conditions have to be chosen that favor the desired reactions and inhibit the undesired ones. Desirable reactions in a catalytic reformer all lead to the formation of aromatics and isoparaffins as follows:

1. Paraffins are isomerized and to some extent converted to naphthenes. The naphthenes are subsequently converted to aromatics.
2. Olefins are saturated to form paraffins which then react as in (1).
3. Naphthenes are converted to aromatics.
4. Aromatics are left essentially unchanged.

Reactions leading to the formation of undesirable products include:

1. Dealkylation of side chains on naphthenes and aromatics to form butane and lighter paraffins.
2. Cracking of paraffins and naphthenes to form butane and lighter paraffins.

As the catalyst ages, it is necessary to change the process operating conditions to maintain the reaction severity and to suppress undesired reactions (Table 6.1).

There are four major reactions that take place during reforming. They are: (1) dehydrogenation of naphthenes to aromatics, (2) dehydrocyclization of paraffins to aromatics, (3) isomerization, and (4) hydrocracking. The first two of these reactions involve dehydrogenation and will be discussed together.

Dehydrogenation Reactions

The dehydrogenation reactions are highly endothermic and cause a decrease in temperature as the reaction progresses. In addition, the dehydrogenation reactions have the highest reaction rates of the reforming reactions which necessitates the use of the interheaters between catalyst beds to keep the mixture at sufficiently high temperatures for the reactions to proceed at practical rates (see Figure 6.1).

The major dehydrogenation reactions are:

TABLE 6.1

Some Basic Relationships in Catalytic Reforming [5]

Reaction	Reaction rate	Heat effect	Effect of high pressure	Effect of high temperature	Effect of high space velocity	Effect on hydrogen production	Effect on RVP	Effect on density	Effect on volumetric yield	Effect on octane
Hydro-cracking	Slowest	Exothermic	Aids	Aids	Hinders	Absorb	Increase	Decrease	Varies	Increase
Isomeri-zation	Rapid	Mildly exothermic	None	Aids	Hinders	None	Increase	Slight decrease	Slight increase	Increase
Cycliza-tion	Slow	Mildly exothermic	Hinders	Aids	Hinders	Evolves	Decrease	Increase	Decrease	Increase
Naphthene isomeri-zation	Rapid	Midly exothermic	None	Aids	Hinders	None	Decrease	Slight increase	Slight increase	Slight decrease
Naphthene dehydro-genation	Very fast	Quite endothermic	Hinders	Aids	Hinders	Evolves	Decrease	Increase	Decrease	Increase

a. Dehydrogenation of alkylcyclohexanes to aromatics.

$$
\begin{array}{ccc}
\underset{\substack{|\\ C \\ \diagup \diagdown \\ H_2-C \quad C-H_2 \\ | \quad | \diagup CH_3 \\ H_2-C \quad C \diagdown \\ \diagdown \diagup \quad H \\ C}}{H_2}
& \longrightarrow &
\underset{\substack{H \\ | \\ C \\ \diagup \diagdown \\ HC \quad CH \\ | \quad \| \\ HC \quad C-CH_3 \\ \diagdown \diagup \\ C \\ | \\ H}}{}
\quad + \; 3\,H_2
\end{array}
$$

Methylcyclohexane   Toulene

b. Dehydroisomerization of alkylcyclopentanes to aromatics.

$$
\begin{array}{ccccc}
\underset{\substack{H_2 \\ | \\ C \\ \diagup \diagdown \; CH_3 \\ H_2C \quad C \diagup \\ | \quad | \diagdown H \\ H_2C - CH_2}}{}
& \longrightarrow &
\underset{\substack{H_2 \\ | \\ C \\ \diagup \diagdown \\ H_2C \quad CH_2 \\ | \quad | \\ H_2C \quad CH_2 \\ \diagdown \diagup \\ C \\ | \\ H_2}}{}
& \longrightarrow &
\underset{\substack{H \\ | \\ C \\ \diagup \diagdown \\ H-C \quad C-H \\ \| \quad | \\ H-C \quad C-H \\ \diagdown \diagup \\ C \\ | \\ H}}{}
\quad + \; 3\,H_2
\end{array}
$$

c. Dehydrocyclization of paraffins to aromatics.

$$
n-C_7H_{16} \longrightarrow
\underset{\substack{H \\ | \\ C \\ \diagup \diagdown \\ H-C \quad C-CH_3 \\ \| \quad | \\ H-C \quad C-H \\ \diagdown \diagup \\ C \\ | \\ H}}{}
\quad + \; 4\,H_2
$$

The dehydrogenation of cyclohexane derivatives is a much faster reaction than either the dehydroisomerization of alkylcyclopentanes or the dehydrocyclization of paraffins, however all three reactions are necessary to obtain the high aromatic concentration needed in the product to produce a high octane.

Aromatics have a higher liquid density than paraffins or naphthenes with the same number of carbon atoms so one volume of paraffins produces only 0.77 volumes of aromatics and one volume of naphthenes about 0.87 volume. In addition, conversion to aromatics increases the gasoline end point because the boiling points of aromatics are higher than the boiling points of paraffins and naphthenes with the corresponding number of carbons.

The yield of aromatics is increased by:

1. High temperature (increases reaction rate but adversely affects chemical equilibrium)
2. Low pressure (shifts chemical equilibrium "to the right")
3. Low space velocity (promotes approach to equilibrium)
4. Low hydrogen-to-hydrocarbon mole ratios (shifts chemical equilibrium "to the right," however, a sufficient hydrogen partial pressure must be maintained to avoid excessive coke formation)

Isomerization Reactions

Isomerization of paraffins and cyclopentanes usually results in a lower octane product than does conversion to aromatics. However, there is a substantial increase over that of the unisomerized materials. These are fairly rapid reactions with small heat effects.

a. Isomerization of normal paraffins to isoparaffins.

$$CH_3-CH_2-CH_2-CH_2-CH_3 \rightarrow CH_3-\underset{\underset{CH_3}{|}}{CH}-CH_2-CH_2-CH_3$$

n-hexane   isohexane

b. Isomerization of alkylcyclopentanes to cyclohexanes, plus subsequent conversion to benzene.

methylcyclopentane   cyclohexane   benzene
91 RON   83 RON   >100 RON

Isomerization yield is increased by:

1. High temperature (which increases reaction rate)
2. Low space velocity
3. Low pressure

There is no isomerization effect due to the hydrogen-to-hydrocarbon mole ratios but high hydrogen-to-hydrocarbon ratios reduce the hydrocarbon partial pressure and thus favor the formation of isomers.

Hydrocracking Reactions

The hydrocracking reactions are exothermic and result in the production of lighter liquid and gas products. They are relatively slow reactions and therefore most of the hydrocracking occurs in the last section of the reactor.

The major hydrocracking reactions involve the cracking and saturation of paraffins.

$$
\begin{array}{ccc}
& & \underset{\text{Decane}}{\text{H—C—C—C—C—C—H}} \\
C_{10}H_{22} + H_2 & \longrightarrow & \\
\end{array}
$$

| Decane | Isohexane | n-butane |

The concentration of paraffins in the charge stock determines the extent of the hydrocracking reaction but the relative fraction of isomers produced in any molecular weight group is independent of the charge stock.

Hydrocracking yields are increased by:

1. High temperature
2. High pressure
3. Low space velocity

In order to obtain high product quality and yields, it is necessary to carefully control the hydrocracking and aromatization reactions. Reactor temperatures are carefully monitored to observe the extent of each of these reactions.

Low pressure reforming is generally used for aromatics production and the following generalizations hold for feedstocks in the 155-345°F (68-175°C) TBP boiling range:

1. One a mole basis, naphthene conversion to aromatics is about
    98% with the number of carbon atoms in the precursor being
    retained in the product as follows:

    methylcyclopentane   produces   benzene

    cyclohexane   produces   benzene

    dimethylcyclopentane   produces   toluene

    dimethylcyclohexane   produces   xylene

    cycloheptane   produces   toluene

    methylcycloheptane   produces   xylene

2. For paraffins the following moles of aromatics are produced from
    one mole of paraffins having the indicated number of carbon atoms:

$$1 \text{ mole } P_6 \text{ yields } 0.05 \text{ moles } A_6$$

$$1 \text{ mole } P_7 \text{ yields } 0.10 \text{ moles } A_7$$

$$1 \text{ mole } P_8 \text{ yields } 0.25 \text{ moles } A_8$$

$$1 \text{ mole } P_9 \text{ yields } 0.45 \text{ moles } A_9$$

$$1 \text{ mole } P_{10} \text{ yields } 0.45 \text{ moles } A_{10}$$

6.1 FEED PREPARATION

The active material in most catalytic reforming catalysts in platinum. Certain
metals, hydrogen sulfide, ammonia, and organic nitrogen and sulfur compounds will deactivate the catalyst [2]. Feed pretreating, in the form of hydrotreating, is usually employed to remove these materials. The hydrotreater
employs a cobalt-molybdenum catalyst to convert organic sulfur and nitrogen
compounds to hydrogen sulfide and ammonia which then are removed from
the system with the unreacted hydrogen. The metals in the feed are retained
by the hydrotreater catalyst. Hydrogen needed for the hydrotreater is obtained from the catalytic reformer. If the boiling range of the charge stock
must be changed, the feed is redistilled before being charged to the catalytic
reformer.

6.2 CATALYTIC REFORMING PROCESSES

There are several major reforming processes in use today. These include the
Platforming process licensed by UOP (Photo 4, Appendix F), Powerforming
(Exxon) (Photo 5, Appendix F), Ultraforming (Std Oil, Ind.), Houdriforming
and Iso-Plus Houdriforming (Houdry), Catalytic Reforming (Engelhard), and
Rheniforming (Chevron). There are several other processes in use at some

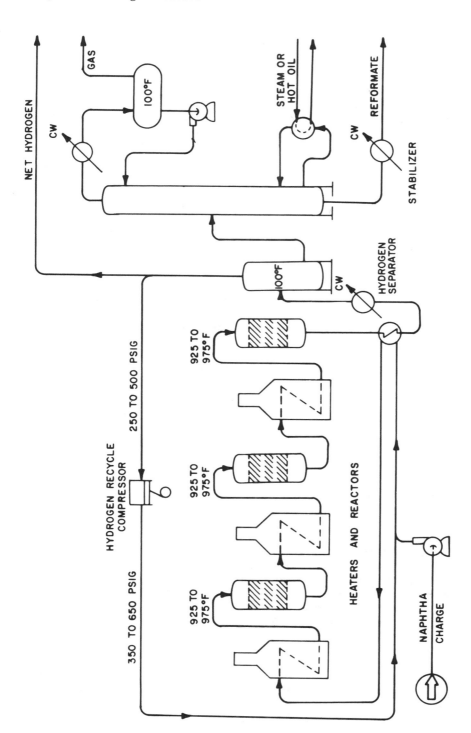

FIG. 6.1. Catalytic reforming, semiregenerative.

refineries but these are limited to a few installations and are not of general interest.

Reforming processes are classified as continuous, cyclic, or semiregenerative depending upon the frequency of catalyst regeneration. The equipment for the continuous process is designed to permit the removal and replacement of catalyst during normal operation. As a result, the catalyst can be regenerated continously and maintained at a high activity. As coke laydown and thermodynamic equilibrum yields of reformate are both favored by low pressure operation, the ability to maintain high catalyst activities by continous catalyst regeneration is the major advantage of the continuous type of unit. This advantage has to be evaluated with respect to the higher capital costs and possible lower operating costs due to lower hydrogen recycle rates and pressures needed to keep coke laydown at an acceptable level.

The semiregenerative unit is at the other end of the spectrum and has the advantage of minimum capital costs. Regeneration requires the unit to be taken off-stream. Depending upon severity of operation, regeneration is required at intervals of 3 to 24 months. High hydrogen recycle rates and operating pressures are utilized to minimize coke laydown and consequent loss of catalyst activity.

The cyclic process is a compromise between these extremes and is characterized by having a swing reactor in addition to those on-stream in which the catalyst can be regenerated without shutting the unit down. When the activity of the catalyst in one of the on-stream reactors drops below the desired level, this reactor is isolated from the system and replaced by the swing reactor. The catalyst in the replaced reactor is then regenerated by admitting hot air into the reactor to burn the carbon off the catalyst. After regeneration it is used to replace the next reactor needing regeneration.

The Platforming process can be obtained as a continuous or semiregenerative operation and other processes as either cyclic or semiregenerative. The Platforming, semiregenerative process is typical of reforming operations and will be described here.

A simplified process flow diagram of the Platforming process is given in Fig. 6.1 [3]. The pretreated feed and recycle hydrogen are heated to 925 to 975°F before entering the first reactor. In the first reactor, the major reaction is the dehydrogenation of naphthenes to aromatics and, as this is strongly endothermic, a large drop in temperature occurs. To maintain the reaction rate, the gases are reheated before being passed over the catalyst in the second reactor. As the charge proceeds through the reactors, the reaction rates decrease and the reactors become larger, and the reheat needed becomes less. Usually three reactors are sufficient to provide the desired degree of reaction and heaters are needed before each reactor to bring the mixture up to reaction temperature. In practice, either separate heaters can be used or one heater can contain several separate coils. The reaction mixture from the last reactor is cooled and the liquid products condensed. The hydrogen-rich gases are separated from the liquid phase in a drum separator and the liquid from the separator is sent to a fractionator to be debutanized.

The hydrogen-rich gas stream is split into a hydrogen recycle stream and a net hydrogen byproduct which is used in hydrotreating or hydrocracking operations or as fuel.

The reformer operating pressure and the hydrogen/feed ratio are compromises among obtaining maximum yields, long operating times between regenerations, and stable operation. It is usually necessary to operate at pressures from 100 to 350 psig and at hydrogen charge ratios of 4,000 to

8,000 scf/bbl fresh feed. Liquid hourly space velocities in the area of 2 to 3 are in general use.

   The Platforming process is classified as a semiregenerative type because catalyst regeneration is infrequent and runs of 3 to 6 months between regenerations are common. In the cyclic processes, regeneration is typically performed on a 24- or 48-hour cycle and a spare reactor is provided so that regeneration can be accomplished while the unit is still on-stream. Because of these extra facilities, the cyclic processes are more expensive but offer the advantages of low pressure operation and higher yields of reformate at the same severity.

   Example Problem

   Calculate the length of time between regeneration of catalyst in a reformer operating at the following conditions:

   Liquid hourly space velocity (LHSV) = 3.0 v/hr/v

   Feed rate = 5,000 BPSD

   Feed gravity = 55.0°API

   Catalyst bulk density = 50 lb/ft^3

   Hydrogen-to-feed ratio = 8,000 scf/bbl

   Number of reactors = 3

   Catalyst deactivates after processing 90 barrels of feed per pound of catalyst. If the catalyst bed is 6 feet deep in each reactor, what are the reactor inside diameters? Assume an equal volume of catalyst in each reactor.

   Solution:

   5,000 BPD = 1,170 ft^3/hr

   Total catalyst = $\dfrac{1,170}{3}$ = 390 ft^3

   (390 ft^3)(50 lb/ft^3) = 19,500 lb

   Time between regenerations:

   $\dfrac{(19,500 \text{ lb})(90 \text{ bbl/lb})}{5,000 \text{ bbl/day}}$ = 351 days

   Volume of catalyst per reactor = 390/3 = 130 ft^3

   Inside area = (130 ft^3)/(6 ft) = 21.67 ft^2

   Inside diameter = 5.25 ft

6.3 REFORMING CATALYST

All of the reforming catalyst in general use today contains platinum supported on a silica or silica-aluminum base. In most cases rhenium is combined with platinum to form a more stable catalyst which permits operation at lower pressures. Platinum is thought to serve as a catalytic site for hydrogenation and dehydrogenation reactions and chlorinated alumina provides an acid site for isomerization, cyclization, and hydrocracking reactions [2]. Reforming catalyst activity is a function of surface area, pore volume, and active platinum and chlorine content. Catalyst activity is reduced during operation by coke deposition and chloride loss. In a high pressure process, up to 200 barrels of charge can be processed per pound of catalyst before regeneration is needed. The activity of the catalyst can be restored by high temperature oxidation of the carbon followed by chlorination. This type of process is referred to as semiregenerative and is able to operate for 6- to 24-month periods between regenerations. The activity of the catalyst decreases during the on-stream period and the reaction temperature is increased as the catalyst ages to maintain the desired operating severity. Normally the catalyst can be regenerated at least three times before it has to be replaced and returned to the manufacturer for reclamation.

6.4 REACTOR DESIGN

Reactors used for catalytic reforming vary in size and mechanical details, but all have basic features as shown on Figure 6.2. Very similar reactors are used for hydrotreating, isomerization, and hydrocracking.

The reactors have an internal refractory lining which is provided to insulate the shell from the high reaction temperatures and thus reduce the required metal thickness. Metal parts exposed to the high temperature hy-

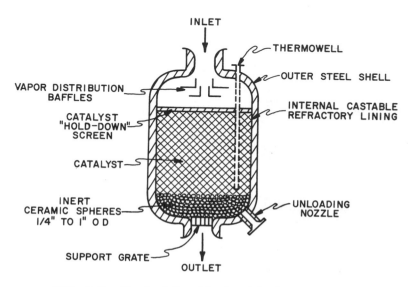

FIG. 6.2. Typical fixed bed catalytic reactor.

drogen atmosphere are constructed from steel containing at least 5% chromium and 0.5% molybdenum to resist hydrogen embrittlement.

Proper distribution of the inlet vapor is necessary to make maximum use of the available catalyst volume. Some reactor designs provide for radial vapor flow rather than the simpler straight-through type shown here. The important feature of vapor distribution is to provide maximum contact time with minimum pressure drop.

Temperature measurement at a minimum of three elevations in the catalyst bed is considered essential to determine catalyst activity and as an aid in coke burn-off operations.

The catalyst pellets are generally supported on a bed of ceramic spheres about 12 to 16 inches deep. The spheres vary in size from about 1 inch in diameter on the bottom to about 0.35 inch in diameter on the top.

6.5 YIELDS AND COSTS

Catalytic reforming yields can be estimated using Figures 6.3 through 6.6. Capital and operating costs can be obtained from Figure 6.7 and its accompanying descriptive material (Table 6.2).

6.6 ISOMERIZATION

The octane number of the LSR gasoline (C_5-160°F) can be improved by the use of an isomerization process to convert normal paraffins into their isomers. This results in a significant octane increase as n-pentane has an unleaded RON of 61.7 and isopentane has a rating of 92.3. In once-through isomerization the unleaded RON of LSR gasoline can be increased from 70 to about 82.

Reaction temperatures of about 300 to 400°F are preferred to higher temperatures because the equilibrium conversion to isomers is enhanced at the lower temperatures. At these relatively low temperatures a very active catalyst is necessary to provide a reasonable reaction rate. The available catalysts used for isomerization contain platinum on various bases. Some

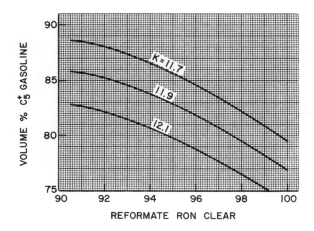

FIG. 6.3. Catalytic reforming yield correlations.

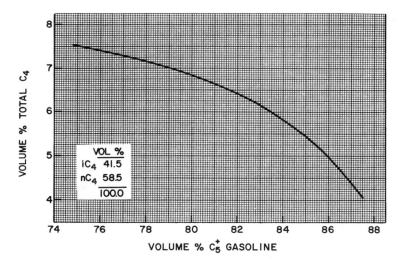

FIG. 6.4. Catalytic reforming yield correlations.

types of catalysts require the continuous addition of very small amounts of organic chlorides to maintain high catalyst activities. This is converted to hydrogen chloride in the reactor and consequently the feed to these units must be free of water and other oxygen sources in order to avoid catalyst deactivation and potential corrosion problems. A second type of catalyst uses a molecular sieve base and is reported to tolerate feeds saturated with water at ambient temperature [4]. This type of catalyst can be regenerated in place. Catalyst life is usually about two to three years or more with either type of catalyst.

An atmosphere of hydrogen is used to minimize carbon deposits on the catalyst but hydrogen consumption is negligible [5].

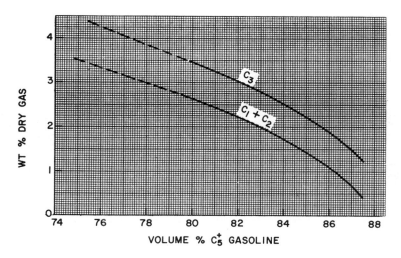

FIG. 6.5. Catalytic reforming yield correlations.

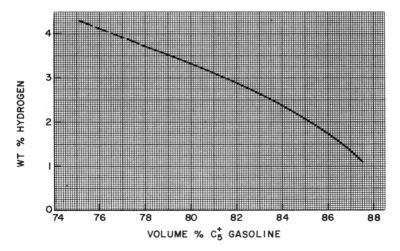

FIG. 6.6. Catalytic reforming yield correlations.

TABLE 6.2 *

Catalytic Reforming Units

Costs Included:
1. All battery limit facilities required for producing 100 RON unleaded reformate from a HSR naphtha sulfur-free feed.
2. Product stabilizer.
3. All necessary control and instrumentation.
4. Preheat and product cooling facilities to accept feed and release products at ambient temperatures.

Costs Not Included:
1. Cooling water, steam, and power supply.
2. Initial catalyst charge.
3. Royalty.
4. Feed fractionation or desulfurization.

Catalyst Charge:
Initial catalyst charge cost is approximately $200/BPD of feed.

Royalties:
Running royalty is about $0.05 to 0.10/bbl
Paid-up royalty is about $50 to $100/BPD

Utility Data (per bbl feed):
Steam, lb	30
Power, kWh	3
Cooling water, gal	600
Fuel gas (LHV), MMBtu	0.3
Catalyst replacement, $	0.10

---

*See Fig. 6.7, page 90.

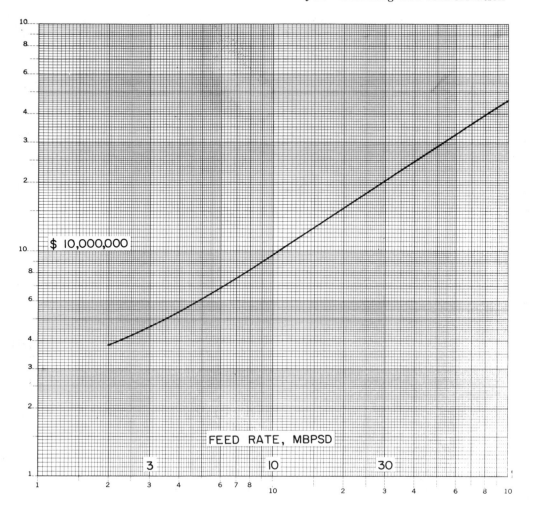

FIG. 6.7   Catalytic reforming units investment cost—1982 U.S. Gulf Coast.
*See Table 6.2, page 89.

The composition of the reactor products can closely approach chemical equilibrium. The actual product distribution is dependent upon the type and age of the catalyst, the space velocity, and the reactor temperature. The pentane fraction of the reactor product is about 70 to 75 weight % iso-pentane and the hexane fraction is about 82 to 87 weight % hexane isomers.

Following is a simplified conversion summary for a typical LSR cut. The values are on a relative weight basis and do not account for the weight loss resulting from hydrocracking to molecules lighter than pentane.

LSR component	Feed weight	Product weight	RON (unleaded)
Isopentane	22	40	92
Normal Pentane	33	15	62
2,2-Dimethybutane	1	13	96
2,3-Dimethybutane	2	5	84
2-Methylpentane	12	14	74
3-Methylpentane	10	7	74
Normal Hexane	20	6	26
Total	100	100	

If the normal pentane in the reactor product is separated and recycled the product RON can be increased by about 3 numbers (83 to 85 RON) [6]. If both normal pentane and normal hexane are recycled the product clear RON can be improved to about 87 to 89. Separation of the normals from the isomers can be accomplished by fractionation or by vapor phase adsorption of the normals on a molecular sieve bed. The adsorption process is well developed in several large units [4].

Some hydrocracking occurs during the reactions resulting in a loss of gasoline and the production of light gas. The amount of gas formed varies with the catalyst type and age and is sometimes a significant economic factor. The light gas produced is typically in the range of 1.0 to 4.0 weight % of the hydrocarbon feed to the reactor. For preliminary estimates the composition of the gas produced can be assumed to be 95 weight % methane and 5 weight % ethane.

For refineries that do not have hydrocracking facilities to supply isobutane for alkylation unit feed, the necessary isobutane can be made from n-butane by isomerization. The process is very similar to that of LSR gasoline isomerization but a feed deisobutanizer is used to concentrate the n-butane in the reactor charge. The reactor product is about 50 to 54 weight % isobutane.

A representative flow scheme for an isomerization unit is shown in Figure 6.8. Typical operating conditions are [5]:

Reactor temperature, °F	300-400
Pressure, psg	250-400
Hydrogen/hydrocarbon mole ratio	2:1
Single-pass LHSV, v/hr/v	1.5-2.5
Liquid product yield, wt %	>98

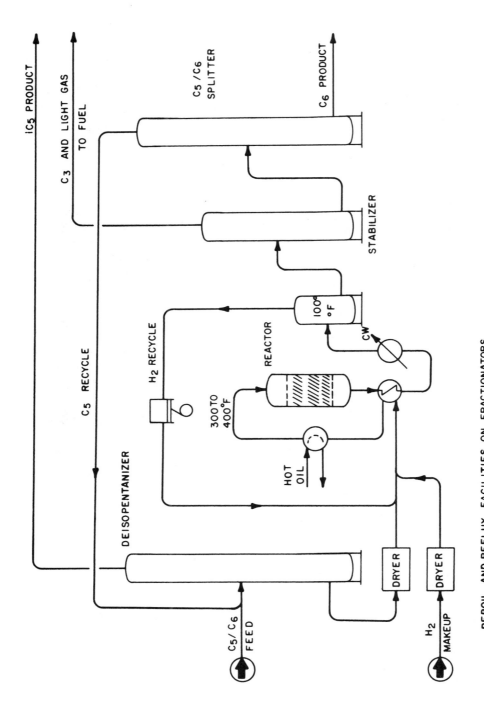

FIG. 6.8. Isomerization.

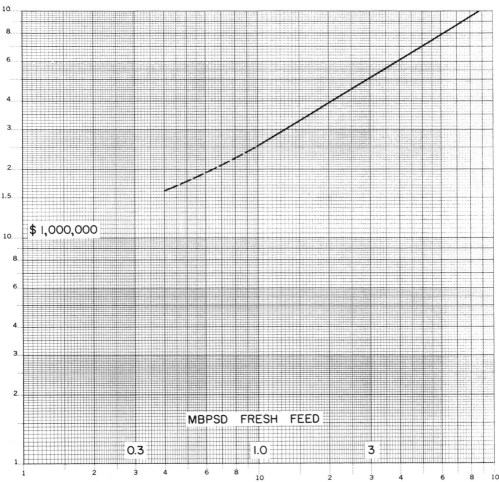

FIG. 6.9*. Paraffin isomerization units (platinum catalyst type) investment costs —
1982 U.S. Gulf Coast.
*See Table 6.3, page 94.

6.7 CAPITAL AND OPERATING COSTS

The capital construction and operating costs for an isomerization unit can be
estimated from Figure 6.9 and Table 6.3.

6.8 ISOMERIZATION YIELDS

Yields vary with feedstock properties and operating severity. A typical
product yield is given in Table 6.4 for a 13 number improvement in both RON
and MON clear for a 13 psi RVP C_5^+ isomerate product.

TABLE 6.3*

Paraffin Isomerization with Platinum Catalysts

Costs Included:

1. Feed drying.
2. Drying of hydrogen makeup.
3. Complete preheat, reaction, and hydrogen circulation facilities.
4. Product stabilization.
5. Sufficient heat exchange to cool products to ambient temperatures.
6. Central controls.
7. Fractionation to produce i-C_5 as a separate stream.
8. Paid up royalty.

Costs Not Included:

1. Hydrogen source.
2. Cooling water, steam, and power supply.
3. Feed desulfurization.
4. Initial catalyst charge is about $150 per BPD of reactor feed.

Utility Data (per bbl "fresh" feed):

Power, kWh	1.0
Cooling water, gal (30°FΔt)	600-1000
Fuel (LHV), MMBtu	0.20
Catalyst replacement, $	0.05
Hydrogen makeup in scf	40

*See Fig. 6.9, page 93.

TABLE 6.4

Isomerization Yields

Component	vol % on feed
C_3	0.5
i-C_4	0.8
n-C_4	2.2
C_5^+	98.4

6.9 CASE-STUDY PROBLEM: NAPHTHA HYDROTREATER AND CATALYTIC REFORMING

In this case, the feed to the catalytic reformer consists of the heavy straight-run (HSR) gasoline (190 to 380°F) from the crude unit and the coker gasoline. In practice the coker gasoline would probably be separated into a C_5-180°F fraction, which would not be reformed, and a 180 to 380° fraction which would be used as reformer feed. As sufficient information is not available to estimate the quantities and properties of these fractions, it is necessary to send all of the coker gasoline to the reformer. The C_5-80° fraction will undergo little octane improvement but its gasoline quality will be helped by saturation of the olefins. The required severity of reforming is not known until after the gasoline blending calculations are made. Therefore it is necessary to assume a value for the first time through. A severity of 94 RON clear is used for this calculation.

Procedure:

1. Calculate the characterization factor (K_w) of the feed.
2. Determine C_5^+ gasoline volume yield from Figure 6.3.
3. Determine weight or volume yields of H_2, $C_1 + C_2$, C_3, i-C_4 and n-C_4 from Figures 6.4, 6.5, and 6.6.
4. Calculate weight yield of all product streams except C_5^+ gasoline.
5. Determine weight yield of C_5^+ gasoline by difference.
6. Calculate API gravity of C_5^+ gasoline.
7. Make sulfur and/or nitrogen balance if needed to determine H_2S and NH_3 made and net hydrogen produced.
8. Estimate utility requirements from Table 6.2.

The results are tabulated in Table 6.5 and 6.6.

PROBLEMS

1. Reactor pressure is an important process variable in catalytic reforming. A common reaction in reforming is the conversion of methylcyclopentane to benzene. Calculate the barrels of benzene formed from one barrel of methylcyclopentane at the following reactor outlet conditions:

    a. 900°F, 600 psia
    b. 900°F, 300 psia

    The hydrogen feed rate to the reactor is 10,000 scf/bbl of methylcyclopentane. Assume the reaction is a single ideal gas-phase reaction and thermodynamic equilibrium is obtained. The National Bureau of Standards values for free energies of formation at 900°F are +66.09 kcal/g mol for methylcyclopentane and +50.78 kcal/g mol for benzene.

2. Determine the yield of n-butane and C_5^+ gasoline when reforming 4,500 BPD of HRS gasoline, K_w = 11.9, to a 92 clear RON.

3. A 180 to 380°F virgin naphtha stream with a mean average boiling point of 275°F and 50.2°API is reformed to a 96 RON clear gasoline blending stock. Make an overall material balance around the reformer for a 10,000 BPD feed rate.

TABLE 6.5

Hydrotreater and Catalytic Reformer Material Balance

Basis: 100,000 BPCD   North Slope, Alaska, Crude Oil
Severity: 94 RON clear K_W = 11.7

Component	vol %	BPD	°API	(lb/hr)/BPD	lb/hr	wt % S	lb/hr S
Feed:							
190-380° HSR	69.3	12,500	48.5	11.47	143,380	0.1	140
Coker							
gasoline	30.7	5,540	54.6	11.09	61,420	0.65	400
	100.0	18,040			204,800		540
Products:							
H_2 wt %,							
total	1.7				3,480		
C_1 + C_2,							
wt %	1.0				2,050		
C_3, wt %	1.8	500	146.5	7.42	3,690		
i-C_4	2.0	360	119.4	8.22	2,960		
n-C_4	2.8	505	110.8	8.51	4,300		
C_5^+ reformate	86.5	15,610	39.7	12.06	188,320		
Hydrogen:							
H_2S					574		540
H_2, net					3,446		

$$H_2S = \frac{540}{32.06} = 16.84 \text{ lb-mol/hr}$$

$$H_2 \text{ in } H_2S = (16.84)(2) = 34 \text{ lb/hr}$$

TABLE 6.6

Hydrotreater and Catalytic Reformer Chemical and
Utility Requirements

Requirements (per bbl feed)				
Utility	Treater	Reformer	Total	Total required
Steam, lb	6	30	36	27,060 lb/hr
Power, kWh	2	3	5	90,200 kWh/day
C.W, gal	300	600	900	11,275 gpm
Fuel, MMBtu	0.1	0.3	0.4	7,216 MMBtu/day
Catalyst, lb	0.002	0.004		36 lb/day for hydrotreater
				72 lb/day for reformer

4. Estimate the installed cost for a 6,200 BPSD isomerization unit to increase the RON of a LSR naphtha by 13 numbers. Determine the utility requirements and estimate the direct operating costs per barrel of feed if two operators are required per shift at an hourly rate of $17.37 per operator.

5. Estimate the direct operating cost per barrel of feed for an 8,400 BPSD catalytic reformer upgrading an HSR naphtha to a 96 RON clear product. The 50.2°API naphtha feed has a MABP of 275°F. Assume the reformer requires two operators per shift at an hourly rate of $17.37 per operator. Include royalty costs as a direct operating cost.

6. For problem 5, express the direct operating cost on a per barrel of C_5^+ reformate basis and compare with the cost of producing a barrel of 90 RON clear reformate from the same feed.

7. For the reforming units in problems 5 and 6, what is the single largest operating expense per barrel? What percentage of the cost is this?

8. Draw a flow diagram for an isomerization unit which increases by 20 research numbers the octane of the feed consisting of 45% (vol) n-pentane and 55% n-hexane.

9. Calculate the product gas composition of a reformer operated to maximize aromatics production with the following feedstock of saturated hydrocarbons:

	Vol. %
C_6H_{12}	10.1
C_6H_{14}	18.9
C_7H_{14}	12.8
C_7H_{16}	21.2
C_8H_{16}	17.7
C_8H_{18}	19.3
	100.0

10. Calculate the total operating costs per gallon of reformate for producing (a) a 92 clear RON product from 10,000 BPCD of the HSR naphtha from the assigned crude oil; (b) a 98 clear RON product. Use heating value of comparable products to obtain values of gases produced.

NOTES

1. R. Varela-Villegas, M. E. Thesis, Colorado School of Mines, 1971.
2. D. P. Thornton, Petro/Chem. Eng. *41*(5), 21-29 (1969).
3. Hydrocarbon Process. *49*(9), 189 (1970).
4. H. W. Kouwenhoven and W. C. Van Z. Langhout, Chem. Eng. Prog. *67*(4), 65-70 (1971).

5. Hydrocarbon Process, *49*(9), 195-197 (1970).
6. G. Bour, C. P. Schwoerev, and G. F. Asselin, Oil Gas J. *68*(43), 57-61 (1970).

ADDITIONAL READING

M. J. Fowle, R. D. Bent, B. E. Milner, and G. P. Masologites, Petrol. Refiner. *31*(4), 156-159 (1952).

Chapter 7
CATALYTIC CRACKING

Catalytic cracking is the most important and widely used refinery process for converting heavy oils into more valuable gasoline and lighter products, with almost 1 million tons of oil processed per day in the United States [1]. Originally cracking was accomplished thermally but the catalytic process has almost completely replaced thermal cracking because more gasoline having a higher octane and less heavy fuel oils and light gases are produced [2]. The light gases produced by catalytic cracking contain more olefins than those produced by thermal cracking (Table 7.1.).

The cracking process produces carbon (coke) which remains on the catalyst particle and rapidly lowers its activity. To maintain the catalyst activity at a useful level, it is necessary to regenerate the catalyst by burning off this coke with air. As a result, the catalyst is continuously moved from reactor to regenerator and back to reactor. The cracking reaction is endothermic and the regeneration reaction exothermic. Some units are designed to use the regeneration heat to supply that needed for the reaction and to heat the feed up to reaction temperature. These are known as "heat balance" units.

Average reactor temperatures are in the range 900 to 1000°F, with oil feed temperatures from 500 to 800°F and regenerator exit temperatures for catalyst from 1,200 to 1,450°F.

The catalytic-cracking processes in use today can all be classified as either moving-bed or fluidized-bed units. There are several modifications under each of the classes depending upon the designer or builder but within a class the basic operation is very similar. The Thermafor catalytic-cracking process (TCC) is representative of the moving-bed units and the fluid catalytic cracker (FCC) of the fluidized-bed units. There are very few TCC units in operation today and the FCC unit has taken over the field. The FCC units can be classified as either bed or riser (transfer line) cracking units depending upon where the major fraction of the cracking reaction occurs.

TABLE 7.1

Thermal vs. Catalytic Cracking Yields on Similar
Topped Crude Feed

	Thermal Cracking		Catalytic Cracking	
	wt %	vol %	wt %	vol %
Fresh feed	100.0	100.0	100.0	100.0
Gas	6.6		4.5	
Propane	2.1	3.7	1.3	2.2
Propylene	1.0	1.8	2.0	3.4
Isobutane	.8	1.3	2.6	4.0
nButane	1.9	2.9	0.9	1.4
Butylene	1.8	2.6	2.6	3.8
C_5-430 Gasoline	26.9	32.1	40.2	46.7
Light cycle oil	1.9	1.9	33.2	32.0
Decant oil			7.7	8.7
Residual oil	57.0	50.2		
Coke	-0-		5.0	
	100.0	96.5	100.0	102.2

The process flows of both types of processes are similar. The hot oil
feed is contacted with the catalyst in either the feed riser line or the reactor.
As the cracking reaction progresses, the catalyst is progressively deactivated
by the formation of coke on the surface of the catalyst. The catalyst and
hydrocarbon vapors are separated mechanically and oil remaining on the
catalyst is removed by steam stripping before the catalyst enters the regenerator. The oil vapors are taken overhead to a fractionation tower for
separation into streams having the desired boiling ranges.

The spent catalyst flows into the regenerator and is reactivated by
burning off the coke deposits with air. Regenerator temperatures are carefully controlled to prevent catalyst deactivation by overheating and to provide the desired amount of carbon burn-off. This is done by controlling
the air flow to give a desired CO_2/CO ratio in the exit flue gases or the
desired temperature in the regenerator. The flue gas and catalyst are separated by cyclone separators and electrostatic precipitators. The catalyst
in some units is steam-stripped as it leaves the regenerator to remove adsorbed oxygen before the catalyst is contacted with the oil feed.

7.1 FLUIDIZED-BED CATALYTIC CRACKING

The FCC process employs a catalyst in the form of very fine particles [average
particle size about 70 micrometers (microns)] which behave as a fluid when
aerated with a vapor. The fluidized catalyst is circulated continuously
between the reaction zone and the regeneration zone and acts as a vehicle to
transfer heat from the regenerator to the oil feed and reactor. Two basic
types of FCC units in use today are the "side-by-side" type, where the
reactor and regenerator are separate vessels adjacent to each other, and the

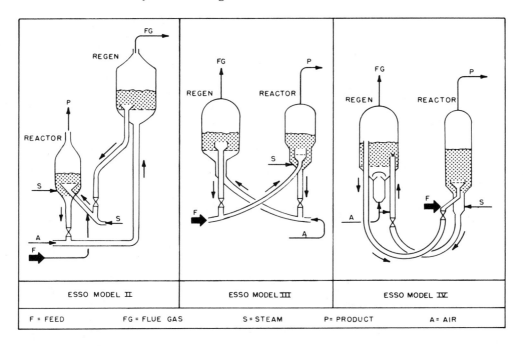

FIG. 7.1a. Fluid catalytic cracking (FCC) unit configurations.

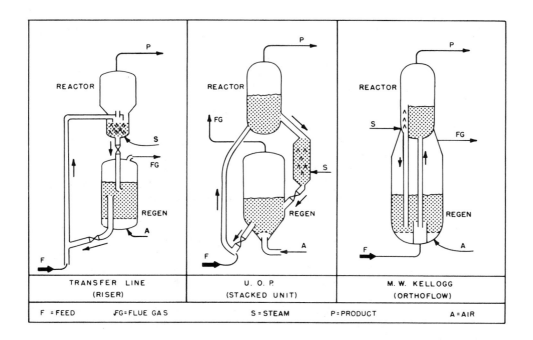

FIG. 7.1b. Fluid catalytic cracking (FCC) unit configurations.

Orthoflow or stacked type, where the reactor is mounted on top of the regenerator. Typical FCC unit configurations are shown in Fig. 7.1 (a and b)
and photos 6 and 7, Appendix F.

    One of the most important process differences in FCC units relates to
the location and control of the cracking reaction. Until about 1965, most
units were designed with a discrete dense-phase fluidized-catalyst bed in
the reactor vessel. The units were operated so most of the cracking occurred
in the reactor bed. The extent of cracking was controlled by varying reactor
bed depth (time) and temperature. Although it was recognized that cracking
occurred in the riser feeding the reactor because the catalyst activity and
temperature were at their highest there, no significant attempt was made to
regulate the reaction by controlling riser conditions. After the more reactive
zeolite catalysts were adopted by refineries, the amount of cracking occurring
in the riser (or transfer line) increased to levels requiring operational changes
in existing units. As a result, most recently constructed units have been
designed to operate with a minimum bed level in the reactor and with control
of the reaction being maintained by varying catalyst circulation rate. Many
older units have been modified to maximize and control riser cracking. Units
are also operating with different combinations of feed-riser and dense-bed
reactors, including feed-riser followed by dense-bed, feed-riser in parallel
with dense-bed, and parallel feed-riser lines (one for fresh feed and the other
for recycle) [3].

    The fresh feed and recycle streams are preheated by heat exchangers
or a furnace and enter the unit at the base of the feed riser where they are
mixed with the hot regenerated catalyst. The heat from the catalyst vaporizes
the feed and brings it up to the desired reaction temperature. The mixture of
catalyst and hydrocarbon vapor travels up the riser into the reactors. The
cracking reactions start when the feed contacts the hot catalyst in the riser
and continues until the oil vapors are separated from the catalyst in the
reactor. The hydrocarbon vapors are sent to the synthetic crude fractionator
for separation into liquid and gaseous products.

    The catalyst leaving the reactor is called spent catalyst and contains
hydrocarbons adsorbed on its internal and external surfaces as well as the
coke deposited by the cracking. Some of the adsorbed hydrocarbons are
removed by steam stripping before the catalyst enters the regenerator. In
the regenerator, coke is burned from the catalyst with air. The regenerator
temperature and coke burnoff are controlled by varying the air flow rate. The
heat of combustion raises the catalyst temperature to 1,150 to 1,350°F and
most of this heat is transferred by the catalyst to the oil feed in the feed
riser. The regenerated catalyst contains from 0.05 to 0.4 wt % residual coke
depending upon the type of combustion (burning to CO or CO_2) in the regenerator.

    The regenerator can be designed and operated to burn the coke on the
catalyst to either a mixture of carbon monoxide and carbon dioxide or
completely to carbon dioxide. Older units were designed to burn to carbon
monoxide to minimize blower capital and operating costs because only about
half as much air had to be compressed to burn to carbon monoxide. Newer
units are designed and operated to burn the coke to carbon dioxide in the
regenerator because they can burn to a much lower residual carbon level on
the regenerated catalyst. This gives a more reactive and selective catalyst
in the riser and a better product distribution results at the same equilibrium
catalyst activity and conversion level.

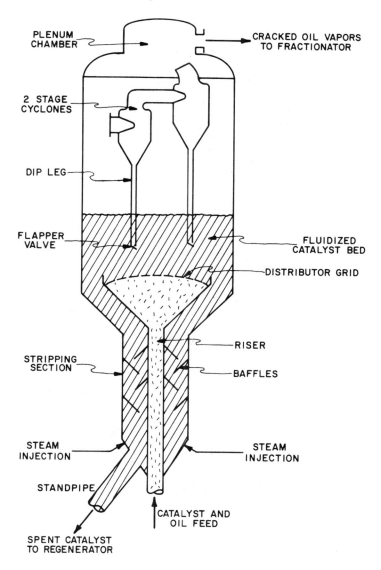

FIG. 7.2. FCC reactor for reactor cracking.

For units burning to carbon monoxide, the flue gas leaving the regenerator contains a large quantity of carbon monoxide which is burned to carbon dioxide in a CO furnace (waste heat boiler) to recover the available fuel energy. The hot gases can be used to generate steam or to power expansion turbines to compress the regeneration air and generate electric power. Schematic diagrams of a typical FCC reactor and regenerator are shown in Figures 7.2 and 7.3 and a flow diagram for a Model III FCC unit is given in Figure 7.4.

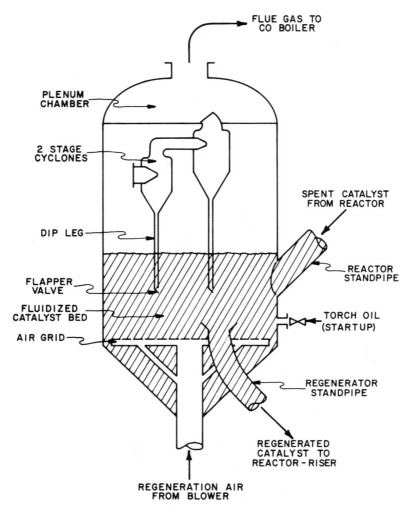

FIG. 7.3. FCC regenerator.

7.2 NEW DESIGNS FOR FLUIDIZED-BED CATALYTIC CRACKING UNITS

Zeolite catalysts have a much higher cracking activity than amorphous catalysts and shorter reaction times are required to prevent overcracking of gasoline to gas and coke. This has resulted in units which have a catalyst-oil separator in place of the fluidized-bed reactor to achieve maximum gasoline yields at a given conversion level. Because of the relatively short reaction time in the riser, this design prevents the flexibility of using amorphous catalysts to maximize middle distillate yields rather than gasoline yields. Some FCC units are designed to permit both riser and bed cracking to give maximum flexibility. Schematic drawings of Exxon Research and Engineering's FLEXICRACKING pure riser-reactor and riser + bed designs are shown in Figure 7.5 [4, 5]. These units incorporate a high height-to-diameter ratio

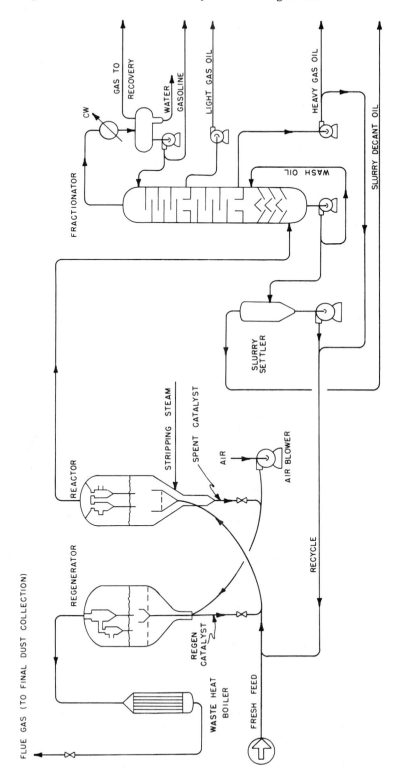

FIG. 7.4. FCC unit, Model III.

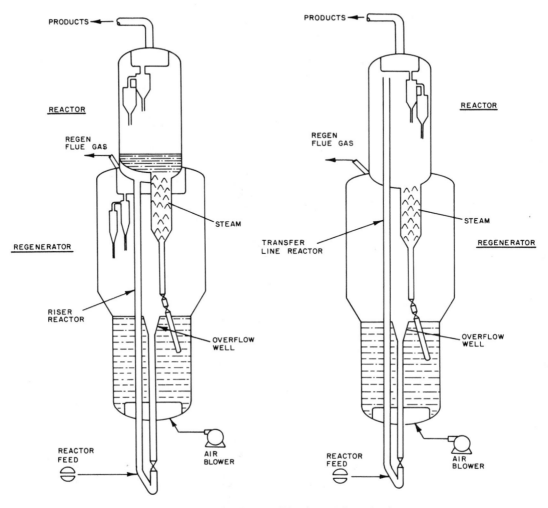

FIG. 7.5.   Latest Exxon Flexicracking designs.

lower regenerator section for more efficient single-stage regeneration and an offset side-by-side or stacked vessel design. UOP designed units utilize high velocity, low inventory regenerators (Figure 7.6a) and Kellogg designed units use plug valves to minimize erosion by catalyst containing streams (Figure 7.6b). All units are now designed for both complete combustion to CO_2 and CO combustion control.

The large differential value between residual fuel and other catalytic cracking feed stocks has caused refiners to blend atmospheric and vacuum tower bottoms into the FCC feed. Residual feed stocks have orders of magnitude higher metals contents (especially nickel and vanadium) and greater coke forming potential (Ramsbottom and Conradson carbon values) than distillate feeds. These contaminants reduce catalyst activity, promote coke and hydrogen formation, and decrease gasoline yield. It has been shown that catalyst activity loss due to metals is caused primarily by vanadium deposition

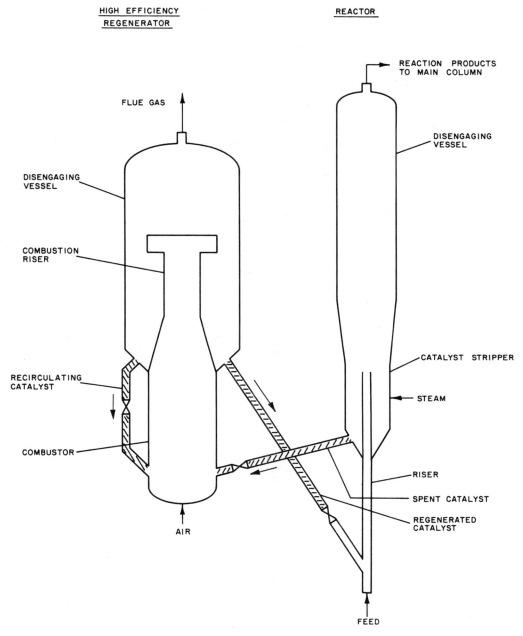

FIG. 7.6a. UOP design, riser FCC unit.

and increased coke and hydrogen formation is due to nickel deposited on the catalyst [6, 7]. The high coke laydown creates problems because of the increased coke burning requirement with the resulting increased air or oxygen demand, higher regenerator temperatures, and greater heat removal. For

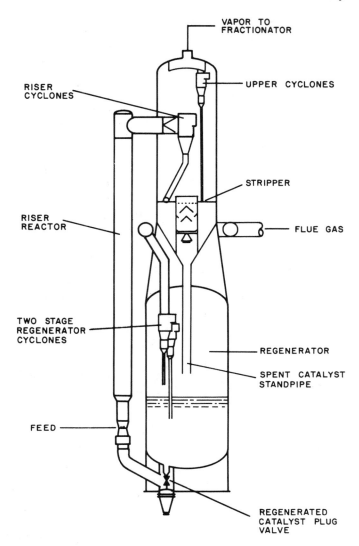

FIG. 7.6b. Kellogg design, riser FCC unit.

feeds containing up to 15 ppm Ni + V, passivation of nickel catalyst activity
can be achieved by the addition of antimony compounds to the feed [8] with
reductions in coke and hydrogen yields up to 15% and 45%, respectively.
For feeds containing >15 ppm metals and having Conradson carbon contents
above 3%, hydroprocessing of the FCC feed may be required to remove
metals, reduce the carbon forming potential, and increase the yield of gasoline and middle distillates by increasing the hydrogen content of the feed.

Designs of FCC units have been modified for higher carbon burning, heat
removal, and gas product rates by larger equipment sizing and introduction
of steam generation coils. Kellogg's Heavy Oil Cracking (HOC) FCC unit developed with Phillips Petroleum Company is shown in Figure 7.7.

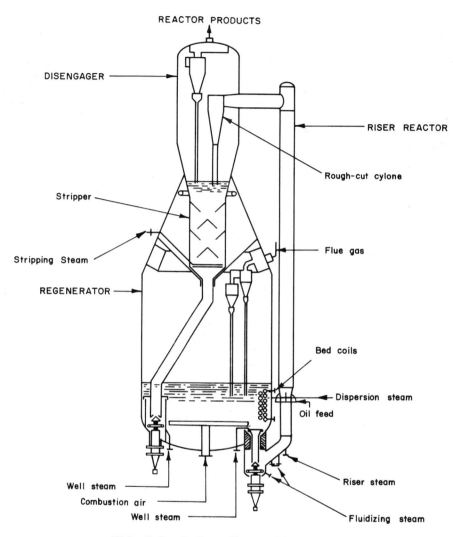

FIG. 7.7. Kellogg Heavy Oil Cracker.

7.3 CRACKING REACTIONS

The products formed in catalytic cracking are the result of both primary and secondary reactions [9]. Primary reactions are designated as those involving the initial carbon-carbon bond scission and the immediate neutralization of the carbonium ion [10]. The primary reactions can be represented as follows:

Paraffin ⟶ paraffin + olefin

Alkyl naphthene ⟶ naphthene + olefin

Alkyl aromatic ⟶ aromatic + olefin

Thomas [11] suggested the mechanism that carbonium ions are formed initially by a small amount of thermal cracking of n-paraffins to form olefins. These olefins add a proton from the catalyst to form large carbonium ions which decompose according to the beta rule (carbon-carbon bond scission takes place at the carbon in the position beta to the carbonium ions and olefins) to form small carbonium ions and olefins. The small carbonium ions propagate the chain reaction by transferring a hydrogen ion from a n-paraffin to form a small paraffin molecule and a new large carbonium ion [7, 12].

As an example of a typical n-paraffin hydrocarbon cracking reaction, we may look at the following sequence for n-octane (where $R = CH_3CH_2CH_2CH_2CH_2-$).

Step 1: Mild thermal cracking initiation reaction.

$$nC_8H_{18} \longrightarrow CH_4 + R-CH=CH_2$$

Step 2: Proton shift.

$$R-CH=CH_2 + H_2O + \begin{bmatrix} O \\ | \\ Al-O-Si \\ | \\ O \end{bmatrix} \longrightarrow$$

$$\overset{+}{R-CH}-CH_3 + \begin{bmatrix} O \\ | \\ HO-Al-O-Si \\ | \\ O \end{bmatrix}^{-}$$

Step 3: Beta scission.

$$\overset{+}{R-CH}-CH_3 \longrightarrow CH_3CH_2=CH_2 + \overset{+}{CH_2}CH_2CH_2CH_3$$

Step 4: Rearrangement toward more stable structure. The order of carbonium ion stability is tertiary > secondary > primary.

$$\overset{+}{CH_2}CH_2CH_2CH_3 \rightleftarrows CH_3\overset{+}{CH}CH_2CH_3 \rightleftarrows CH_3-\overset{\overset{\displaystyle CH_3}{|}}{\underset{}{CH}}-\overset{+}{CH_2} \rightleftarrows CH_3-\overset{\overset{\displaystyle CH_3}{|}}{\underset{+}{C}}-CH_3$$

Step 5: Hydrogen ion transfer.

$$CH_3-\overset{\overset{\displaystyle CH_3}{\displaystyle |}}{\underset{\displaystyle +}{C}}-CH_3 + C_8H_{18} \longrightarrow i\text{-}C_4H_{10} + CH_3\overset{+}{C}HCH_2R$$

Thus another large carbonium ion is formed and the chain is ready to repeat itself.

Even though the basic mechanism is essentially the same, the manner and extent of response to catalytic cracking differs greatly among the various hydrocarbon types.

7.4 CRACKING OF PARAFFINS

The catalytic cracking of paraffins is characterized by: high production of C_3 and C_4 hydrocarbons in the cracked gases, reaction rates and products determined by size and structure of paraffins, and isomerization to branched structures and aromatic hydrocarbons formation resulting from secondary reactions involving olefins [13]. In respect to reaction rates, the effect of the catalyst is more pronounced as the number of carbon atoms in the molecule increases, but the effect is not appreciable until the number of carbon atoms is at least six.

The cracking rate is also influenced by the structure of the molecule, with those containing tertiary carbon atoms cracking most readily while quaternary carbon atoms are most resistant. Compounds containing both types of carbon atoms tend to neutralize each other on a one-to-one basis. For example, 2,2,4-trimethylpentane (one tertiary and one quaternary) cracks only slightly faster than n-octane while 2,2,4,6,6-pentamethylheptane (one tertiary and two quaternary) cracks at a slower rate than does n-dodecane.

7.5 OLEFIN CRACKING

The catalytic cracking rates of olefinic hydrocarbons are much higher than those of the corresponding paraffins. The main reactions are [12]:

1. Carbon-carbon bond scissions
2. Isomerization
3. Polymerization
4. Saturation, aromatization, and carbon formation

Olefin isomerization followed by saturation and aromatization are responsible for the high octane number and lead susceptibility of catalytically cracked gasolines. The higher velocity of hydrogen transfer reactions for branched olefins results in ratios of iso- to normal paraffins higher than the equilibrium ratios of the parent olefins. In addition, naphthenes act as hydrogen donors in transfer reactions with olefins to yield isoparaffins and aromatics.

7.6 CRACKING OF NAPHTHENIC HYDROCARBONS

The most important cracking reaction of naphthenes in the presence of silica-alumina is dehydrogenation to aromatics. There is also carbon-carbon bond scission in both the ring and attached side chains but at temperatures below 1,000°F the dehydrogenation reaction is considerably greater. Dehydrogenation is very extensive for C_9 and larger naphthenes and a high-octane gasoline results. The nonring liquid products and cracked gases resulting from naphthenic hydrocarbon cracking are more saturated than those resulting from cracking paraffins.

7.7 AROMATIC HYDROCARBON CRACKING

Aromatic hydrocarbons with alkyl groups containing less than three carbon atoms are not very reactive. The predominant reaction for aromatics with long alkyl chains is the clean splitting off of side chains without breaking the ring. The carbon-carbon bond ruptured is that adjacent to the ring and benzene compounds containing alkyl groups can be cracked with nearly quantitative recovery of benzene [12].

7.8 CRACKING CATALYSTS

Commercial cracking catalysts can be divided into three classes: (1) acid-treated natural aluminosilicates, (2) amorphous synthetic silica-alumina combinations, and (3) crystalline synthetic silica-alumina catalysts called zeolites or molecular sieves [1]. Most catalysts used in commercial units today are either class (3) or mixtures of classes (2) and (3) catalysts [14]. The advantages of the zeolite catalysts over the natural and synthetic amorphous catalysts are:

1. Higher activity
2. Higher gasoline yields at a given conversion
3. Production of gasolines containing a larger percentage of paraffinic and aromatic hydrocarbons
4. Lower coke yield (and therefore usually a larger throughput at a given conversion level)
5. Increased isobutane production
6. Ability to go to higher conversions per pass without overcracking

The high activity of zeolitic cracking catalyst permits short residence time cracking and has resulted in many cracking units being adapted to riser cracking operations [15]. Here the adverse effects of carbon deposits on catalyst activity and selectivity are minimized because of the negligible amount of catalyst backmixing in the riser. In addition, separate risers can be used for cracking the recycle stream and the fresh feed so that each can be cracked at their own optimum conditions.

The catalytic effects of zeolitic catalysts can be achieved with only 10 to 20% of the circulating catalyst as zeolites [16] and the remainder amorphous silica-alumina cracking catalyst. Amorphous catalysts have higher attrition

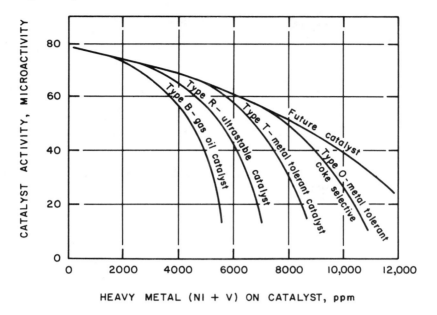

FIG. 7.8a. Catalyst activity loss from heavy metals [20].

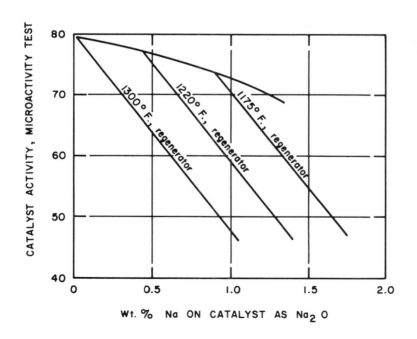

FIG. 7.8b. Catalyst activity loss from sodium contamination [20].

resistance and are less costly than zeolitic catalysts. Most commercial catalysts contain approximately 15% zeolites and thus obtain the benefits of the higher activity and gasoline selectivity of the zeolites and the lower costs and make-up rates of the amorphous catalysts. Lower attrition rates also greatly improve particulate emission rates.

Basic nitrogen compounds, iron, nickel, vanadium and copper in the oil act as poisons to cracking catalysts [1]. The nitrogen reacts with the acid centers on the catalyst and lowers the catalyst activity. The metals deposit and accumulate on the catalyst and cause a reduction in throughput by increasing coke formation and decreasing the amount of coke burnoff per unit of air by catalyzing coke combustion to CO_2 rather than to CO.

It is generally accepted that nickel has about four times as great an effect on catalyst activity and selectivity as vanadium and some companies use a factor of 4Ni + V (expressed in ppm) to correlate the effects of metals loading while other use Ni + V/4 or Ni + V/5 [17, 18, 19].

Although the deposition of nickel and vanadium reduces catalyst activity by occupying active catalytic sites, the major effects are to promote the formation of gas and coke and reduce the gasoline yield at a given conversion level. Tolen [20] and others [21] have discussed the effects of nickel and vanadium deposits on equilibrium catalysts (Figure 7.8a and b). Metals loadings on equilibrium catalysts are now as high as 10,000 ppm and, in 1982, the mean was over 1000 ppm of nickel plus vanadium [22].

Phillips Petroleum Company [8] has developed a metals passivation process which adds an organic antimony compound to the feed. The antimony is deposited on the catalyst with the nickel and forms a compound having less catalytic activity than the nickel alone.

A range of catalysts is available from catalyst manufacturing companies which are compounded to give high gasoline octanes (0.5 to 1.5 RON improvement), resist deactivation due to sulfur or metals in the feed, and to transfer sulfur from the regenerator to the reactor (to reduce SO_4 in the regenerator flue gas). For each operation and feed, economic and/or environmental considerations affect the catalyst used.

In his book "Catalytic Processes and Proven Catalysts," C. L. Thomas lists the chemical composition, physical characteristics, and suppliers of craking catalysts [1].

7.9 PROCESS VARIABLES

In addition to the nature of the charge stock, the major operating variables affecting the conversion and product distribution are the cracking temperature, catalyst/oil ratio, space velocity, catalyst type and activity, and recycle ratio. For a better understanding of the process, several terms should be defined.

Activity: Ability to crack a gas oil to lower boiling fractions.

Catalyst/oil ratio = C/O = lb catalyst/lb feed.

$$Conversion = 100 \left(\frac{Volume\ of\ feed - volume\ of\ cycle\ stock}{Volume\ of\ feed} \right).$$

Cycle stock: Portion of catalytic-cracker effluent not converted to naphtha and lighter products (generally the material boiling above 430°F).

$$Efficiency = \left(\frac{\% \text{ gasoline}}{\% \text{ conversion}}\right) \times 100.$$

Recycle ratio = Volume recycle/volume fresh feed.

Selectivity: The ratio of the yield of desirable products to the yield of undesirable products (coke and gas).

Space velocity: Space velocity may be defined on either a volume (LHSV) or a weight (WHSV) basis. In a fluidized-bed reactor, the LHSV has little meaning because it is difficult to establish the volume of the bed. The weight of the catalyst in the reactor can be easily determined or calculated from the residence time and C/O ratio.

LHSV = liquid hour space velocity in volume feed/(volume catalyst) (hr).

WHSV = weight hour space velocity in lb feed/(lb catalyst) (hr). If t is the catalyst residence time in hours, then WHSV = 1/(t)(C/O).

Within the limits of normal operations, increasing

1. Reaction temperature
2. Catalyst/oil ratio
3. Catalyst activity
4. Contact time

results in an increase in conversion while a decrease in space velocity increases conversion. It should be noted that an increase in conversion does not necessarily mean an increase in gasoline yield, as an increase in temperature above a certain level can increase conversion, coke and gas yields, and octane number of the gasoline but decrease gasoline yield [12, 23].

In many FCC units, conversion and capacity are limited by the regenerator coke burning ability. This limitation can be due to either air compression limitations or by the afterburning temperatures in the last stage regenerator cyclones. In either case FCC units are generally operated at the maximum practical regenerator temperature with the reactor temperature and throughput ratio selected to minimize the secondary cracking of gasoline to gas and coke. With the trend to heavier feedstocks, the carbon forming potential of catalytic cracker feeds are increasing and some units limited in carbon burning ability because of limited blower capacity are adding oxygen to the air to the regenerator to overcome this limitation. Oxygen contents of the gases to the regenerator are being increased to 24 to 30% by volume and are limited by regenerator temperature capability and heat removal capacity [24, 25].

TABLE 7.2

Comparison of Fluid, Thermafor, and Houdry
Catalytic Cracking Units

	FCC	TCC	HCC
Reactor space vel	1.1–13.4[a]	1–3[b]	1.5–4[b]
C/O	5–16[c]	2–7[d]	3–7[d]
Recycle/fresh feed, vol	0–0.5	0–0.5	0–0.5
Catalyst requirement, lb/bbl feed	0.15–0.25	0.06–0.13	0.06–0.13
Cat crclt rate, ton cat/bbl total feed	0.9–1.5	0.4–0.6	0.4–0.6
On-stream efficiency, %	96–98		
Reactor temp, °F	885–950[e]	840–950	875–950
Regenerator temp, °F	1200–1500	1100–1200	1100–1200
Reactor press, psig	8–30[e]	8–12	9–10
Regenerator press, psig	15–30		
Turndown ratio			2:1
Gasoline octane, clear			
RON	92–99	88–94	88–94
MON	80–85		

[a]lb/hr/lb
[b]v/hr/v
[c]wt
[d]vol
[e]One company has operated at 990°F and 40 psig to produce a 98 RON (clear) gasoline with a C_3-650°F liquid yield of 120 vol % on feed (once-through); there was approximately 90% yield of the C_5-650°F product.

In fluidized-bed units, the reactor pressure is generally limited to 15 to 20 psig by the design of the unit and is therefore not widely used as an operating variable. Increasing pressure increases coke yield and the degree of saturation of the gasoline but decreases the gasoline octane. It has little effect on the conversion.

The initial catalyst charge to a FCC unit using riser cracking is about 1 ton of catalyst per 100 BPSD charge rate (catalyst in circulation is about 0.3 ton per BPSD charge rate with approximately twice that in storage for makeup, etc.).

Catalyst circulation rate is approximately 1 ton/min per MBPD charge rate.

Typical operations of these units are given in Table 7.2.

7.10 HEAT RECOVERY

Fuel and energy costs are a major fraction of the direct costs of refining crude oils and as a result of large increases in crude oil and natural gas prices there is a great incentive to conserve fuel by the efficient utilization of the energy in the off-gases from the catalytic cracker regenerator. Tem-

perature control in the regenerator is easier if the carbon on the catalyst
is burned to carbon dioxide rather than carbon monoxide but much more heat
is evolved and regenerator temperature limits many be exceeded. A better
yield structure with lower coke laydown and higher gasoline yield is obtained
at a given conversion level when burning to carbon dioxide to obtain a lower
residual carbon on the catalyst. In either case, the hot gases are at high
temperature (1,200 to 1,450°F in the former case and 1,250 to 1,500°F in the
latter) and at pressures of 15 to 25 psig. Many catalytic crackers include
waste heat boilers which recover the sensible heat by steam generation and
others use power recovery turbines to generate electric power or compress
the air used in the catlytic cracker regenerator. Some refineries recover
the heat of combustion of the carbon monoxide in the flue gas by installing
CO-burning waste heat boilers in place of those utilizing only the sensible
heat of the gases. An even higher rate of energy recovery can be achieved
by using a power recovery turbine prior to the CO or waste heat boiler although when regenerator pressures are less than 15 psig, power recovery
turbines usually are not economic.

Assuming a regenerator flue gas discharge pressure of 20 psig and
1,000°F, the available horsepower per pound per second of gas flow for the
various schemes are:

1. Waste heat boiler only   45 hp
2. Power recovery only   78 hp
3. Power recovery plus waste heat boiler   106 hp
4. CO burning waste heat boiler   145 hp
5. Power recovery plus CO burning
    waste heat boiler   206 hp

7.11 YIELD ESTIMATION

Correlations have been developed to estimate the product quality and yields
from the catalytic cracking of virgin gas oils. The correlations are very useful for estimating typical yields for preliminary studies and to determine
yield trends when changes are made in conversion levels. The yield structure
is very dependent upon catalyst type. Figures 7.9 through 7.14 are for
silica-alumina catalyst and Figures 7.15 through 7.21 are for zeolitic or molecular-sieve catalyst. If the feedstocks contain sulfur, Figure 7.22 can be
used to estimate the distribution of sulfur in the product streams.

It is necessary to make a weight balance in order to obtain product
distribution and properties as the cycle gas oil yields are obtained by difference. When using molecular-sieve catalyst, the procedure is as follows:
(Gravities shown on curves are those of the fresh feed).

1. Calculate weight of feed.
2. Determine yields and weights of all products except cycle gas oils
    from Figures 7.15 through 7.19.
3. Determine weight of cycle gas oils by difference.
4. Using Figure 7.21 to estimate total cycle gas oil gravity, calculate
    volume of total cycle gas oils produced.

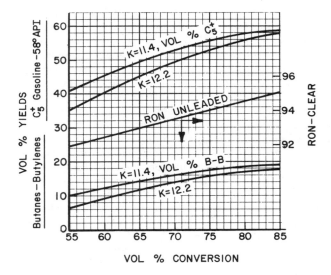

FIG. 7.9. Catalytic cracking yields. Silica-alumina catalyst (butanes, butylenes, C_5^+ gasoline). The butane-butylene fraction typically contains about 40 vol % isobutane, 12 vol % normal butane, and 48 vol
% butylenes.

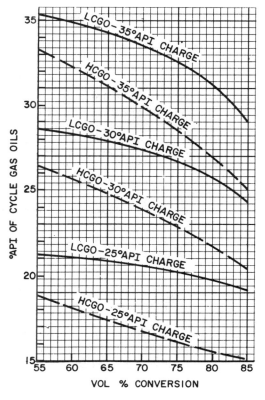

FIG. 7.10. Catalytic cracking yields. Silica-alumina catalyst (cycle gas oils).

5. From Figure 7.20 and feed rate, calculate the volume of heavy cycle gas oil (HCGO).
6. Subtract volume of HCGO from volume of total cycle gas oil (Step 4) to obtain volume yield of light cycle gas oil (LCGO).
7. Use Figure 7.21 to estimate HCGO gravity and calculate weight yield of HCGO.

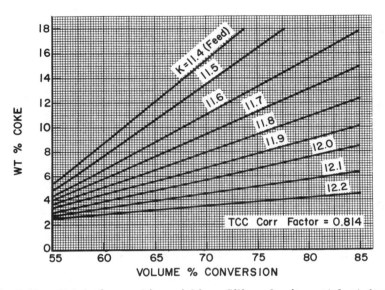

FIG. 7.11. Catalytic cracking yields. Silica-alumina catalyst (coke).

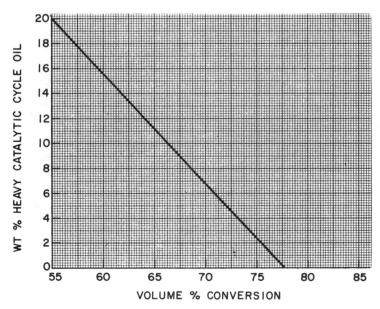

FIG. 7.12. Catalytic cracking yields. Silica-alumina catalyst (heavy catalytic cycle oil).

8. Subtract weight yield of HCGO from weight of total cycle gas oil (Step 3) to obtain weight yield of LCGO.

9. Divide weight yield of LCGO by volume yield of LCGO (Step 6) to obtain density of LCGO. Use tables to find API gravity.

Product yields and properties are obtained in a similar manner when silica-alumina catalyst is used in the reactor, except the gravity of the LCGO can be obtained from Figure 7.10 and the yield of HCGO from Figure 7.12. It is necessary to find the weight yield of LCGO by difference.

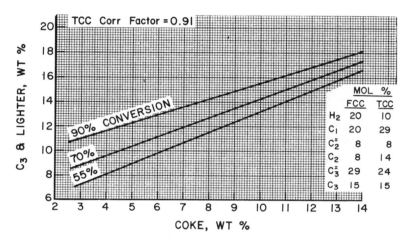

FIG. 7.13. Catalytic cracking yields. Silica-alumina catalyst (C_3 and lighter).

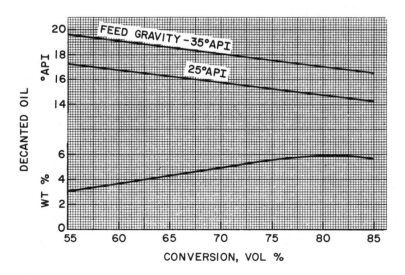

FIG. 7.14. Catalytic cracking yields. Silica-alumina catalyst (decanted oil).

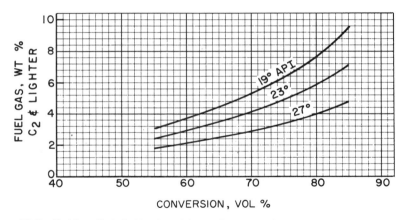

FIG. 7.15. Catalytic cracking yields. Zeolite catalyst (fuel gas).

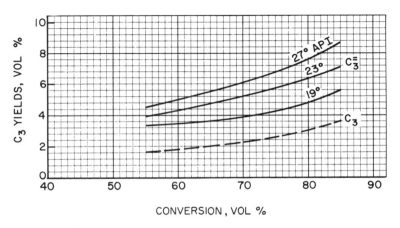

FIG. 7.16. Catalytic cracking yields. Zeolite catalyst (C_3).

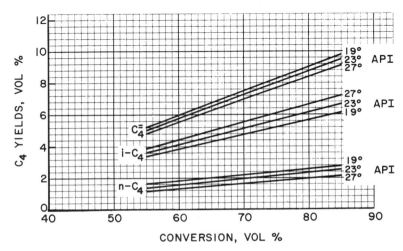

FIG. 7.17. Catalytic cracking yields. Zeolite catalyst (C_4).

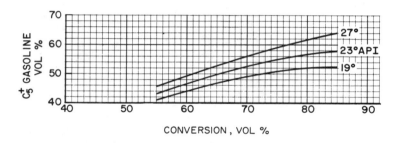

FIG. 7.18. Catalytic cracking yields. Zeolite catalyst (C_5^+ gasoline).

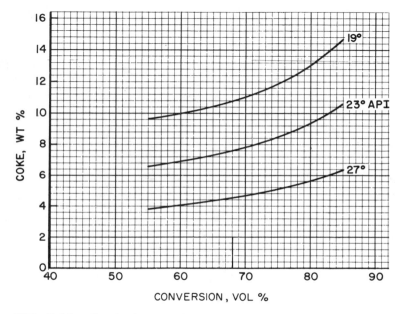

FIG. 7.19. Catalytic cracking yields. Zeolite catalyst (coke).

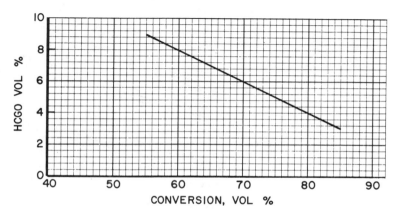

FIG. 7.20. Catalytic cracking yields. Zeolite catalyst (heavy cycle gas oil).

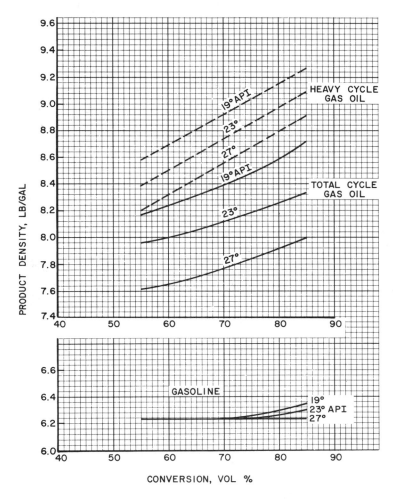

FIG. 7.21. FCC product gravity. Zeolite catalyst.

7.12 CAPITAL AND OPERATING COSTS

Capital construction and operating costs for a fluid catalytic-cracking unit can be estimated using Figure 7.23 and its accompanying descriptive material (Table 7.3). Multiplying factors can be found in Chapter 15 (Cost Estimation).

7.13 CASE-STUDY PROBLEM: CATALYTIC CRACKER

The choice of feedstocks for a catalytic cracking unit should be based on an economic evaluation of alternative uses of the hydrocarbon streams. This is especially important in the case of a refinery having both catalytic cracking and hydrocracking units. For this example, both the heavy gas oil from the atmospheric pipe still and the vacuum gas oil from the vacuum unit are used as FCC feedstocks, although it is quite possible the vacuum gas oil might give a better return if used as hydrocracker feed.

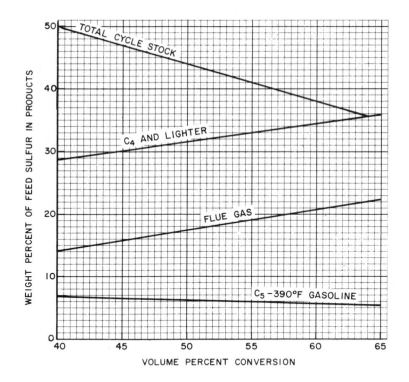

FIG. 7.22. Distribution of sulfur in catalytic cracking products [26].

TABLE 7.3*

Fluid Catalytic Cracking

Costs Included:
1. Product fractionation.
2. Gas compression and concentration for recovery of 95% of C_4's and 80% of C_3.
3. Complete reactor-regenerator section.
4. Sufficient heat exchange to cool products to ambient temperatures.
5. Central control system.

Costs Not Included:
1. Feed fractionation.
2. Off-gas and product treating.
3. Cooling water, steam, and power supply.
4. Initial catalyst charge.[c]

Royalty, $/bbl feed: 0.04

Utility Data (per bbl feed):

Steam, lb[a]	—
Power, kWh[b]	6.0
Cooling water, gal (30FΔt)	500
Fuel (LHV), MMBtu	0.1
Catalyst replacement, $[c]	0.15

---

*See Fig. 7.23, page 125.

[a]Waste heat steam production usually is in excess of consumption by approximately 30 pounds of steam per barrel of fresh feed.

[b]Includes electric drive for air blower and off-gas compressor.

[c]Refer to Appendix C.

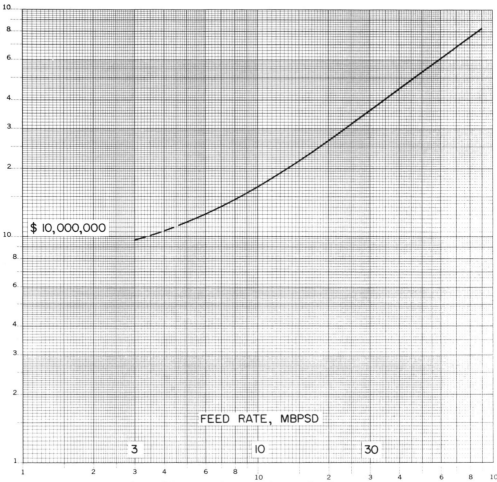

FIG. 7.23*. Fluid catalytic cracking units investment cost.
*See Table 7.3, page 124.

        The selection of a 75% conversion level was made because of the high
boiling characteristics of the feed. Checks should be made at higher and
lower conversion levels to determine the level of conversion giving the best
economic return.

        The gravity of the combined charge stock is obtained by dividing the
total pounds of charge by the volume. If the gravity is to be used for the
calculation of a characterization factor (K_w) for yield prediction, the weight
of sulfur and nitrogen should be subtracted from the weight of feed before
dividing by the volume. If this is not done, the characterization factor will
not be representative of the types of hydrocarbon compounds present in the
feedstocks.

        The yields for the various product streams are obtained from Figures
7.9 through 7.14 for silica-alumina catalyst and from Figures 7.15 through
7.21 for zeolitic catalyst. The sulfur distribution in the products is estimated
from Figure 7.22. In the case of the C_4 and lighter stream shown on Figure
7.22, it is assumed the sulfur is in the form of hydrogen sulfide (H_2S).

The weight of the total cycle gas oil (TCGO) product stream is determined by difference, the gravity from Figure 7.21 and the volume by dividing
the weight by the gravity. This volume is then used to check the conversion.
In this example, the conversion obtained is 75.2% versus the 75% used as the
base. This is a satisfactory check. If it does not check within ±1%, some
other method of determining product yields may better fit the characteristics
of the feed.

TABLE 7.4

FCC Material Balance

Basis: 100,000 BPCD North Slope, Alaska, Crude Oil
Severity: 75% Conversion, Zeolite Catalyst

Component	vol %	BPD	°API	(lb/hr)/BPD	lb/hr	wt % S	lb/hr S
Feed:							
650–800°	41.8	13,500	23.0	13.36	180,360	0.8	1,440
800–1,000°	58.2	18,800	16.8	13.92	261,700	1.5	3,930
K_W = 11.7	100.0	32,300			442,060		
Products:							
Coke, wt %	11.8				52,160		1,290*
C_2 & ltr, wt %	6.4				28,290		1,990
$C_3^=$	4.4	1,420		7.61	10,810		
C_3	2.8	900		7.42	6,680		
$C_4^=$	8.4	2,710		8.76	23,740		
i-C_4	5.3	1,710		8.22	14,060		
n-C_4	2.4	780		8.51	6,640		
C_5^+ gasoline	51.0	16,470	57.0	10.95	180,350		270
TCGO	(25.0)	(8,080)	8.3	14.77	(119,330)		1,820
LCGO		(6,460)	10.7	(14.52)	(93,780)		
HCGO	5.0	1,620	-1.0	15.82	25,550		
					442,060		5,370

*in flue gas as SO_2

TABLE 7.5

FCC Catalyst and Utility Requirements

Power, kWh/day	194,000
Cooling water, gpm	11,200
Fuel, MMBtu/day	3,230
Catalyst, lb/day	6,460

The material balance is shown in Table 7.4.

Utility requirements are estimated from table 7.3 and are listed in Table Table 7.5.

PROBLEMS

1. For a 27.0°API catalytic cracker feedstock with a boiling range of 650 to 900°F and a sulfur content of 1.2% by weight, make an overall weight and volume material balance for 10,000 BPD feed rate when operating at a 65% conversion level and a once-through operation with:

   a. Zeolite catalyst
   b. Silica-alumina catalyst

2. If the catalytic gas oil produced in Problem 1 is recycled to extinction (no products heavier than kerosine or No. 1 fuel oil are withdrawn from the unit), how much will gasoline production be increased? Coke laydown?

3. Estimate the direct operating cost, including royalty, per barrel of feed for a 20,000 BPD catalytic cracking unit if labor costs are $900/day, electric power is $0.015/kWh, steam is $1.55/Mlb, fuel is $1.20/MMBtu, silica-alumina catalyst is $400/ton, a mixture of silica-alumina and zeolite catalysts is $550/ton and zeolite catalyst is $800/ton.

4. Using today's construction costs for a 20,000 BPD catalytic cracking unit and double-rate declining-balance depreciation, estimate the cost per barrel of feed added by depreciation costs. Assume a 20-year life and a salvage value equal to dismantling costs. Calculate for the first and fifth years of operation. Compare with depreciation costs per barrel when straight-line depreciation is used.

5. For the following conditions, calculate the (a) wt % hydrogen in coke, (b) coke yield, and (c) catalyst-to-oil ratio.

Carbon on spent catalyst:	1.50 wt %
Carbon on regenerated catalyst:	0.80 wt %
Air from blower:	155,000 lb/hr
Hydrocarbon feed to reactor:	295,000 lb/hr
Flue gas analysis (Orsat), vol %:	

CO_2	12.0
CO	6.0
O_2	0.7
N_2	81.3
	100.0

6. Estimate the cost, including royalty, per gallon of gasoline boiling range product for a 20,000 BPSD catalytic cracking unit if labor costs are $1150/day, electric power is $0.07/kwh, steam is $4.25/M lb, fuel is $4.49/MM Btu, silica-alumina catalyst is $750/ton, zeolite catalyst is $1700/ton, and depreciation is $625,000/yr. The on-stream factor is 96.5%. The C_5^+ gasoline yield is 58% at 85% conversion level.

7. For a 24.0°API catalytic cracker feedstock with a boiling range of 325°C (617°F) to 510°C (950°F) and a sulfur content of 0.45% by weight, make an overall weight and volume material balance for 25,000 BPSD feed rate when operating at an 85% conversion level and a once-through operation with zeolite catalyst.

8. For the feedstock and conditions in Problem 7, compare the relative cost per barrel of C_5^+ gasoline produced at 75% conversion with that at 85% conversion assuming constant direct operating costs. Use Oil and Gas Journal values for other fuels.

9. For a 23.0°API catalytic cracker feedstock with a boiling range of 600-900°F and containing 1.5 wt % sulfur, make an overall weight and volume material balance for 10,000 BPCD feed rate when operating at 70% conversion using a zeolite catalyst.

NOTES

1. C. L. Thomas, *Catalytic Processes and Proven Catalysts* (Academic Press, New York, 1970), pp. 26-35.
2. W. L. Nelson, *Petroleum Refinery Engineering*, 4th ed. (McGraw-Hill Book Company, New York, 1958), pp. 759-810.
3. A. L. Saxton and A. C. Worley, Oil Gas J., pp. 82-99 (18 May 1970).
4. J. J. Blazek, Davison Catalagram, No. 63, pp. 2-10, 1981.
5. Hydrocarbon Process., *61*(9), 155-158 (1982).
6. E. T. Habib, Jr., and D. S. Henderson, Davison Catalagram, No. 65, 3-5, 1982.
7. R. E. Ritter, L. Rheaume, W. A. Welch, and J. S. Magee, Oil Gas J. *79*(27), 103-106 (1981).
8. G. H. Dale and D. L. McKay, Hydrocarbon Process. *56*(9), 97-99 (1979).
9. L. F. Hatch, Hydrocarbon Process. *48*(2), 77-88 (1969).
10. J. J. McKetta, ed., *Advances in Petroleum Chemistry and Refining*, vol. 5 (Interscience Pub., New York, 1962), pp. 211-254.
11. C. L. Thomas, Ind. Eng. Chem. *41*, 2564 (1949).
12. W. A. Gruse and D. R. Stevens, *Chemical Technology of Petroleum*, 3rd ed. (McGraw-Hill Book Company, New York, 1960), pp. 375-387.
13. S. B. Zdonik, E. J. Green, and L. P. Hallee, Oil Gas J., pp. 98-101 (11 Sept. 1967).
14. H. W. Knowlton, R. R. Beck, and J. J. Melnyk, Chevron Reports on Cat-Cracker Runs, Oil Gas J., pp. 57-61 (9 Nov. 1970).
15. Anon, Fluid Catalytic Cracking with Molecular Sieve Catalysts, Petro/Chem. Eng., p. 12-15 (May 1969).
16. J. S. Magee, J. J. Blazek, and R. E. Ritter, Preprints, Division of

Petroleum Chemistry, pp. B52-B65, Americal Chemical Society, New York Meeting, Aug. 27-Sept. 1, 1982.

17. E. T. Habib, H. Owen, P. W. Snyder, C. W. Streed, and P. B. Venuto, Ind. Eng. Chem. Prod. Res. Dev. *16*, 291 (1977).
18. B. R. Mitchell, Ind. Eng. Chem. Prod. Res. Dev. *19*, 209 (1980).
19. P. G. Thiel, Davison Catalagram, No. 62, 18-22 (1981).
20. D. F. Tolen, Oil and Gas J. *79*(13), 90-93 (1981).
21. C. L. Hemler, C. W. Strother, B. E. McKay, and G. D. Myers, "Catalytic Conversion of Residual Stocks," 1979 NPRA Annual Meeting, San Antonio, TX, Mar. 25-27, 1979.
22. H. V. Stuart, Davison Catalagram, No. 65, 1982.
23. M. D. Edgar, Oil and Gas. J., pp. 166-169 (3 June 1974).
24. F. Macerato and S. Anderson, Oil Gas. J. *79*(9), 101-106 (1981).
25. D. P. Bhasin, A. I. Dalton, A. J. F. Paterson, and G. Bohl, Air Products and Chemicals, Inc. 1982.
26. M. J. Fowle and R. D. Bent, Petrol. Refiner. *26*(11), 719-727 (1947).

ADDITIONAL READING

R. J. Hengstebeck, *Petroleum Processing* (McGraw-Hill Book Company, New York, 1959), pp. 148-178.

E. E. Winfree and R. Newirth, Petro/Chem. Eng., pp. 16-20 (May 1969).

D. F. Tolen, Personal Communication, 1983.

R. J. Fahrig, *Ultracat Regeneration*, National Petroleum Refiners Association, 72nd Annual Meeting, Mar. 31-Apr. 2, 1974.

R. R. Dean, J. L. Mauleon, and W. S. Letzsch, Oil Gas. J. *80*(40), 75-80; *80*(41), 168-176 (1982).

B. P. Castiglioni, Hydrocarbon Process. *62*(2), 35-38 (1983).

Chapter 8
HYDROTREATING

The terms hydrotreating, hydroprocessing, hydrocracking, and hydrodesulfurization are used rather loosely in the industry because, in the
processes hydrocracking, hydrodesulfurization, and other operations occur
simultaneously and it is relative as to which predominates. In this text,
hydrotreating refers to a relatively mild operation whose primary purpose
is to saturate olefins and/or reduce the sulfur and/or nitrogen content (and
not to change the boiling range) of the feed. Hydrocracking refers to
processes whose primary purpose is to reduce the boiling range and in
which 100% of the feed is converted to products with boiling ranges lower
than that of the feed. Hydrotreating and hydrocracking set the two ends
of the spectrum and those processes with a substantial amount of sulfur
and/or nitrogen removal and a significant change in boiling range of the
products versus the feed are called hydroprocessing.

Hydrotreating is a process to catalytically stabilize petroleum products and/or remove objectionable elements from products or feedstocks
by reacting them with hydrogen. Stabilization involves converting unsaturated hydrocarbons such as olefins and gum-forming unstable diolefins to
paraffins. Objectionable elements removed by hydrotreating include sulfur,
nitrogen, oxygen, halides, and trace metals. Hydrotreating is applied
to a wide range of feedstocks from naphtha to reduced crude. When the
process is employed specifically for sulfur removal it is usually called hydrodesulfurization or HDS.

Although there are about 30 hydrotreating processes available for
licensing [1], most of them have essentially the same process flow for a
given application. Figure 8.1 illustrates a typical hydrotreating unit.

The oil feed is mixed with hydrogen-rich gas either before or after
it is preheated to the proper reactor inlet temperature. Most hydrotreating
reactions are carried out below 800°F to minimize cracking and the feed is
usually heated to between 500 and 800°F. The oil feed combined with the
hydrogen-rich gas enters the top of the fixed-bed reactor. In the presence

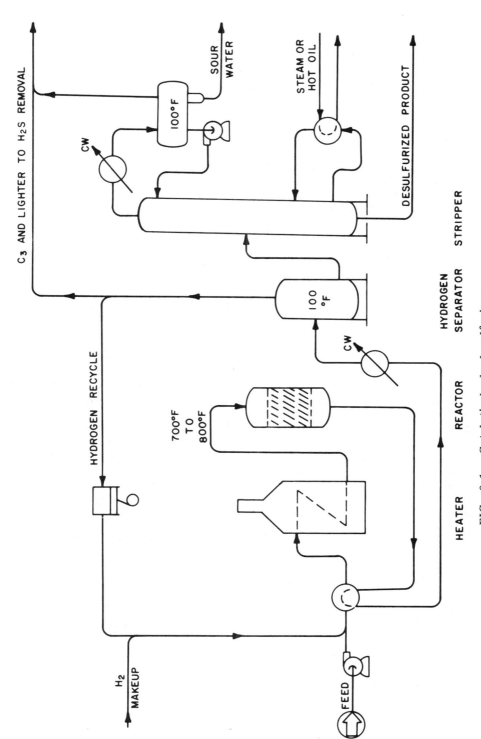

FIG. 8.1. Catalytic hydrodesulfurizer.

of the metal-oxide catalyst, the hydrogen reacts with the oil to produce
hydrogen sulfide, ammonia, saturated hydrocarbons, and free metals. The
metals remain on the surface of the catalyst and other products leave the
reactor with the oil-hydrogen stream. The reactor effluent is cooled before
separating the oil from the hydrogen-rich gas. The oil is stripped of any
remaining H_2S and light ends in a stripper. The gas is treated to remove
hydrogen sulfide and recycled to the reactor.

8.1 HYDROTREATING CATALYSTS

Catalysts developed for hydrotreating include cobalt and molybdenum oxides
on alumina, nickel oxide, nickel thiomolybdate, tungsten and nickel sulfides,
and vanadium oxide [2]. The cobalt and molybdenum oxides on alumina
calalysts are in most general use today because they have proven to be highly
selective, easy to regenerate, and resistant to poisons.

If, however, the removal of nitrogen is a significant consideration,
catalysts composed of nickel-cobalt-molybdenum or nickel-molybdenum compounds supported on alumina are more efficient. Nitrogen is usually more
difficult to remove than sulfur from hydrocarbon streams and any treatment
which reduces excess nitrogen concentration to a satisfactory level will ec-
fectively remove excess sulfur. Nickel-containing catalysts generally require
activation by presulfiding with carbon disulfide, mercaptans, or dimethyl
sulfide before bringing up to reaction temperature; however, some refiners
activate these catalysts by injecting the sulfiding chemical into the oil feed
during startup [3]. The sulfiding reaction is highly exothermic and care
must be taken to prevent excessive temperatures during activation.

Cobalt-molybdenum catalysts are selective for sulfur removal and nickel-
molybdenum catalysts are selective for nitrogen removal although both
catalysts will remove both sulfur and nitrogen [4]. Nickel-molybdenum catalysts have a higher hydrogenation activity than cobalt-molybdenum which
results, at the same operating conditions, in a greater saturation of aromatic
rings. Simply stated, if sulfur reduction is the primary objective, then a
cobalt-molybdenum catalyst will reduce the sulfur a given amount at less
severe operating conditions with a lower hydrogen consumption than nickel-
molybdenum catalyst. If nitrogen reduction or aromatic ring saturation is
desired, nickel-molybdenum catalyst is the preferred catalyst.

Catalyst consumption varies from 0.001 to 0.007 lb/bbl feed depending
upon the severity of operation and the gravity and metals content of the feed.

8.2 REACTIONS

The main hydrotreating reaction is that of desulfurization but many others
take place to a degree proportional to the severity of the operation. Typical
reactions are:

1. Desulfurization:
   a. Mercaptans: $RSH + H_2 \rightarrow RH + H_2S$
   b. Sulfides: $R_2S + 2H_2 \rightarrow 2RH + H_2S$
   c. Disulfides: $(RS)_2 + 3H_2 \rightarrow 2RH + 2H_2S$

d. Thiophenes:

$$HC{-\!\!\!-}CH$$
$$\|\quad\;\|$$
$$HC\quad CH + 4H_2 \rightarrow C_4H_{10} + H_2S$$
$$\diagdown_S\diagup$$

2. Denitrogenation:
   a. Pyrrole: $\quad C_4H_4NH + 4H_2 \rightarrow C_4H_{10} + NH_3$
   b. Pyridine: $\quad C_5H_5N + 5H_2 \rightarrow C_5H_{12} + NH_3$

3. Deoxidation:
   a. Phenol. $\quad\;\; C_6H_5OH + H_2 \rightarrow C_6H_6 + H_2O$
   b. Peroxides: $\quad C_7H_{13}OOH + 3H_2 \rightarrow C_7H_{16} + 2H_2O$

4. Dehalogenation:
   Chlorides: $\quad RCl + H_2 \rightarrow RH + HCl$

5. Hydrogenation:
   Pentene: $\quad C_5H_{10} + H_2 \rightarrow C_5H_{12}$

6. Hydrocracking:
   $C_{10}H_{22} \rightarrow C_4H_8 + C_6H_{14}$

Nitrogen removal requires more severe operating conditions than does desulfurization. The ease of desulfurization is dependent upon the type of compound. Lower-boiling compounds are desulfurized more easily than higher-boiling ones. The difficulty of sulfur removal increases in the order paraffins, naphthenes, aromatics [5].

Hydrogen consumption is about 70 scf/bbl of feed per percent sulfur, about 320 scf/bbl oil feed per percent nitrogen and 180 scf/bbl per percent oxygen removed. If operating conditions are severe enough that an appreciable amount of cracking occurs, hydrogen consumption increases rapidly. It is important to note that actual hydrogen makeup requirements are from two to ten times the amount of stoichiometric hydrogen required. This is due to the solubility loss in the oil leaving the reactor effluent separator.

All reactions are exothermic and, depending on the specific conditions, a temperature rise through the reactor of 5 to 20°F is usually observed.

8.3 PROCESS VARIABLES

The principal operating variables are temperature, hydrogen partial pressure, and space velocity.

Increasing temperature and hydrogen partial pressure increases sulfur and nitrogen removal, and hydrogen consumption. Increasing pressure also increases hydrogen saturation and reduces coke formation. Increasing space velocity reduces conversion, hydrogen consumption, and coke formation. Although increasing temperature improves sulfur and nitrogen removal, excessive temperatures must be avoided because of the increased coke formation.

Typical ranges of process variables in hydrotreating operations are [1]:

Temperature, °F	600-800
Pressure, psig	100-3,000
Hydrogen, scf/bbl charge	
Recycle	2,000
Consumption	200-800
Space velocity (LHSV)	1.5-8.0

8.4 CONSTRUCTION AND OPERATING COSTS

Hydrogen consumption for sulfur, nitrogen, and oxygen removal can be estimated from a nomograph prepared by Nelson [6] or from Section 8.2. Assume

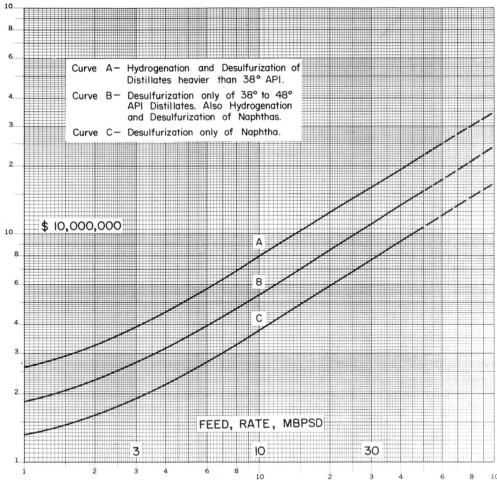

FIG. 8.2*. Catalytic desulfurization and hydrogenation units investment cost — 1982
U.S. Gulf Coast.
*See Table 8.1, page 135.

TABLE 8.1*

Catalytic Desulfurization and Hydrogenation

Costs Included:
1. Product fractionation.
2. Complete preheat, reaction, and hydrogen circulation facilities.
3. Sufficient heat exchange to cool products to ambient temperature.
4. Central control system.
5. Initial catalyst charge.

Costs Not Included:
1. Feed fractionation.
2. Makeup hydrogen generation.
3. Sulfur recovery from off-gas.
4. Cooling water, steam, and power supply.

Royalty, $/bbl feed: 0.015

Utility Data (per bbl feed):

	Type operation		
	"A"	"B"	"C"
Steam, lb	10	8	6
Power, kWh	6.0	3.0	2.0
Cooling water, gal crclt (30°FΔt)	500	400	300
Fuel (LHV), MMBtu	0.2	0.15	0.1
Hydrogen makeup, scf	400-800	150-400	100-150
Catalyst replacement, $	0.04	0.03	0.02

*See Fig. 8.2, page 134.

hydrogen loss by solution in products is approximately one pound per barrel of feed (190 scf/bbl).

Construction and operating costs can be estimated from Figure 8.2 and Table 8.1.

PROBLEMS

1. Estimate the hydrogen consumption required to completely remove the sulfur from a hydrotreater feedstock and to reduce the nitrogen content of the product to 15 ppm by weight. The 48.5°API naphtha feed to the unit contains 0.62% sulfur, 0.15% nitrogen, and 0.09% oxygen by weight.

2. The hydrogen required for hydrotreating in a refinery is usually obtained from catalytic reforming operations. Calculate the minimum barrels per day of reformer feed required to provide 120% of the hydrogen necessary to completely desulfurize 1,000 bbl/day of naphtha having the following properties:

°API	55.0
N, wt %	Nil
S, as mercaptans (RSH), wt %	0.5
S, as sulfides (R_2S), wt %	0.5
S, total, wt %	1.0

The reformer feed has the following properties:

Component	vol %	gal/lb mol
C_6 naphthenes	20.0	13.1
C_7 naphthenes	10.0	15.4
C_8 naphthenes	10.0	18.3
Paraffins	50.0	18.5
Aromatics	10.0	13.0

Make the following simplifying assumptions:

1. Conversion of naphthenes is 90% for each type.
2. C_6 naphthenes convert only to benzene, C_7 naphthenes convert only to toluene, and C_8 naphthenes convert only to xylene.
3. Aromatics and paraffins in the reformer feed do not react.

Express the actual hydrogen consumption in the hydrotreater and the hydrogen yield from the reformer as scf/bbl of feed.

NOTES

1. R. J. Hengstebeck, *Petroleum Processing* (McGraw-Hill Book Company, New York, 1959), pp. 272-279.
2. W. A. Gruse and D. R. Stevens, *Chemical Technology of Petroleum*, 3rd ed. (McGraw-Hill Book Company, New York, 1960), pp. 117-121, 306-309.
3. Oil Gas J. *66*(12), 114-120 (1968).
4. T. F. Kellett, C A. Trevino, and A. F. Sartor, Oil Gas J. *78*(18), (1980).
5. Hydrocarbon Process. *49*(9), 204-232 (1970).
6. W. L. Nelson, Oil Gas J. *69*(9), 64 (1971).

Chapter 9
CATALYTIC HYDROCRACKING AND HYDROPROCESSING

Although hydrogenation is one of the oldest catalytic processes used in refining petroleum, it has only been in recent years that catalytic hydrocracking has developed to any great extent in this country. This interest in the use of hydrocracking has been caused by several factors; among them: (1) the demand for petroleum products has shifted to high ratios of gasoline compared with the usages of middle distillates, and (2) byproduct hydrogen at low cost and in large amounts has become available from catalytic reforming operations in recent years.

The hydrocracking process was commercially developed by I. G. Farben Industrie in 1927 for converting lignite into gasoline and was brought to this country by Esso Research and Engineering Company in the early 1930's for use in upgrading petroleum feedstocks and products. Improved catalysts have been developed which permit operations at relatively low pressures, and the demand for high octane gasoline and a decrease in demand for distillate fuels has caused a necessity for converting higher boiling petroleum materials to gasoline and jet fuels.

Product balance is of major importance to any petroleum refiner. There are a number of things that can be done to balance the products made with the demand, but there are relatively few operations that offer the versatility of catalytic hydrocracking. Some of the advantages of hydrocracking are:

1. Better balance of gasoline and distillate production
2. Greater gasoline yield
3. Improved gasoline pool octane quality and sensitivity
4. Production of relatively high amounts of isobutane in the butane fraction
5. Supplements catalytic cracking to upgrade heavy cracking stocks, aromatics, cycle oils, and coker oils to gasoline, jet fuels, and light fuel oils.

137

In a modern refinery catalytic cracking and hydrocracking work as a team. The catalytic cracker takes the more easily cracked paraffinic atmospheric and vacuum gas oils as charge stocks, while the hydrocracker uses more aromatic cycle oils and coker distillates as feed [1]. These streams are very refractory and resist catalytic cracking while the higher pressures and hydrogen atmosphere make them relatively easy to hydrocrack. The new zeolite cracking catalysts help improve the gasoline yields and octanes from catalytic crackers as well as reducing the cycle stock and gas make. However, the cycle oil still represents a difficult fraction to crack catalytically to extinction. One alternative is to use the cycle stock as a component for fuel oil blending, but this is limited as it is a relatively poor burning stock and burns with a smoky flame. For this reason a limit is placed on the percentage that can be blended into distillate fuel oils. The cycle oils that result from cracking operations with zeolite catalysts tend to be highly aromatic and therefore make excellent feedstocks for hydrocracking.

In addition to middle distillates and cycle oils used as feed for hydrocracking units, it is also possible to process residual fuel oils and reduced crude by hydrocracking. This usually requires a different technology and for the purposes of our discussion the hydrocracking operation are broken into two general types of processes; those which operate on distilled feed (hydrocracking) and those which process residual materilas (hydroprocessing). These processes are similar and some the licensed processes can be adapted to operate on both types of feedstocks. There is a major difference, however, between the two processes in regard to the type of catalyst and operating conditions. During the design stages of the hydrocracker the process can be tailored to convert heavy residue into lighter oils or to change straight-run naphthas into liquefied petroleum gases. This is difficult to do after the unit is built as the processing of residual oil requires special consideration with respect to such factors as asphaltenes, ash, and metal contents of the feedstocks [2, 3].

9.1 HYDROCRACKING REACTIONS

Although there are hundreds of simultaneous chemical reactions occurring in hydrocracking, it is the general opinion that the mechanism of hydrocracking is that of catalytic cracking with hydrogenation superimposed (see Figure 9.1). Catalytic cracking is the scission of a carbon-carbon single bond and hydrogenation is the addition of hydrogen to a carbon-carbon double bond. An example of the scission of a carbon-carbon single bond followed by hydrogenation is the following:

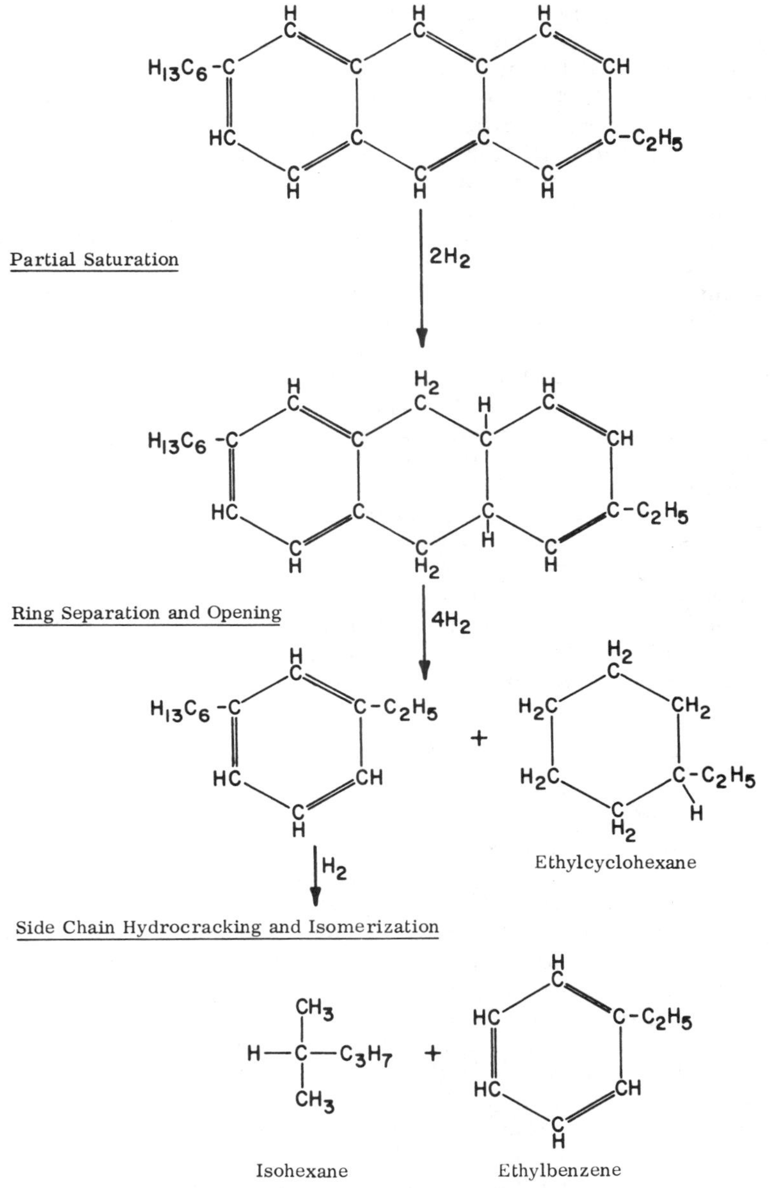

FIG. 9.1. Typical hydrocracking reaction.

This shows that cracking and hydrogenation are complementary, for cracking provides olefins for hydrogenation, while hydrogenation in turn provides heat for cracking. The cracking reaction is endothermic and the hydrogenation reaction is exothermic. The reaction actually provides an excess of heat because the amount of heat released by the exothermic hydrogenation reactions is much greater than the amount of heat consumed by the endothermic cracking reactions. This surplus of heat causes the reactor temperature to increase and accelerate the reaction rate. This is controlled by injecting cold hydrogen as quench into the reactors to absorb the excess heat of reaction.

Another reaction that occurs and illustrates the complementary operation of the hydrogenation and cracking reactions is the initial hydrogenation of a condensed aromatic compound to a cycloparaffin. This allows subsequent cracking to proceed to a greater extent and thus converts a low-value component of catalytic cycle oils to a useful product.

Isomerization is another reaction type that occurs in hydrocracking and accompanies the cracking reaction. The olefinic products formed are rapidly hydrogenated, thus preventing the reverse reaction back to straight-chain molecules and maintaining a high concentration of high octane isoparaffins. An interesting point in connection with the hydrocracking of these compounds is the relatively small amounts of propane and lighter materials that are produced as compared with normal cracking processes. The volumetric yield of liquid products can be as high as 125% of the feed because the hydrogenated products have a higher API gravity than the feed.

Hydrocracking reactions are normally carried out at average catalyst temperatures between 550 and 750°F and at reactor pressures between 1,200 and 2,000 psig. The circulation of large quantities of hydrogen with the feedstock prevents excessive catalyst fouling and permits long runs without catalyst regeneration. Careful preparation of the feed is also necessary in order to remove catalyst poisons and to give long catalyst life. Frequently the feedstock is hydrotreated to remove sulfur and nitrogen compounds as well as metals before it is sent to the first hydrocracking stage or, sometimes, the first reactor in the reactor train can be used for this purpose.

9.2 FEED PREPARATION

Hydrocracking catalyst is susceptible to poisoning by metallic salts, oxygen, organic nitrogen compounds, and sulfur in the feedstocks. The feedstock is hydrotreated to saturate the olefins and remove the sulfur, nitrogen, and oxygen compounds. Certain of the metals are retained on the catalyst. The nitrogen and sulfur compounds are removed by conversion to ammonia and hydrogen sulfide. Although organic nitrogen compounds are thought to act as permanent poisons to the catalyst, the ammonia produced by reaction of the organic nitrogen compounds with hydrogen does not affect the catalyst permanently. For some types of hydrocracking catalysts, the presence of hydrogen sulfide in low concentrations acts as a catalyst to inhibit the saturation of aromatic rings. This is a beneficial effect when maximizing gasoline production as it conserves hydrogen and produces a higher octane product.

In the hydrotreater a number of hydrogenation reactions, such as olefin saturation and aromatic ring saturation, take place but cracking is almost insignificant at the operating conditions used. The exothermic heats of the desulfurization and denitrogenation reactions are high (about 65 to 75 Btu/scf of hydrogen consumed). If the nitrogen and sulfur contents of the feedstock are high this effect contributes appreciably to the total heat of reaction.

Other reactions contributing to high heat release in the hydrotreating process are the saturation of olefins, as the heat of reaction for olefin saturation is about 140 Btu/scf of hydrogen consumed. For cracked feedstocks the olefin content is very high and olefin saturation is responsible for a large portion of the total heat of reaction. For virgin stocks, however, the olefin content is negligible and this is not an important contribution to heat of reaction. The overall heat of reaction for most hydrotreating reactors used for the preparation of hydrocracker feed is approximately 25,000 to 35,000 Btu per barrel of raw feed.

In addition to the removal of nitrogen and sulfur compounds and metals, it is also necessary to reduce the water content of the feed streams to less than 25 ppm because, at the temperatures required for hydrocracking, steam causes the crystalline structure of the catalyst to collapse and the dispersed rare-earth atoms to agglomerate. Water removal is accomplished by passing the feed stream through a silica gel or molecular sieve dryer. Exceptions to this are the Unicracking and GOfining processes, which can tolerate water contents as high as 400 to 500 ppm and it is necessary only to remove free water from the feed.

On the average, the hydrogen treating process requires approximately 150 to 300 cubic feed of hydrogen per barrel of feed.

9.3 THE HYDROCRACKING PROCESS

There are a number of hydrocracking processes available for licensing and some of these are given in Table 9.1. With the exception of the H-Oil and LC-Fining processes all hydrocracking processes in use today are fixed-bed catalytic processes with liquid downflow. The hydrocracking process may require either one or two stages, depending upon the process and the feedstocks used. The process flows of most of the fixed-bed processes are similar and the GOFining process will be described as a typical fixed-bed hydrocracking process.

The GOFining process is a fixed-bed regenerative process employing a molecular-sieve catalyst impregnated with a rare-earth metal. The process employs either single-stage or two-stage cracking with typical operating conditions ranging from 500 to 800°F and from 1,000 to 2,000 psig. The temperature and pressure vary with the age of the catalyst, the product desired, and the properties of the feedstock.

TABLE 9.1

Hydrocracking Processes Available for License

Process	Company
Isomax [4]	Standard Oil Co. (Calif.) and UOP
Unicracking	Union Oil Co.
GOFining	EXXON Research and Engineering
H-G hydrocracking [5]	Gulf Oil Co. and Houdry Process and Chemical Co.
Ultracracking [6]	Standard Oil Co. (Indiana)
H-Oil [7]	Hydrocarbon Research, Inc. and
LC-Fining	Texaco Development
	C-E Lummus
Shell	Shell Development Co.
BASF-IFP hydrocracking	Badische Anilin und Soda Fabrik, and Institut Francais Petrole
Unibon	UOP

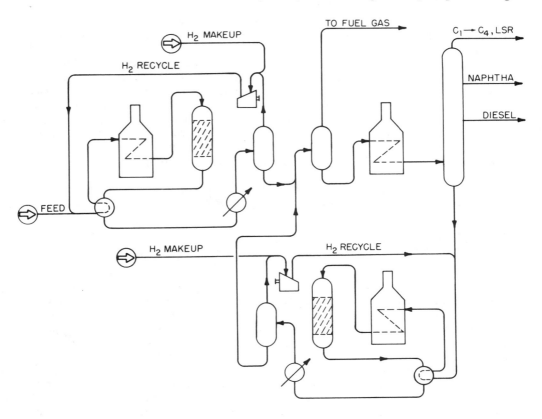

FIG. 9.2.   Two-stage hydrocracker.

The decision to use a single- or two-stage system is dependent upon the size of the unit and the product desired. For most feedstocks the use of a single stage will permit the total conversion of the feed material to gasoline and lighter products by recycling the heavier material back to the reactor. The process flow for a two-stage reactor system is shown in Figure 9.2. If only one stage is used, the process flow is the same as for the second stage of the two-stage plant.

The fresh feed is mixed with makeup hydrogen and recycle gas (high in hydrogen content) and passed through a heater to the first reactor. The first stage reactor is operated at a sufficiently high temperature to convert 40 to 50 vol % of the reactor effluent to material boiling below 400°F. The reactor effluent goes through heat exchangers to a high-pressure separator where the hydrogen-rich gases are separated and recycled to the first stage for mixing both makeup hydrogen and fresh feed. The liquid product from the separator is sent to a distillation column where the gasoline and lighter fractions are taken overhead and the bottoms used as feed to the second stage reactor. If jet fuel or diesel fuel is one of the products desired, then the distillation column separation is made with the jet fuel or diesel fuel going overhead. The bottoms from the distillation column are mixed with recycle hydrogen and sent through a furnace to the second stage reactor. Here the temperature is maintained to bring the total conversion of the unconverted oil from the first stage and second stage recycle to 50 to 70 vol % per pass.

The second stage product is combined with the first stage product prior to fractionation.

Both the first and second stage reactors contain several beds of catalysts. The major reason for having separate beds is to provide locations for injecting cold recycled hydrogen into the reactors for temperature control. In addition, redistribution of the feed and hydrogen between the beds helps to maintain a more uniform utilization of the catalyst.

When operating hydrocrackers for total conversion of distillate feeds to gasoline, the butane-and-heavier liquid yields are generally from 120 to 125 vol % of fresh feed.

9.4 HYDROCRACKING CATALYST

There are a number of hydrocracking catalysts available and the actual composition is tailored to the process, feed material, and the products desired. Most of the Hydrocracking catalysts consist of a crystalline mixture of silica-alumina with a small uniformly-distributed amount of rare earths contained within the crystalline lattice. The silica-alumina portion of the catalyst provides cracking activity while the rare-earth metals promote hydrogenation. Catalyst activity decreases with use and reactor temperatures are raised during a run to increase reaction rate and maintain conversion. The catalyst selectivity also changes with age and more gas is made and less naphtha produced as the catalyst temperature is raised to maintain conversion. With typical feedstocks it will take from two to four years for catalyst activity to decrease from the accumulation of coke and other deposits to a level which will require regeneration. Regeneration is accomplished by burning off the catalyst deposits and catalyst activity is restored to close to its original level. The catalyst can undergo several regenerations before it is necessary to replace it.

Almost all hydrocracking catalysts use silica-alumina as the cracking base but the rare-earth metals vary according to the manufacturer. Those in most common use are platinum, palladium, tungsten, and nickel.

9.5 MOVING-BED HYDROCRACKING PROCESSES

There are two moving-bed processes available for license today. These are the H-Oil and LC-fining processes which were developed by Hydrocarbon Research Incorporated (HRI) and Cities Service and C-E Lummus. The LC-fining and H-Oil processes are designed to process heavy feeds such as atmospheric tower bottoms and use metals removal and hydrotreating catalysts. A simplified process flow diagram for the LC-fining process is shown in Figure 9.3a and b. (See also Photo 9, Appendix F.)

The preheated feed, recycle, and makeup hydrogen are charged to the first reactor of the unit. The liquid passes upward through the catalyst, which is maintained as an ebullient bed. The first stage reactor effluent is sent to the second stage reactor for additional conversion. The product from the second reactor passes through a heat exchanger to a high-pressure separator where the recycle gas is removed. The liquid from the high-pressure separator is sent to a low-pressure flash drum to remove additional gases. The liquid stream at low pressure then goes to a rectification column for sepa-

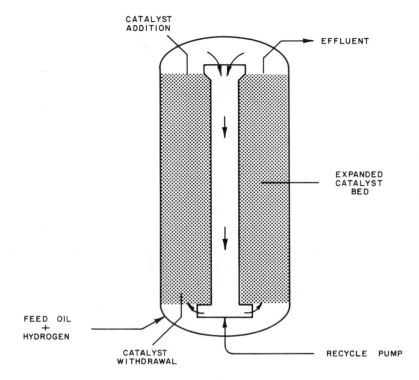

FIG. 9.3a. Expanded or ebullated bed reactor [8].

ration into products. The operating pressure of the units is a function
of feed boiling point with operating pressures up to 3,000 psig used when
charging vacuum tower residuum. The operating temperature is a function
of charge stock and conversion but is normally in the range of 800 to 850°F.

One of the main advantages of the moving-bed reactor process is the
ability to add and remove catalyst during operation. This permits operators
to regenerate catalyst while remaining on-stream and to maintain catalyst activity by either regeneration or the addition of fresh catalyst.

Typical product yields from LC-fining cracking are given in Table 9.2.

9.6 PROCESS VARIABLES

The severity of the hydrocracking reaction is measured by the degree of
conversion of the feed to lighter products. Conversion is defined as the
volume percent of the feed which disappears to form products boiling below
the desired product end point. In order to compare operation severities it is
necessary to equate conversions to the same product end point. A given
percent conversion at a low product end point represents a more severe operation than does the same percent conversion at a higher product end point.

The primary reaction variables are reactor temperature and pressure,
space velocity, hydrogen consumption, nitrogen content of feed, and hydrogen sulfide content of the gases. The effects of these are as follows:

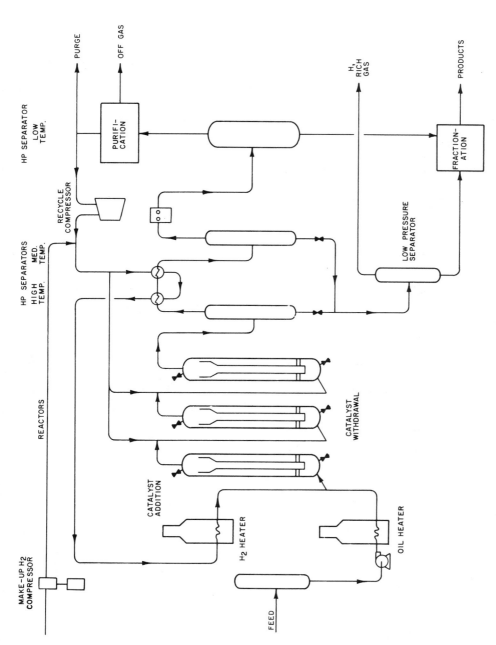

FIG. 9.3b. LC-fining expanded bed hydroprocessing unit [8].

TABLE 9.2

LC-Fining Yields [8]

	Long Resid	Short Resid
Feed stocks		
Gravity, °API	15.7	10.7
Sulfur, wt %	2.7	3.2
RCR, wt %	9.4	
Metals:		
Vanadium, ppm	110	147
Nickel, ppm	27	35
IBP-1050°F, vol %	55	33.2
1050°F+, vol %	45	66.8
Yields		
Gas, C_3^-, scf/bbl	350	590
Naphtha, C_4-400°F, vol %	17.8	13.1
Gravity, °API	61.0	64.0
Sulfur, wt %	<0.1	0.04
Kerosine,		
400-500°F, vol %	10.2	8.9
Gravity, °API	38.8	37.2
Sulfur, wt %	0.1	0.1
AGO, 500-600°F, vol %	19.7	14.3
Gravity, °API	31.8	30.1
Sulfur, wt %	0.25	0.18
VGO, 650-1050°F, vol %	37.1	36.1
Gravity, °API	22.4	23.0
Sulfur, wt %	0.6	0.6
Pitch, 1050°F+, vol %	20.0	32.7
Gravity, °API	7.7	7.0
Sulfur, wt %	1.3	2.3
H_2 consumption, scf/bbl	985	1310
Catalyst consumption,		
lb/bbl	0.15	0.12

Temperature

Reactor temperature is the primary means of conversion control. At normal reactor conditions a 20°F increase in temperature almost doubles the reaction rate but does not affect the conversion level as much because a portion of the reaction rate involves material that has already been converted to materials boiling below the product end point. As the run progresses it is necessary to raise the average temperature from 0.1 to 0.2°F per day to compensate for the loss in catalyst activity.

Reactor Pressure

The primary effect of reactor pressure is in its effects on the partial pressures of hydrogen and ammonia. An increase in total pressure increases the partial pressures of both hydrogen and ammonia. Conversion increases with increasing hydrogen partial pressure and decreases with increasing ammonia partial pressure. The hydrogen effect is greater, however, and the net effect of raising total pressure is to increase conversion.

Space Velocity

The volumetric space velocity is the ratio of liquid flow rate in barrels per hour to catalyst volume in barrels. The catalyst volume is constant, therefore the space velocity varies directly with feed rate. As the feed rate increases, the time of catalyst contact for each barrel of feed is decreased and conversion is lowered. In order to maintain conversion at the proper level when the feed rate is increased, it is necessary to increase the temperature.

Nitrogen Content

The organic nitrogen content of the feed is of great importance as the hydrocracking catalyst is deactivated by contact with organic nitrogen compounds. As increase in organic introgen content of the feed causes a decrease in conversion.

Hydrogen Sulfide

At low concentrations the presence of hydrogen sulfide acts as a catalyst to inhibit the saturation of aromatic rings. This conserves hydrogen and produces a product with a higher octane number because the aromatic naphtha has a higher octane than does its naphthenic counterpart. However, hydrocracking in the presence of a small amount of hydrogen sulfide normally produces a very low smoke point jet fuel. At high hydrogen sulfide levels corrosion of the equipment becomes important and the cracking activity of the catalyst is also affected adversely.

9.7 HYDROCRACKING YIELDS

The yields for hydrocracking to produce gasoline as the primary product can be calculated from charts and equations developed by W. L. Nelson [9-11]. The data needed to start the calculations are the Watson characterization factor (K_w) of the feed and the hydrogen consumption in scf/bbl feed. With this information proceed as follows:

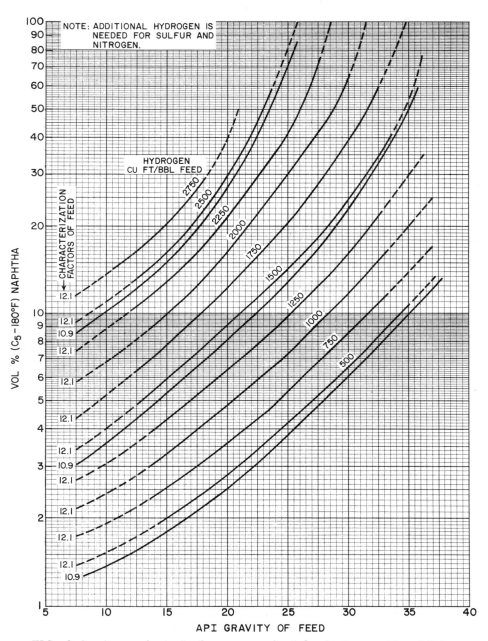

FIG. 9.4. Approximate hydrogen required for hydrocracking [12] .

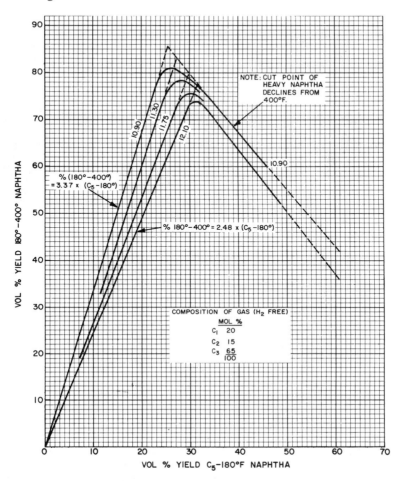

FIG. 9.5. Relationship between the yields of (C_5-180°F) and (180-400°F) hydrocrackates [12].

1. Use Figure 9.4 to determine the volume % (C_5-180°F) naphtha.
2. Enter Figure 9.5 with the volume % (C_5-180°F) naphtha and feed K to obtain volume % (180 to 400°F) naphtha.
3. Calculate the liquid volume % butanes formed from:
   LV% i-C_4 = 0.377 [LV% (C_5-180°)]
   LV% n-C_4 = 0.186 [LV% (C_5=180°)]
4. Calculate the weight % of propane and lighter from:
   wt % C_3 and lighter = 1.0 + 0.09 [LV% (C_5-180°)].

It is necessary to make both weight and hydrogen balances on the unit. The gravities of the product streams can be calculated using the K factors of the product streams obtained from Figure 9.6 and average mid-boiling points of 131°, 281°, and 460°F for the (C_5-180°F) naphtha, (180 to 400°F) naphtha, and +400°F streams, respectively. The weight of the 400°F+ stream is obtained by difference. The chemical hydrogen consumed should be included with the total weight of the feed.

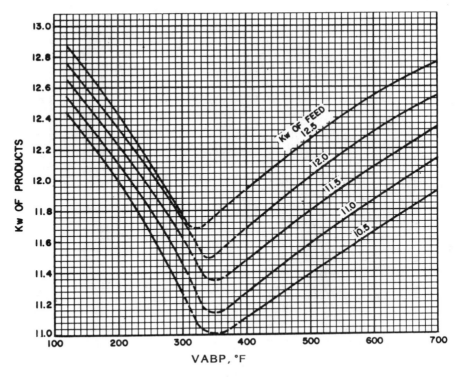

FIG. 9.6. Characterization factor of hydrocracker products [10].

Hydrogen contents of the streams can be estimated using the weight % hydrogen for each stream, except the heavy hydrocrackate (180 to 400°F), obtained from Figure 9.7 [13]. The heavy hydrocrackate is highly naphthenic and contains from 13.3 to 14.5 wt % hydrogen (avg 13.9%). Assume hydrogen loss by solution in products of 1 lb/bbl feed (range 0.8 to 1.3 lb H_2/bbl feed) [9].

It should be noted that if the yield of C_5-180°F naphtha is greater than 25 to 30 volume % the yield of heavy hydrocrackate (180 to 400°F naphtha) will be determined from a curve having a negative slope. This indicates an economically unattractive situation in that heavy hydrocrackate is being cracked to lighter materials. A less severe operation should be used.

The composition of the C_3 and lighter stream will vary depending upon feedstock properties and operating conditions. For the purpose of preliminary studies, the following composition can be assumed:

	mol %	wt %
C_1	20	8.8
C_2	15	12.4
C_3	65	78.7
	100	100.0

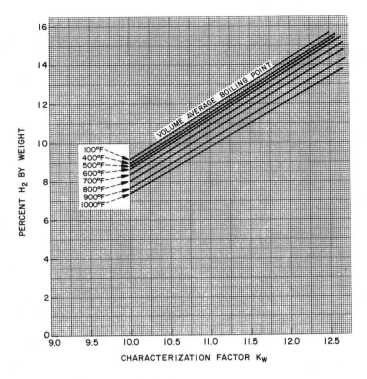

FIG. 9.7.   Hydrogen content of hydrocarbons. From "Industrial Chemical
Calculations," D. A. Hogan and K. M. Watson, copyright 1938,
John Wiley and Sons. Reprinted by permission of John Wiley and
Sons, Inc.

These values were determined by averaging the compositions obtained when
processing thirteen feedstocks ranging from virgin gas oil to coker and fluid
catalytic cracker gas oil. Within the precision of the data, the same average
composition was found for the C_3 and lighter streams when operating to
obtain all gasoline or maximum jet fuel liquid products.

9.8   INVESTMENT AND OPERATING COSTS

Capital investment costs for catalytic hydrocracking units can be estimated
from Figure 9.8. Table 9.3 lists the items included in the investment cost
obtained from Figure 9.8 and also the utility requirements for operation.

9.9   HYDROPROCESSING

The term hydroprocessing is used to denote those processes used to reduce
the boiling range of the feedstock as well as to remove substantial amounts
of impurities such as metals, sulfur, nitrogen, and high carbon-forming compounds. As presently practiced, H-Oil and LC-fining fall into this category

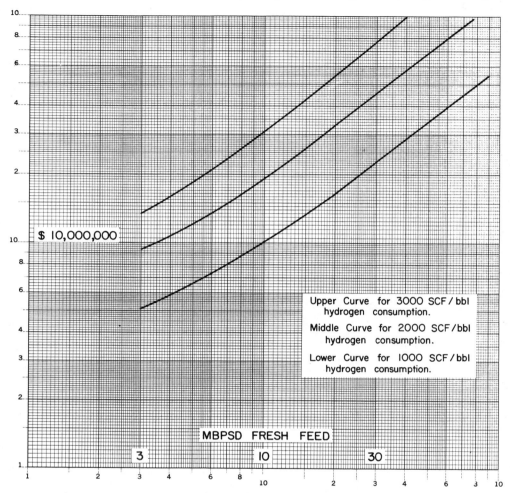

FIG. 9.8* Catalytic hydrocracking units investment cost—1982 U.S. Gulf
             Coast.
*See Table 9.3, page 153.

rather than hydrocracking. In hydroprocessing processes, feed conversion
levels of 25 to 65% can be attained. Other names applied to this operation are
hydroconversion, hydrorefining, and resid HDS.

    In U.S. refineries, hydroprocessing units are used to prepare residual
stream feedstocks for cracking and coking units. Although vacuum resids
can be used as feedstocks, most units use atmospheric resids as feeds because
the lower viscosities and impurity levels give better overall operations and
greater impurity reductions in the 1050°F+ (566°C+) fractions. Typically the
heavy naphtha fraction of the products will be catalytically reformed to improve octanes, the atmospheric gas oil fraction hydrotreated to reduce aromatic content and improve cetane number, the vacuum gas oil fraction used as
conventional FCC unit feed, and the vacuum tower bottoms sent to a heavy oil
cracker or coker.

TABLE 9.3*

Catalytic Hydrocracking Units

Costs Included:

1. Stabilization of gasoline.
2. Fractionation into two products.
3. Complete preheat, reaction, and hydrogen circulation facilities.
4. Hydrogen sulfide removal from hydrogen recycle.
5. Sufficient heat exchange to cool products to ambient temperature.
6. Central control system.
7. Electric-motor driven hydrogen recycle compressors.

Costs Not Included:

1. Initial catalyst charge. This is approximately $150/BPD of feed.
2. Hydrogen generation and supply facilities.
3. Spare hydrogen recycle compressors.
4. Recovery of butanes, propane, etc., from gas.
5. Feed fractionation.
6. Conversion of hydrogen sulfide to sulfur.
7. Cooling water, steam, and power supply.
8. Paid up royalty.

Royalty ($/bbl feed):   0.05 to 0.10

Utility Data (per bbl feed):

Hydrogen consumption, scf	1,000	2,000	3,000
Steam, lb	50	75	100
Power, kWh†	8	13	18
Cooling Water, gal crclt (30° FΔt)	300	450	600
Fuel (LHV), MMBtu	0.1	0.2	0.3
Catalyst Replacement, $	0.05	0.10	0.20

---

*See Fig. 9.8, page 152.
† Includes electric drive for hydrogen recycle compressors.

With the exception of H-Oil and LC-fining, the processes have fixed-bed reactors and usually require the units to be shut down to have the catalyst changed when catalyst activity declines below the accepted level of activity. A typical process flow diagram is shown in Figure. 9.9.

All units operate at very high pressures, usually above 2000 psig (135 bars), and low space velocities of 0.2 to 0.5 v/hr/v. The low space velocities and high pressures limit charge rates to 30,000 to 40,000 BPSD per train of reactors. Typically each train will have a guard reactor to reduce the metals contents and carbon-forming potential of the feed, followed by three to four hydroprocessing reactors in series. The guard reactor's catalyst is a large pore-size (150-200 A°) silica-alumina catalyst with a low-level loading of hydrogenation metals such as cobalt and molybedenum. The catalysts in the other reactors are tailor-made for the feedstocks and conversion levels desired and may contain catalysts with a range of pore and particle sizes as well as different catalytic metal loadings and types (i.e.

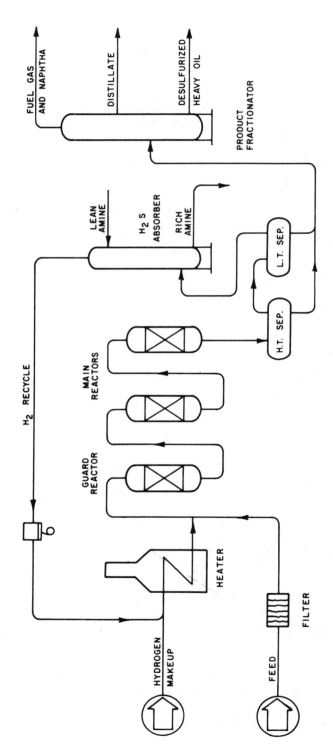

FIG. 9.9. Exxon RESIDfining hydroprocessing unit.

cobalt and molybdenum or nickel and molybdenum). Typical pore sizes will be in the 80 to 100 A° range.

The process flow, as shown on Figure 9.9, is very similar to that of a conventional hydrocracking unit except for the amine absorption unit to remove hydrogen sulfide from the recycle hydrogen stream and the guard reactor to protect the catalyst in the reactor train. The heavy crude oil feed to the atmospheric distillation unit is desalted in a two- or three-stage desalting unit to remove as much of the inorganic salts and suspended solids as possible because these will be concentrated in the resids. The atmospheric resids are filtered before being fed to the hydroprocessing unit to remove solids greater than 25 A° in size, mixed with recycle hydrogen, heated to reaction temperature and charged into the top of the guard reactor. Suspended solids in the feed will be deposited in the top section of the guard reactor and most of the metals will be deposited on the catalyst. There is a substantian reduction in the Conradson and Ramsbottom carbons in the guard reactor and the feed to the following reactors is low in metals and carbon forming precursors.

The three or four reactors following the guard reactor are operated to remove sulfur and nitrogen and to crack the 1050°F+ (566°C+) material to lower boiling compounds. Recycle hydrogen is separated and the hydrocarbon liquid stream fractionated in atmospheric and vacuum distillation columns. Results of hydroprocessing a heavy Venezuelan crude oil (Jobo) are shown in Table 9.4.

TABLE 9.4

Results from Hydroprocessing Jobo Crude [14]

	Feed	Product
Gravity, °API	8.5	22.7
Sulfur, wt %	4.0	.8
Nickel, ppm	89	5
Vanadium, ppm	440	19
Residue, wt %	13.8	2.8

9.10 CASE-STUDY PROBLEM: HYDROCRACKER

Table 9.5 shows the hydrocracker material balance. Hydrocracker catalyst and utility requirements are shown in Table 9.6.

TABLE 9.5

Hydrocracker Material Balance

Basis: 100,000 BPCD, North Slope, Alaska, Crude Oil
Hydrocracker: 17,200 BPCD Fresh Feed
Severity: 2,750 scf H_2/bbl

Component	vol %	BPD	°API	(lb/hr)/BPD	lb/hr	wt % S	lb/hr S	K_w
Feed:								
Coker GO	61.8	10,740	24.0	13.27	142,530	1.95	2,780	11.3
FCC LCGO	38.2	6,460	10.7	14.52	93,780	1.2	1,130	10.0
	100.0	17,200	18.9	13.73	236,310		3,910	10.8
Hydrogen, scfb	2,750				10,240			
					246,550			
Products:								
H_2S					4,150		3,910	
C_3 and ltr (wt %)	2.9				6,850			
i-C_4	8.1	1,370	119.4	8.22	11,260			
n-C_4	4.0	680	110.8	8.51	5,790			
C_5-180°	21.5	3,640	77.5	9.88	35,960			12.44
180-400°	74.0	12,530	46.0	11.63	145,720			11.45
400-520°	16.9	2,870	29.3	12.85	36,820			11.38
	124.5	21,090			246,550			

Hydrogen Balance	wt % H_2	lb/hr		lb/hr H_2
Out:				
H_2S		4,150		240
C_3 and ltr	20.0	6,850		1,370
C_4	17.2	17,050		2,930
C_5-180°	15.4	35,960		5,540
180-400°	13.8	145,720		20,110
400-520°	12.5	36,820		4,600
H_2 in soln				
(1 lb/bbl feed)				710
				35,500
In:				
Coker GO	11.9	139,750		16,630
FCC LCGO	8.9	92,650		8,250
Hydrogen	100.0	10,240		10,240
				35,120
Added H_2*				380
				35,500

*Increase H_2 consumption by 380 lb/hr to give 2,850 scfb H_2 used.

TABLE 9.6

Hydrocracker Catalyst and Utility Requirements

Steam, lb/hr	66,310
Power, MkWh/day	245
Cooling water, gpm	6,620
Fuel, MMBtu/day	4,655
Catalyst, $/day	550

PROBLEMS

1. A hydrocracker feedstock has a boiling range of 650 to 920°F, an API gravity of 23.7° and contains 1.7 wt % sulfur. If the hydrocracking hydrogen consumption is 1,500 scf/bbl of feed and the feed rate is 7,500 BPSD, determine (a) total hydrogen consumption, (b) barrels of gasoline, and (c) barrels of jet fuel produced per day.

2. For the feed of Problem 1, calculate the feed rate in barrels per day needed to produce sufficient isobutane for an alkylation unit producing 3,500 BPD of alkylate. Assume the hydrocracking hydrogen consumption is 1,750 scf/bbl of feed and 0.65 bbl of isobutane is needed to produce 1 bbl of alkylate.

3. For the feedstock and under the conditions of Problem 1, estimate the characterization factors of the gasoline and jet fuel fractions produced by the hydrocracker.

4. Make an overall material balance, including a hydrogen balance for a 10,000 BPSD hydrocracker with a 26.5°API feedstock having a characterization factor of 12.1 and containing 0.7% sulfur, 0.3% nitrogen, and 0.15% oxygen by weight. The hydrocracking hydrogen consumption is 2,000 scf/bbl of feed.

5. Calculate direct operating costs, excluding labor, per barrel of feed for a 10,000 BPSD hydrocracker that has a total hydrogen consumption of 1,780 scf/bbl of feed and hydrogen has a value of $3.00/Mscf.

6. Calculate the total hydrogen consumption for hydrocracking 10,000 BPSD of 24.0°API, 325°C (617°F) to 510°C (950°F) boiling range feedstock containing 0.45% sulfur, 0.18% nitrogen, and 0.11% oxygen by weight to a total gasoline liquid product.

7. Using the cost figures in Problem 7.6 and assuming labor costs are equal on the two units, compare the costs per barrel of gasoline product for hydrocracking and fluid catalytic cracking the feedstocks in Problems 7.7 and 9.5. The cost of hydrogen is $3.00/Mscf.

8. Calculate the hydrogen consumption and jet fuel production for the feedstock in Problem 9.6 if the hydrocracker was operated to maximize jet fuel production (45°API fraction). How much gasoline is produced?

9. Using the cost figures in Problem 7.6 and product prices in a current
    issue of the Oil and Gas Journal, estimate the hydrocracking processing
    costs per barrel of liquid product and the value added per barrel of
    product.

NOTES

1. J. R. Murphy, M. R. Smith, and C. H. Viens, Oil Gas J. *68*(23),
    108-112 (1970).
2. Hydrocarbon Process. *49*(9), 167-173 (1970).
3. Petro/Chem. Eng. *41*(5), 30-52 (1969).
4. C. H. Watkins and W. L. Jacobs, Hydrocarbon Process. *45*(5), 159-
    164 (1966).
5. R. G. Craig, E. A. White, A M. Henke, and S. J. Kwolek, Hydro-
    carbon Process. *45*(5), 159-164 (1966).
6. C. G. Frye, D. L. Muffat, and H. W. McAninch, Oil Gas J. *68*(20),
    69-71 (1970).
7. W. McFatter, E. Meaux, W. Mounce, and R. Van Driesen, Oil and
    Gas J. *67*(27), 119-122 (1969).
8. J. A. Celestinos, R. G. Zermeno, R. P. Van Driesen, and E. D.
    Wysoki, Oil Gas. J. *46*(48), 127-134 (1975).
9. W. L. Nelson, Oil Gas J. *65*(26), 84-85 (1967).
10. W. L. Nelson, Oil Gas J. *69*(9), 64-65 (1971).
11. W. L. Nelson, Oil Gas J. *71*(44), 108 (1973).
12. W. L. Nelson, Oil Gas J. *65*(26), 85 (1967).
13. O. A. Hougen and K. M. Watson, *Chemical Process Principles*, vol. 1
    (John Wiley and Sons, New York, 1943), p. 333.
14. Exxon Research and Engineering Company, *RESIDfining Information*,
    Florham Park, New Jersey, May, 1982.

ADDITIONAL READING

J. W. Ward, R. C. Hansford, A. D. Reichle, and J. Sosnowski, Oil
Gas J. *71*(22), 69-73 (1973).
W. L. Nelson, Oil Gas J. *65*(26), 84 (1967).
R. P. Van Briesen, *Altering Today's Refinery, Energy Bureau Con-
ference On Petroleum Refining*, Houston, Texas, Sept. 28, 1981.

Chapter 10
ALKYLATION AND POLYMERIZATION

The addition of an alkyl group to any compound is an alklation reaction but in petroleum refining terminology the term alkylation is used for the reaction of low-molecular-weight olefins with an isoparaffin to form higher-molecular-weight isoparaffins [1]. Although this reaction is simply the reverse of cracking, the belief that paraffin hydrocarbons are chemically inert delayed its discovery until about 1935 [2]. The need for high-octane aviation fuels during World War II acted as a stimulus to the development of the alkylation process for production of isoparaffinic gasolines of high octane number.

Although alkylation can take place at high temperatures and pressures without catalysts, the only processes of commercial importance involve low-temperature alkylation conducted in the presence of either sulfuric or hydro-fluoric acid. The reactions occurring in both processes are complex and the product has a rather wide boiling range. By proper choice of operating conditions, most of the product can be made to fall within the gasoline boiling range with motor octane numbers from 88 to 94 range [2] and research octane numbers from 94 to 99.

10.1 ALKYLATION REACTIONS

In alkylation processes using hydrofluoric or sulfuric acids as catalysts, only isoparaffins with tertiary carbon atoms, such as isobutane or isopentane, react with the olefins. In practice only isobutane is used because isopentane has a sufficiently high octane number and low vapor pressure to allow it to be effectively blended directly into finished gasolines.

The process using sulfuric acid as a catalyst is much more sensitive to temperature than the hydrofluoric acid process. With sulfuric acid it is necessary to carry out the reactions at 50 to 70°F or lower, to minimize oxidation

reduction reactions which result in the formation of tars and the evolution
of sulfur dioxide. When anhydrous hydrofluoric acid is the catalyst, the
temperature is usually limited to 100°F or below [2]. In both processes, the
volume of acid employed is about equal to that of the liquid hydrocarbon
charge and sufficient pressure is maintained on the system to keep the hydrocarbons and acid in the liquid state. High isoparaffin/olefin ratios (4:1 to
15:1) are used to minimize polymerization and to increase product octane. Efficient agitation to promote contact between the acid and hydrocarbon phases is
essential to high product quality and yields. Contact times from 10 to 40 minutes
are in general use. The yield, volatility, and octane number of the product is
regulated by adjusting the temperature, acid/hydrocarbon ratio, and isopar-
affin/olefin ratio. At the same operating conditions, the products from the
hydrofluoric and sulfuric acid alkylation process are quite similar [3, 4]. In
practice, however, the plants are operated at different conditions and the
products are somewhat different. The effects of variables will be discussed
for each process later, but for both processes the more important variables
are:

1. Reaction temperature
2. Acid strength
3. Isobutane concentration
4. Olefin space velocity

The principal reactions which occur in alkylation are the combinations
of olefins with isoparaffins as follows:

$$CH_3-\underset{\underset{CH_3}{|}}{C}=CH_2 + CH_3-\underset{\underset{CH_3}{|}}{CH}-CH_3 \rightarrow CH_3-\underset{\underset{CH_3}{|}}{\overset{\overset{CH_3}{|}}{C}}-CH_2-\underset{\underset{CH_3}{|}}{CH}-CH_3$$

isobutylene   isobutane   2,2,4-trime+hylpentane
                                                    (isooctane)

$$CH_2{=}CH-CH_3 + CH_3-\underset{\underset{CH_3}{|}}{CH}-CH_3 \rightarrow CH_3-\underset{\underset{CH_3}{|}}{CH}-CH_2-CH_2-CH_3$$

propylene   isobutane   2,2-dimethylpentane
                                                    (isoheptane)

Another significant reaction in propylene alkylation is the combination of propylene with isobutane to form propane plus isobutylene. The isobutylene then
reacts with more isobutane to form 2,2,4-trimethylpentane (isooctane). The
first step involving the formation of propane is referred to as a hydrogen

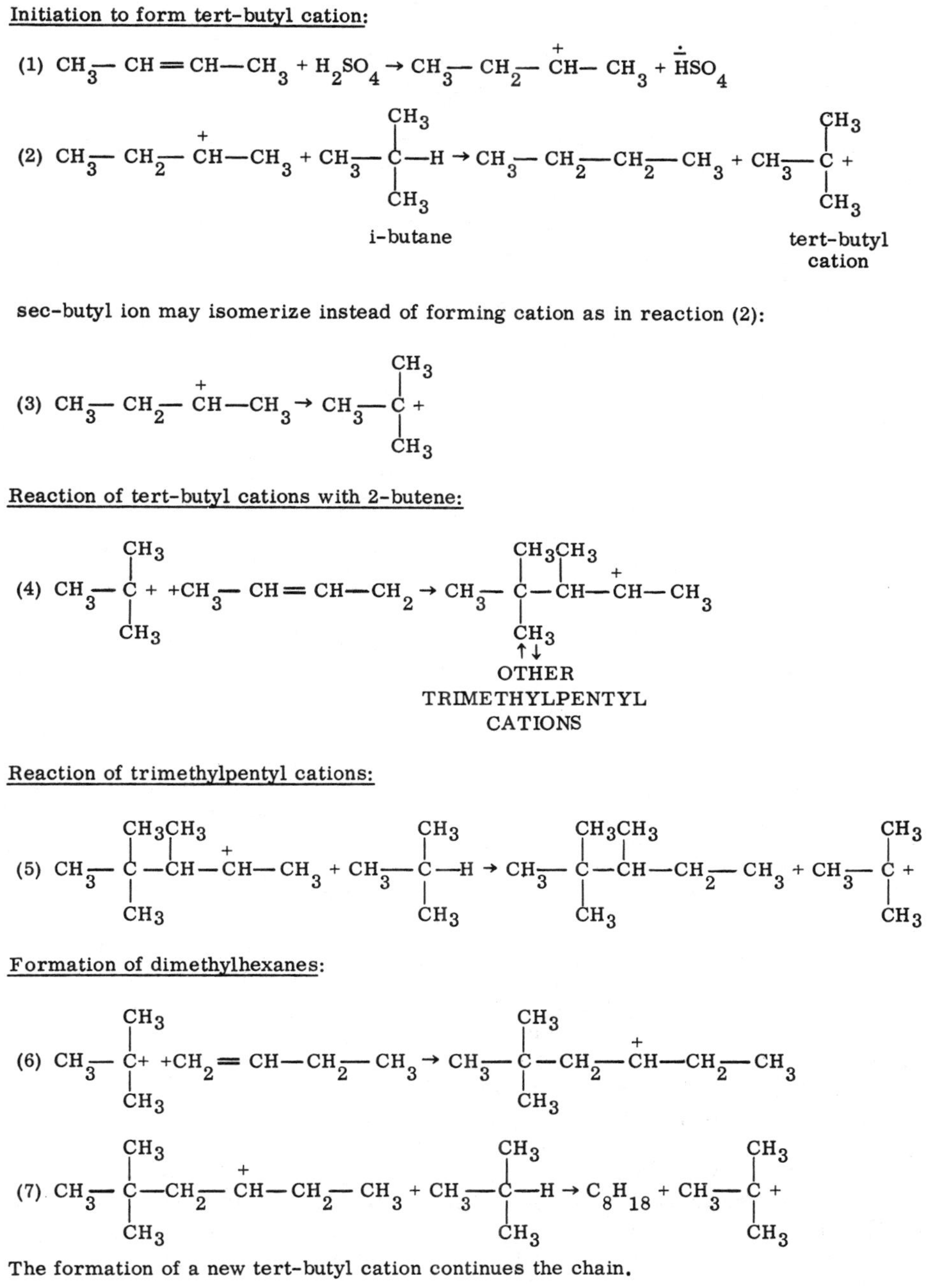

Initiation to form tert-butyl cation:

(1) $CH_3-CH=CH-CH_3 + H_2SO_4 \rightarrow CH_3-CH_2-\overset{+}{C}H-CH_3 + \overset{..}{H}SO_4$

(2) $CH_3-CH_2-\overset{+}{C}H-CH_3 + CH_3-\underset{\underset{CH_3}{|}}{\overset{\overset{CH_3}{|}}{C}}-H \rightarrow CH_3-CH_2-CH_2-CH_3 + CH_3-\underset{\underset{CH_3}{|}}{\overset{\overset{CH_3}{|}}{C}}+$

i-butane $\qquad\qquad$ tert-butyl cation

sec-butyl ion may isomerize instead of forming cation as in reaction (2):

(3) $CH_3-CH_2-\overset{+}{C}H-CH_3 \rightarrow CH_3-\underset{\underset{CH_3}{|}}{\overset{\overset{CH_3}{|}}{C}}+$

Reaction of tert-butyl cations with 2-butene:

(4) $CH_3-\underset{\underset{CH_3}{|}}{\overset{\overset{CH_3}{|}}{C}}+ +CH_3-CH=CH-CH_2 \rightarrow CH_3-\underset{\underset{CH_3}{|}}{\overset{\overset{CH_3\,CH_3}{|\quad|}}{C}}-CH-\overset{+}{C}H-CH_3$

OTHER
TRIMETHYLPENTYL
CATIONS

Reaction of trimethylpentyl cations:

(5) $CH_3-\underset{\underset{CH_3}{|}}{\overset{\overset{CH_3CH_3}{|\quad|}}{C}}-CH-\overset{+}{C}H-CH_3 + CH_3-\underset{\underset{CH_3}{|}}{\overset{\overset{CH_3}{|}}{C}}-H \rightarrow CH_3-\underset{\underset{CH_3}{|}}{\overset{\overset{CH_3CH_3}{|\quad|}}{C}}-CH-CH_2-CH_3 + CH_3-\underset{\underset{CH_3}{|}}{\overset{\overset{CH_3}{|}}{C}}+$

Formation of dimethylhexanes:

(6) $CH_3-\underset{\underset{CH_3}{|}}{\overset{\overset{CH_3}{|}}{C}}+ +CH_2=CH-CH_2-CH_3 \rightarrow CH_3-\underset{\underset{CH_3}{|}}{\overset{\overset{CH_3}{|}}{C}}-CH_2-\overset{+}{C}H-CH_2-CH_3$

(7) $CH_3-\underset{\underset{CH_3}{|}}{\overset{\overset{CH_3}{|}}{C}}-CH_2-\overset{+}{C}H-CH_2-CH_3 + CH_3-\underset{\underset{CH_3}{|}}{\overset{\overset{CH_3}{|}}{C}}-H \rightarrow C_8H_{18} + CH_3-\underset{\underset{CH_3}{|}}{\overset{\overset{CH_3}{|}}{C}}+$

The formation of a new tert-butyl cation continues the chain.

FIG. 10.1. Alkylation chemistry.

transfer reaction. Research on catalyst modifiers is being conducted to promote this step since it produces a higher octane alkylate than is obtained by formation of isoheptanes.

A number of theories have been advanced to explain the mechanisms of catalytic alkylation and these are discussed in detail by Gruse and Stevens [2]. The one most widely accepted involves the formation of carbonium ions by transfer of protons from the acid catalyst to olefin molecules, followed by combination with isobutane to produce tertiary-butyl cations. The tertiary-butyl ion reacts with 2-butene to form C_8 carbonium ions capable of reacting with isobutane to form C_8 paraffins and tertiary-butyl ions. These tertiary-butyl ions then react with other 2-butene molecules to continue the chain. Figure 10.1 illustrates the above sequence using sulfuric acid, 2-butene, and isobutane as the example reaction. The alkylation reaction is highly exothermic, with the liberation of 124,000 to 140,000 Btu per barrel of isobutane reacting [5].

10.2 PROCESS VARIABLES

The most important process variables are reaction temperature, acid strength, isobutane concentration, and olefin space velocity. Changes in these variables affect both product quality and yield.

Reaction temperature has a greater effect in sulfuric acid processes than in those using hydrofluoric acid. Low temperatures mean higher quality and the effect of changing sulfuric acid reactor temperature from 25 to 55°F is to decrease product octane from one to three numbers depending upon the efficiency of mixing in the reactor. In hydrofluoric acid alkylation, increasing the reactor temperature from 60 to 125°F degrades the alkylate quality about three octane numbers [4].

In sulfuric acid alkylation, low temperatures cause the acid viscosity to become so great that good mixing of the reactants and subsequent separation of the emulsion is difficult. At temperatures above 70°F, polymerization of the olefins becomes significant and yields are decreased. For these reasons the normal sulfuric acid reactor temperature is from 40 to 50°F with a maximum of 70°F and a minimum of 30°F.

For hydrofluoric acid alkylation, temperature is less significant and reactor temperatures are usually in the range of 70 to 100°F.

Acid strength has varying effects on alkylate quality depending on the effectiveness of reactor mixing and the water content of the acid. In sulfuric acid alkylation, the best quality and highest yields are obtained with acid strengths of 93 to 95% by weight of acid, 1 to 2% water and the remainder hydrocarbon diluents. The water concentration in the acid lowers its catalytic activity about 3 to 5 times as much as hydrocarbon diluents, thus an 88% acid containing 5% water is a much less effective catalyst than the same strength acid containing 2% water. The poorer the mixing in a reactor the higher the acid strength necessary to keep acid dilution down [4]. Increasing acid strength from 89 to 93% by weight increases alkylate quality by one to two octane numbers.

In hydrofluoric acid alkylation the highest octane number alkylate is attained in the 86 to 90% by weight acidity range. Commercial operations usually have acid concentrations between 83 and 92% hydrofluoric acid and contain less than 1% water.

Isobutane concentration is generally expressed in terms of isobutane/olefin ratio. High isobutane/olefin ratios increase octane number and yield and reduce side reactions and acid consumption. In industrial practice the isobutane/olefin ratio on reactor charge varies from 5:1 to 15:1. In reactors employing internal circulation to augment the reactor feed ratio, internal ratios from 100:1 to 1,000:1 are realized.

Olefin space velocity is defined as the volume of olefin charged per hour divided by the volume of acid in the reactor. Lowering the olefin space velocity reduces the amount of high boiling hydrocarbons produced, increases the product octane and lowers acid consumption. Olefin space velocity is one way of expressing reaction time; another is by using contact time. Contact time is defined as the residence time of the fresh feed and externally recycled isobutane in the reactor. Contact time for hydrofluoric acid alkylation ranges from 5 to 25 minutes and for sulfuric acid alkylation from 5 to 40 minutes [6].

Although the relationship is only approximate, Mrstik, Smith, and Pinkerton [5] developed a correlating factor, F, which is useful in predicting trends in alkylate quality where operating variables are changed.

$$F = \frac{I_E (I/O)_F}{100 (SV)_O}$$

where

I_E = isobutane in reactor effluent, liquid volume %

$(I/O)_F$ = volumetric isobutane/olefin ratio in feed

$(SV)_O$ = olefin space velocity, v/hr/v

The higher the value of F, the better the alkylate quality. Normal values of F range from 10 to 40.

10.3 ALKYLATION FEEDSTOCKS

Olefins and isobutane are used as alkylation unit feedstocks. The chief sources of olefins are catalytic cracking and coking operations. Butenes and propenes are the most common olefins used but ethylene and pentenes are included in some cases. Olefins can be produced by dehydrogenation of paraffins and isobutane is cracked commercially to provide alkylation unit feed.

Hydrocrackers and catalytic crackers produce a great deal of the isobutane used in alkylation but it is also obtained from catalytic reformers, crude distillation, and natural gas processing. In some cases, normal butane is isomerized to produce additional isobutane for alkylation unit feed.

TABLE 10.1

Range of Operating Variables in Alkylation [11]

	HF	H_2SO_4
Isobutane concentrations		
Vol % in reaction zone	30-80	40-80
External ratio to olefins	3-12	3-12
Internal ratio to olefins	–	50-1,000
Olefin concentration		
Total HC contact time, min	8-20	20-30
Olefin space velocity, v/hr/v	–	0.1-0.6
Reactor temperature, °F	60-115	35-60
Reactor acid conc, wt %	80-95	88-95
Acid in emulsion, vol %	25-80	40-60

10.4 ALKYLATION PRODUCTS

In addition to the aklylate stream, the products leaving the alkylation unit include the propane and normal butane that enter with the saturated and unsaturated feed streams as well as a small quantity of tar produced by polymerization reactions.

The product streams leaving an alkylation unit are:

1. LPG grade propane liquid
2. Normal butane liquid
3. C_5^+ alkylate
4. Tar

Only about 0.1% by volume of olefin feed is converted into tar. This is not truly a tar but a thick dark brown oil containing complex mixtures of conjugated cyclopentadienes with side chains [7].

Typical alkylation operating conditions are shown in Table 10.1 and theoretical yields of alkylates and isobutane requirements based on olefin reacted are given in Table 10.2.

TABLE 10.2

Theoretical Yields and Isobutane Requirements
Based on Olefin Reacting [7]

	Alkylate vol %	i-Butane vol %
Ethylene	188	139
Propene	181	128
Butenes (mixed)	172	112
Pentenes (mixed)	165	96

10.5 CATALYSTS

Concentrated sulfuric and hydrofluoric acids are the only catalysts used commerically today for the production of high octane alkylate gasoline but other catalysts are used to produce ethylbenzene, cumene, and long chain (C_{12} to C_{16}) alkylated benzenes [7].

As discussed in Section 10.1, the desirable reactions are the formation of C_8 carbonium ions and the subsequent formation of alkylate. The main undesirable reaction is polymerization of olefins. Only strong acids can catalyze the alkylation reaction but weaker acids can cause polymerization to take place. Therefore, the acid strengths must be kept above 88% by weight H_2SO_4 or HF in order to prevent excessive polymerization. Sulfuric acid containing free SO_3 also causes undesired side reactions and concentrations greater than 99.3% H_2SO_4 are not generally used [7].

Isobutane is soluble in the acid phase only to the extent of about 0.1% by weight in sulfuric acid and about 3% in hydrofluoric acid. Olefins are more soluble in the acid phase and a slight amount of polymerization of the olefins is desirable as the polymerization products dissolve in the acid and increase the solubility of isobutane in the acid phase.

If the concentration of the acid becomes less than 88%, some of the acid must be removed and replaced with stronger acid. In hydrofluoric acid units, the acid removed is redistilled and the polymerization products removed as a thick dark oil. The concentrated HF is recycled in the unit and the net consumption is about 0.3 lb per barrel of alkylate produced [8].

The sulfuric acid removed must be regenerated in a sulfuric acid plant which is generally not a part of the alkylation unit and the acid consumption ranges from 18 to 30 lb per barrel of alkylate produced. Makeup acid is usually 99.3% by weight H_2SO_4.

10.6 HYDROFLUORIC ACID PROCESSES [9, 10, 11]

There are two commercial alkylation processes using hydrofluoric acid as the catalyst. They are designed and licensed by Phillips Petroleum Company, and the UOP Process Division of Universal Oil Products Company. Typical operating conditions are given in Tables 10.3 and 10.4.

The basic flow scheme is the same for both the Phillips and the UOP process (Fig. 10.2). See also Photo 10, Appendix F.

Both the olefin and isobutane feeds are dehydrated by passing the feedstocks through a solid bed desiccant unit. Good dehydration is essential to minimize potential corrosion of process equipment which results from addition of water to hydrofluoric acid.

TABLE 10.3

HF Alkylation Yields, Product Octanes
and Isobutane Requirements

	Vol/vol olefin		Clear octane no.	
	Isobutane	Alkylate	Research	Motor
Propylene	1.33	1.77	93	91
Butylenes	1.16	1.75	96	94

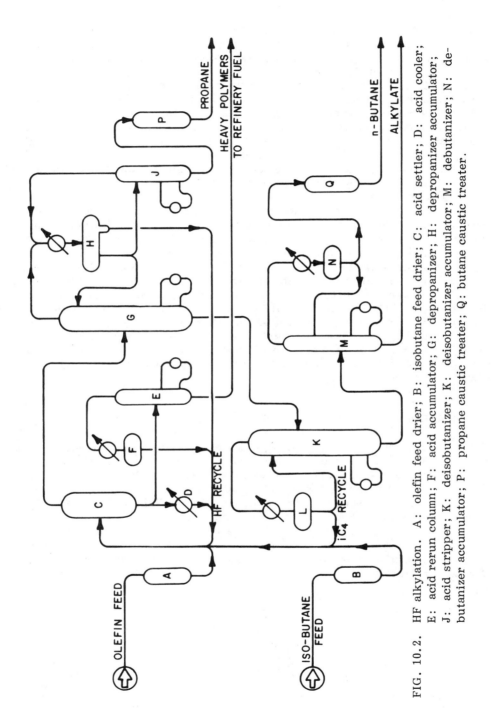

FIG. 10.2. HF alkylation. A: olefin feed drier; B: isobutane feed drier; C: acid settler; D: acid cooler; E: acid rerun column; F: acid accumulator; G: depropanizer; H: depropanizer accumulator; J: acid stripper; K: deisobutanizer; K: deisobutanizer accumulator; M: debutanizer; N: debutanizer accumulator; P: propane caustic treater; Q: butane caustic treater.

TABLE 10.4

HF Alkylate Properties

Gravity, °API	71.4
RVP, psi	4.5
ASTM distillation, °F	
IBP	110
5%	155
10%	172
20%	190
50%	217
70%	222
90%	245
EP	370

After dehydration the olefin and isobutane feeds are mixed with hydro-fluoric acid at sufficient pressure to maintain all components in the liquid phase. The reaction mixture is allowed to settle into two liquid layers. The acid has a higher density than the hydrocarbon mixture and is withdrawn from the bottom of the settler and passed through a cooler to remove the heat gained from the exothermic reaction. The acid is then recycled and mixed with more fresh feed, thus completing the acid circuit.

A small slip-stream of acid is withdrawn from the settler and fed to an acid rerun column to remove dissolved water and polymerized hydrocarbons. The acid rerun column contains about five trays and operates at 150 psig [12]. The overhead product from the rerun column is clear hydrofluoric acid which is condensed and returned to the system.

The bottom product from the rerun column is a mixture of tar and an HF-water azeotrope. These components are separated in a tar settler (not shown on the flow diagram). The tar is used for fuel and the HF-water mixture is neutralized with lime or caustic. This rerun operation is necessary to maintain the activity of the hydrofluoric acid catalyst.

The hydrocarbon layer removed from the top of the acid settler is a mixture of propane, isobutane, normal butane, and alkylate along with small amounts of hydrofluoric acid. These components are separated by fractiona-tion and the isobutane is recycled to the feed. Propane and normal butane products are passed through caustic treaters to remove trace quantities and hydrofluoric acid.

Although the flow sheet (Fig. 10.2) shows the fractionation of propane, isobutane, normal butane, and alkylate to require three separate fractionators, many alkylation plants have a single tower where propane is taken off over-head, a partially purified isobutane recycle is withdrawn as a liquid several trays above the feed tray, a normal butane product is taken off as a vapor several trays below the feed tray and the alkylate is removed from the bottom.

The design of the acid settler-cooler-reactor section is critical to good conversion in a hydrofluoric acid alkylation system. Different reactor sys-tem designs have been made over the years by both UOP and Phillips. Many of the reactor systems designed by UOP are similar to a horizontal shell and tube heat exchanger with cooling water flowing inside the tubes to maintain the reaction temperatures at the desired level. Good mixing is attained in the

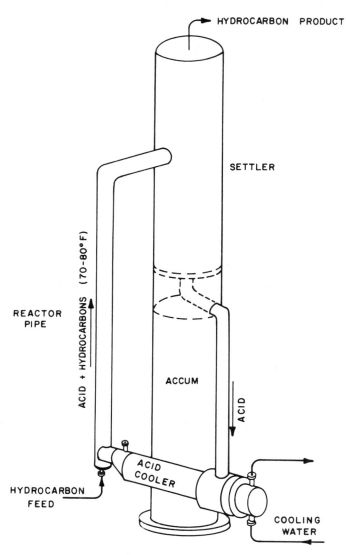

FIG. 10.3. Phillips HF reactor.

reactor by using a recirculating pump to force the mixture through the reactor at a rate about eight to ten times the mixed hydrocarbon feed rate to the reactor.

Reactor systems designed by Phillips usually have been similar to that illustrated in Figure 10.3. Acid circulation in this system is by gravity differential and thus a relatively expensive acid circulation pump is not necessary.

In portions of the process system where it is possible to have HF-water mixtures, the process equipment is fabricated from Monel metal or Monel-clad steel. The other parts of the system are carbon steel.

Special precautions are taken to protect maintenance and operating personnel from injury by accidental contact with acid. These precautions include: special seals on acid containing equipment such as pumps and valve stems; rubber safety jackets, pants, gloves, and boots which must be worn by personnel entering an acid area; safety eyeglasses; caustic tubs for washing all hand tools; safety showers; special acid drain systems; and many others.

Careful attention to engineering design details and extensive operator training combined with the above precautions are necessary to provide safe operations for hydrofluoric acid alkylation units.

10.7 SULFURIC ACID ALKYLATION PROCESSES

The major alkylation processes using sulfuric acid as a catalyst are the Cascade Autorefrigeration process, licensed by M. W. Kellogg Company, and the Effluent Refrigeration process, licensed by Stratford Engineering Corporation. There are also some older units using time-tank reactors but no new units of ths type have been constructed recently (Photo 11, Appendix F).

The major differences between the Cascade and Effluent Refrigeration processes are in the reactor designs and the point in the process at which propane and isobutane are evaporated to induce cooling. A simplified flow diagram for the Cascade Autorefrigeration process is shown in Figure 10.4.

The Cascade Autorefrigeration process uses a multistage cascade reactor with mixers in each stage to emulsify the hydrocarbon-acid mixture. Acid and isobutane enter the first stage of the reactor and pass in series through the remaining stages. The olefin hydrocarbon feed is split and injected in equal quantities into each of the stages. The temperature in the reactor is controlled by vaporizing a portion of the hydrocarbon phase in each stage in the reactor. Each stage operates at the proper pressure to maintain its temperature at the desired level. The gases flashed are primarily propane and isobutane.

These flashed gases are compressed and liquefied. A portion of this liquid is vaporized in an economizer to cool the olefin hydrocarbon feed before it is sent to the reactor. The vapors are returned for recompression. The remainder of the liquefied hydrocarbon is sent to a depropanizer column for removal of the excess propane which accumulates in the system. The liquid isobutane from the bottom of the depropanizer is pumped to the first stage of the reactor.

The acid-hydrocarbon emulsion from the last reactor stage is separated into acid and hydrocarbon phases in a settler. The acid is removed from the system for reclamation and the hydrocarbon phase is pumped through a caustic wash to eliminate trace amounts of acid and sent to a deisobutanizer. The deisobutanizer separates the hydrocarbon feed stream into isobutane (which is returned to the reactor), n-butane, and alkylate product.

The Effluent Refrigeration process uses a single-stage reactor in which the temperature is maintained by cooling coils (Fig. 10.5). The reactor contains an impeller that emulsifies the acid-hydrocarbon mixture and recirculates it in the reactor. Average residence time in the reactor is on the order of 20 to 25 minutes [13, 14].

Emulsion removed from the reactor is sent to a settler for phase separation. The acid is recirculated and the pressure of the hydrocarbon phase is

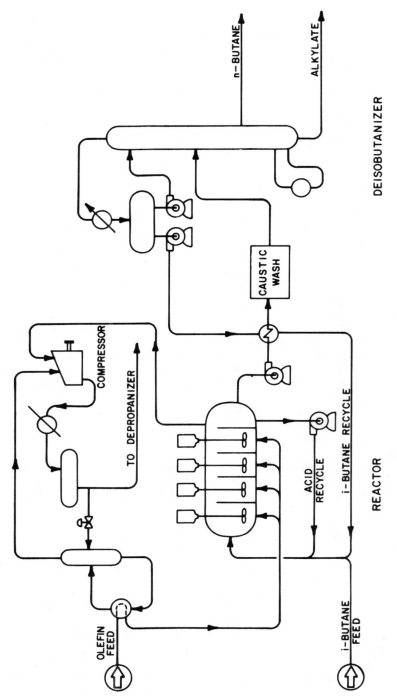

FIG. 10.4. Cascade sulfuric acid alkylation unit.

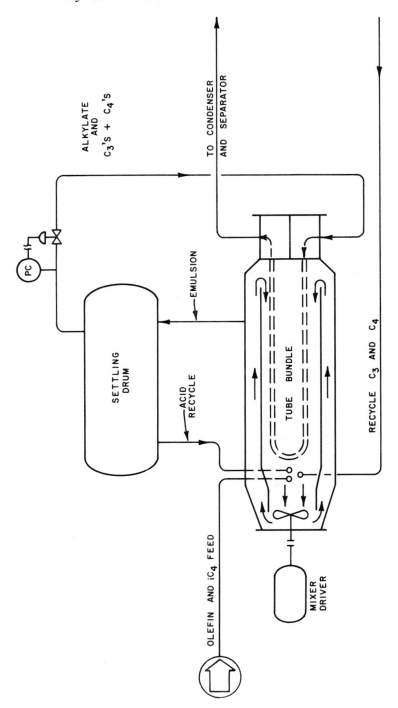

ALKYLATE AND C_3'S + C_4'S

TO CONDENSER AND SEPARATOR

EMULSION

RECYCLE C_3 AND C_4

PC

SETTLING DRUM

ACID RECYCLE

TUBE BUNDLE

OLEFIN AND iC_4 FEED

MIXER DRIVER

FIG. 10.5. Stratco Contactor.

lowered to flash vaporize a portion of the stream and reduce the liquid temperature to about 20°F [14]. The cold liquid is used as coolant in the reactor tube bundle.

The flashed gases are compressed and liquefied, then sent to the depropanizer where LPG grade propane and recycle isobutane are separated. The hydrocarbon liquid from the reactor tube bundle is separated into isobutane, n-butane, and alkylate streams in the deisobutanizer column. The isobutane is recycled and n-butane and alkylate are product streams.

10.8 COMPARISON OF PROCESSES

The most desirable alkylation process for a given refinery is governed by economics. In particular the location of the refinery with respect to acid supply and disposal is very important. If the refinery is at a distance from either sulfuric acid suppliers or purchasers of spent sulfuric acid, the cost of transportation of fresh acid and/or the cost of disposing of the large quantities of spent acid can render the use of sulfuric acid economically unattractive. Only a small amount of makeup hydrofluoric acid is required for the HF process, as facilities are provided to regenerate the spent hydrofluoric acid. As a result, the cost of transporting hydrofluoric acid from a remote supplier is not a major cost. L. F. Albright [15], in a series of articles on alkylation processes, summarized the comparison of the processes as follows:

The important question for a refinery is which alkylation process is best for the production of the desired product. Many factors are important, including total operating expenses, initial capital costs, alkylate quality, flexibility of operation, reactants available, yields and conversion of reactants, maintenance problems, safety, experience with a given process, and patents, licensing arrangements and possible royalties.

Advocates of the hydrofluoric acid process argue that both capital and total operating costs are less than those of sulfuric acid processes for the following reasons:

1. Small and simpler reactor designs are feasible.
2. Cooling water can be used instead of refrigeration.
3. Smaller settling devices are needed for emulsions.
4. Essentially complete regeneration of the hydrofluoric acid catalyst occurs. Hence, hydrofluoric acid consumption and costs are very low. Disposal of spent acids is not necessary.
5. There is increased flexibility of operation relative to temperature, external ratio of isobutane to olefin, etc.
6. There is decreased need for turbulence or agitation when acid and hydrocarbon streams are combined.

Advocates of sulfuric acid processes counter the above arguments for hydrofluoric acid processes with the following:

1. Additional equipment is needed for the hydrofluoric acid process to recover or neutralize the hydrofluoric acid in various streams. Such equipment includes the hydrofluoric acid stripper tower, hydrofluoric

acid regeneration tower, and neutralization facilities for the several product streams. With sulfuric acid, the entire effluent hydrocarbon stream is neutralized.

2. Equipment is required to dry the feed streams to a few ppm in hydrofluoric acid processes. Drying is beneficial but less critical in sulfuric acid processes.

3. Additional equipment at increased cost is required for safety. In some hydrofluoric acid plants, a closed water system is required as a safety measure in the event of hydrofluoric acid leakage into the system. Maintenance costs and the amount of safety equipment in hydrofluoric acid processes are greater.

4. Capital costs for hydrofluoric acid processes are little cheaper (if at all) than those for sulfuric acid processes. In general, the relative costs appear to favor hydrofluoric acid processes in small plants.

5. Royalty and licensing costs of hydrofluoric acid processes are greater.

6. Isobutane is not fully used, since self-alkylation occurs to a higher extent when hydrofluoric acid is used as a catalyst.

7. There are greater limitations on obtaining alkylates with high octane numbers.

10.9 ALKYLATION YIELDS AND COSTS

Typical alkylation yields based on percent of olefin in fresh feed are given in Table 10.5. These do not take into account effects of temperature and are based on isobutane olefin ratios of 10:1 for propylene, 6:1 for butylenes, and 10:1 for amylenes.

Cost curves for construction of alkylation units are shown on Figure 10.6. The costs are average costs and include the items given in Table 10.6. Power and chemical consumption are also given in Table 10.6.

10.10 POLYMERIZATION

Propene and butenes can be polymerized to form a high octane product boiling in the gasoline boiling range. The product is an olefin having unleaded octane numbers of 97 RON and 83 MON. The polymerization process was widely used in the 1930's and 40's to convert low boiling olefins into gasoline blending stocks but was supplanted by the alkylation process after World War II. The mandated reduction in use of lead in gasoline and the increasing proportion of the market demand for unleaded gasolines created a need for low-cost processes to produce high octane gasoline blending components. Polymerization produces about 0.7 barrels of polymer gasoline per barrel of olefin feed as compared with about 1.5 barrels of alkylate by alkylation and the product has a high octane sensitivity but capital and operating costs are much lower than for alkylation. As a result, polymerization processes are being added to many refineries.

Typical polymerization reactions are shown in Table 10.7 but while i-C_4H_8 reacts to give primarily diisobutylene, propene gives mostly trimers and dimers with only about 10% conversion to dimer (16).

TABLE 10.5

Design Yield Factors for Propylene, Butylene, Amylene Alkylation
Expressed as Percent of Olefin

	Propylene					Butylene					Amylene				
	MW	lb/gal	vol %	wt %	mol %	MW	lb/gal	vol %	wt %	mol %	MW	lb/gal	vol %	wt %	mol %
Feeds:															
Olefin	42.0	4.38	100.00	100.00	100.00	56.0	5.00	100.00	100.00	100.00	70.1	5.46	100.00	100.00	100.00
Isobutane	58.1	4.69	160.00	171.32	124.07	58.1	4.69	120.00	112.56	108.76	58.1	4.69	140.00	120.25	145.06
			260.00	271.32	224.07			220.00	212.56	208.76			240.00	220.25	245.06
Products:															
Propane	44.0	4.26	29.80	29.05	27.70	—	—	—	—	—	—	—	—	—	—
n–Butane	—	—	—	—	—	58.1	4.88	12.71	12.40	12.00	—	—	—	—	—
Pentane	72.1	5.21	7.17	8.53	4.98	72.1	5.21	6.80	7.08	5.50	72.1	5.25	45.20	43.48	42.24
Depent rerun alkylate	112.6	5.84	159.40	212.56	79.27	112.6	5.84	151.10	176.56	88.03	110.9	5.86	145.60	156.19	98.72
Alkyl bottoms	165.0	6.31	12.60	18.18	4.61	165	6.31	11.90	15.02	5.12	230	6.67	14.40	17.58	5.37
Tar, 20° API	360	7.78	1.70	3.00	0.35	360	7.78	0.96	1.50	0.23	360	7.78	2.11	3.00	0.58
			210.67	271.32	116.91			183.47	212.56	110.88			207.31	220.25	146.91
(Total feeds) (Total products)		1.234		1.000	1.917		1.199		1.000	1.883		1.158		1.000	1.668

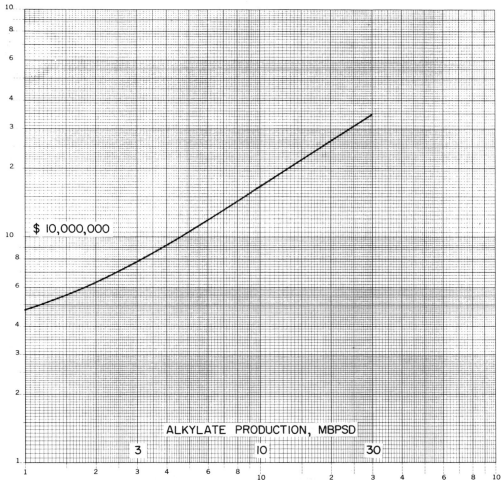

FIG. 10.6*. Alkylation units investment cost--1982 U.S. Gulf Coast.
*See Table 10.6, page 176.

The most widely used catalyst is phosphoric acid on an inert support.
This can be in the form of phosphoric acid mixed with kieselguhr (solid phos-
pheric acid) or a film of liquid phosphoric acid on crushed quartz. Sulfur in
the feed poisons the catalyst and any basic materials neutralize the acid and
increase catalyst consumption. Oxygen dissolved in the feed adversely affects
the reactions and must be removed. Normal catalyst consumption rates are
in the range of one pound of catalyst per 100 to 200 gallons of polymer produced.

The feed, consisting of propane and butane as well as propene and
butene, is caustic washed to remove mercaptans and contacted with an amine
solution to remove hydrogen sulfide. It is then scrubbed with water to remove any caustic and amines then dried by passing through a silica gel or
molecular sieve bed. Finally, a small amount of water (350-400 ppm) is added

TABLE 10.6*

Alkylation Unit Costs

Costs Included:
1. All facilities required for producing alkylate from a feedstream of isobutane and $C_3^=$ to $C_5^=$ unsaturates in proper proportions.
2. All necessary controllers and instrumentation.
3. All BL process facilities.
4. Feed treating (molecular sieve unit to remove moisture in feed).

Costs Not Included:
1. Cooling water, steam and power supply.
2. Feed splitter in lower cost line.
3. Feed and product storage.

	HF	H_2SO_4
Royalty, $/bbl TA†:	0.15	0.10
Utility Data (per bbl TA):		
Steam, lb	11	390
Power, kWh	3.7	4.6
Cooling water, gal crclt	3,700	3,300
Fuel, MMBtu	1.04	0.85
Chemicals (per bbl TA):		
Acid, lb	0.3	30
Caustic, lb	0.2	0.2

*See Fig. 10.6, page 175.
†Total alkylate

to promote ionization of the acid before the olefin feed steam is heated to about 400°F (204°C) and passed over the catalyst bed. Reactor pressures are about 500 psig.

The polymerization reaction is highly exothermic and temperature is controlled either by injecting a cold propane quench or by generating steam. The propane and butane in the feed act as diluents and a heat sink to help control the rate of reaction and the rate of heat release. Propane is also recycled to help control the temperature.

After leaving the reactor the product is fractionated to separate the butane and lighter material from the polymer gasoline. Gasoline boiling range polymer production is normally 90-97 wt % on olefin feed or about 0.7 barrel of polymer per barrel of olefin feed. A simplified process flow diagram for the UOP unit is shown in Figure. 10.7 [17] and ranges of reaction conditions given in Table 10.8 [16].

Insitut Francais du Petrole licenses a process to produce dimate (isohexene) from propene using a homogeneous aluminum alkyl catalyst which is not recovered. The process requires a feed stream that is better than 99% pro-

TABLE 10.7

Polymerization Reactions

$$C=C-C \; + \; C=C-C \; \rightarrow \; C-\overset{\overset{\displaystyle C}{|}}{\underset{\underset{\displaystyle C}{|}}{C}}-C=C-C$$

with methyl groups on the appropriate carbons

$$C=C-C \; + \; C=C-C \; \rightarrow \; C-C-\overset{\overset{\displaystyle C}{|}}{C}=C-C$$

TABLE 10.8

Polymerization Operating Conditions

Temperature: 175-235°C (300-425°F)
    Usually 200-220°C

Pressure: 25-80 Atm (400-1,500 PSI)

Space Velocity: 2.5 1/kg-hr (0.3 gal/lb)

(Higher Pressures Mean Lower
    Temperatures)

TABLE 10.9

Dimersol Feed Impurities [18]

Component	ppm
H_2O	3
Acetylene	15
Sulfur	5
Propadiene	1
Butadiene	10

TABLE 10.10

Yields Per bbl Mixed Olefin Feed

	Alkylate	Polymer	Dimate
C_5^+ gasoline	1.55	0.68	0.68
RON	94	97	97
MON	91	83	82

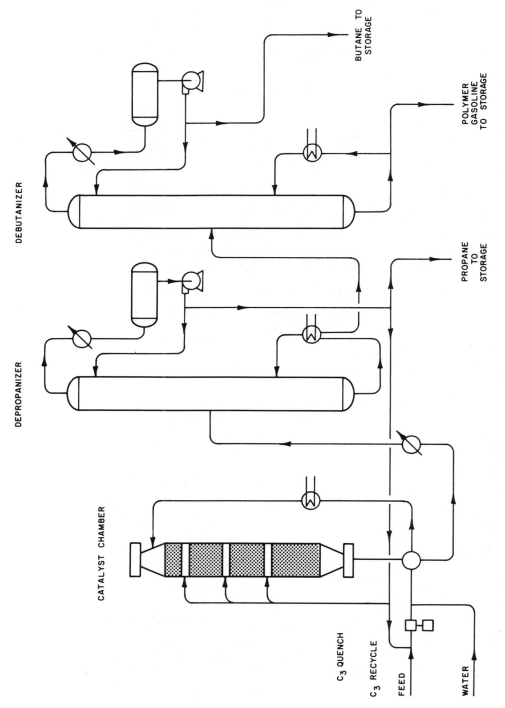

FIG. 10.7. UOP solid phosphoric acid polymerization unit [7].

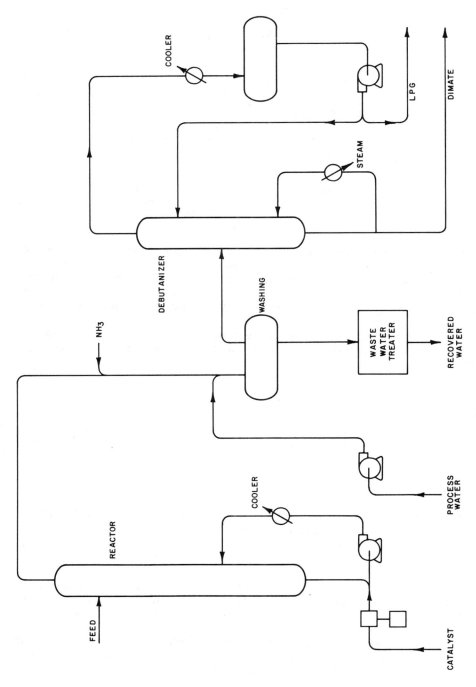

FIG. 10.8. DIMERSOL polymerization unit [18].

TABLE 10.11

Polymerization Unit Costs

Investment Cost—1982 U.S. Gulf Coast	
500 BPD	$ 600,000
2000 BPD	1,400,000
Costs Included:	
Costs are for reactor section only but including preheat heat exchange and product cooling.	
Utility Data (per bbl polymer):	
Fuel, MM Btu	0.036
Cooling water, gal crclt, 30°F Δt	290

propane and propene because C_2's and C_4's poison the catalyst. Dienes and triple bonded hydrocarbons can create problems and in some cases it is necessary to selectively hydrogenate the feed to eliminate these compounds. The major advantage of this process is the low capital cost because it operates at low pressures. A simplified process flow diagram is given in Figure 10.8 [18] and feed impurity limits in Table 10.9.

Product properties and yields are shown in Table 10.10 for alkylate, polymer and dimer blending stocks. Table 10.11 shows Polymerization Unit Costs.

10.11 CASE-STUDY PROBLEM: ALKYLATION

Table 10.12 gives the material balance and Table 10.13 shows the alkylation unit chemical and utility requirements.

PROBLEMS

1. A refinery has available 2,700 BPCD of butylene and 2,350 BPCD of isobutane for possible akylation unit feed. How many barrels of alkylate can be made from these feedstocks? What will be the production of other products?

2. An unsaturated feed stream consisting of 1,750 BPCD of butylene and 1,550 BPCD of propylene is fed to an alkylation unit. How many BPCD of isobutane will be required for the unit? How much alkylate will be made?

3. Make an overall material balance for an alkylation unit with feed rates of 1,710 BPCD of propylene, 3,320 BPCD of butylene, 1,550 BPCD of amylene, and 9,570 BPCD of isobutane.

4. Estimate the 1982 construction costs for building the alkylation unit of problem 3 and the chemical and utility requirements (a) if sulfuric acid catalyst is used and (b) if hydrofluoric acid catalyst is used.

TABLE 10.12

HF Alkylation Unit

Component	BPD	°API	(lb/hr)/BPD	lb/hr
Feed:				
$C_3^=$, coker	260			1,990
$C_3^=$, cat cracker	1,420			10,810
	1,680			
$C_4^=$, coker	240			2,060
$C_4^=$, cat cracker	2,710			23,740
	2,950			
i-C_4, crude unit	200			1,640
i-C_4, coker	110			890
i-C_4, cat cracker	1,710			14,060
i-C_4, HC	1,370			11,260
i-C_4, reformer	360			2,960
	3,750			
i-C_4 purchased	2,480		8.22	20,390
	6,230			89,800
Products:				
C_3	500		7.39	3,730
n-C_4	370		8.51	3,150
Alkylate	7.460	71.5	10.18	75,910
Alkylate btms	560	55.4	11.04	6,190
Tar	60	20.0	13.62	820
				89,800

Alkylate produced: $C_3^=$:   (1.666)(1,680) = 2,800

                      $C_4^=$:   (1.579)(2,950) = 4,660

                                         7,460

i-C_4 required: $C_3^=$:   (1.6)(1,680) = 2,690

                     $C_4^=$:   (1.2)(2,950) = 3,540

                                         6,230

C_3 produced: (0.298)(1,680) = 500

n-C_4 produced: (0.127)(2,950) = 370

Alkylate btms produced: (0.126)(1,680) = 210   tar: (0.017)(1,680) = 30

        6.31 lb/gal   (0.119)(2,950) = 350   (0.0096)(2,950) = 30

                                              560   60

TABLE 10.13

Alkylation Unit Chemical and Utility Requirements

Steam, lb/hr	3,430
Power, MkWh/day	27.6
Cooling water, gpm	5,710
Fuel, MMBtu/day	1,940
HF acid, lb/day	2,240
Caustic, lb/day	1,490

5. Using the construction costs and chemical and utility requirements of problem 4, what are the costs per barrel of alkylate produced for direct operation, including royalties but not labor, and depreciation. Assume 16-year straightline depreciation, with dismantling costs equal to salvage value. Compare total costs per barrel of alkylate for hydrofluoric and sulfuric acid processes.

6. Compare the operating costs per barrel of alkylate produced from a 4000 BPSD sulfuric acid alkylation plant and a 4000 BPSD hydrofluoric acid alkylation plant. Obtain chemical costs from the latest issue of "Chemical Marketing Reporter". Use the following utility costs: steam, $2.25/Mlb; electric power, $0.02/kwh; cooling water: make-up, $0.45/M gal, circulating, see p. 264 fuel, $2.49/MM Btu; labor costs, $150/day.

7. A refinery has 5,560 BPSD of isobutane available for alkylation. There are also 2,110 BPSD of propylene, 2200 BPSD of butylene, and 1400 BPSD of amylene available. What combination of feedstocks to the alkylation unit will maximize alkylate production?

8. Estimate the 1978 construction costs for building an alkylation unit to feed 2400 BPSD of propylene, 4000 BPSD of butylene and 8350 BPSD of isobutane.

9. Assuming 16-year straight-line depreciation and dismantling costs to equal salvage value, calculate the operating costs, including depreciation and running royalties, but not labor, per barrel of alkylate produced for hydrofluoric and sulfuric acid plants producing 10,000 BPSD of alkylate.

10. For Problem 10.6, compare the daily operating costs for a 4000 BPSD feed rate hydrofluoric acid plant when producing alkylate from (1) 98% propylene feedstock with that when (2) using a 97.5% butylene feedstock.

11. Make an overall volume and weight material balance for an alkylation unit with feed rates of 1,550 BPCD of propylene, 3,520 BPCD of butylene, 1,800 BPCD of amylene and 9,510 BPCD of isobutane.

12. For an alkylation plant with a 90% on-stream factor producing 9,000 BPCD of alkylate gasoline having clear octane numbers of 95.9 MON and 97.3 RON, estimate its 1973 U.S. Gulf Coast construction cost and its annual chemical and utility requirements.

13. Make an overall weight and volume material balance for an alkylation unit with a feedrate of 2500 BPCD of butylene and 2800 BPCD of isobutane.

14. A refinery has 8000 BPSD of isobutane available for alkylation. There are also 2120 BPSD of propylene, 3985 BPSD of butylene, and 1880 BPSD of amylene available. What is the maximum yield of pentane-free alkylate than can be produced?

NOTES

1. V. N. Ipatieff and L. Schmerling, *Advances in Catalysis*, vol. I (Academic Press, New York, 1948), pp. 27-63.
2. W. A. Gruse and D. R. Stevens, *Chemical Technology of Petroleum*, 3rd ed. (McGraw-Hill Book Company, New York, 1960), pp. 153-163.
3. A. R. Glasgow, A. J. Streiff, C. B. Willingham, and F. D. Rossini, J. Res. Nat. Bur. Stand. *38*, 537 (1947).
4. R. E. Payne, Petrol. Refiner. *37*(9), 316-329 (1958).
5. A. V. Mrstik, K. A. Smith, and R. D. Pinkerton, Advan, Chem. Ser. *5*(97) (1951).
6. R. J. Hengstebeck, *Petroleum Processing* (McGraw-Hill Book Company, New York, 1959), pp. 218-233.
7. C. L. Thomas, *Catalytic Processes and Proven Catalysts* (McGraw-Hill Book Company, New York, 1970), pp. 87-96.
8. P. C. Templeton and B. H. King, paper presented before the Western Petroleum Refiners Association, June 21-22, 1956.
9. Hydrocarbon Process. *49*(9), 198-203 (1970).
10. Petrol. Process. *12*(5), 146; *12*(8), 79 (1957).
11. Petrol. Refiner. *31*(9), 156-164 (1952).
12. L. F. Albright, Chem. Eng. *73*(19), 205-210 (1966).
13. L. F. Albright, Chem. Eng. *73*(14), 119-126 (1966).
14. L. F. Albright, Chem. Eng. *73*(17), 143-150 (1966).
15. L. F. Albright, Chem. Eng. *73*(21), 209-215 (1966).
16. C. L. Thomas, *Catalytic Processes and Proven Catalysts* (McGraw-Hill Book Company, New York, 1970), pp. 67-69.
17. Petrol. Refiner. *39*(9), 232 (1960).
18. J. W. Andrews and P. L. Bonnifay, Hydrocarbon Process. *56*(4), 161-164 (1977).

Chapter 11
PRODUCT BLENDING

Increased operating flexibility and profits result when refinery operations produce basic intermediate streams that can be blended to produce a variety of on-specification finished products. For example, naphthas can be blended into either gasoline or jet fuel, depending upon the product demand. Aside from lubricating oils, the major refinery products produced by blending are gasolines, jet fuels, heating oils, and diesel fuels. The objective of product blending is to allocate the available blending components in such a way as to meet product demands and specifications at the least cost and to produce incremental products which maximize overall profit. The volumes of products sold, even by a medium-sized refiner, are so large that savings of a fraction of a cent per gallon will produce a substantial increase in profit over the period of one year. For example, if a refiner sells about one billion gallons of gasoline per year (several refiners sell more than that in the United States), a saving of one one-hundredth of a cent per gallon results in an additional profit of $100,000 per year.

Today's trend is to using computer-controlled in-line blending for blending gasolines and other high-volume products. Inventories of blending stocks, together with cost and physical property data are maintained in the computer. When a certain volume of a given quality product is specified, the computer uses linear programming models to optimize the blending operations to select the blending components to produce the required volume of the specified product at the lowest cost.

To ensure that the blended streams meet the desired specifications, stream analyzers, such as boiling point, specific gravity, RVP, and research and motor octane comparators are installed to provide feedback control of additives and blending streams.

Blending components to meet all critical specifications most economically is a trial-and-error procedure and, because of the large number of variables, it is possible to have a number of equivalent solutions that give the same total overall cost or profit.

The same basic techniques are used for calculating the blending components for any of the blended refinery products. Gasoline is the largest volume refinery product and will be used as an example to help clarify the procedures.

For purposes of preliminary cost evaluation studies, calculations generally are not made on the percent distilled specifications at intermediate percentages, even though these are important with respect to such operating characteristics as warm-up, acceleration, and economy. The allowable blending stocks are those with boiling ranges within the product specifications (e.g., C_4-380°F) and the control criteria are to meet RVP and octane requirements.

11.1 REID VAPOR PRESSURE (RVP)

The desired RVP of a gasoline is obtained by blending n-butane with C_5-380°F naphtha. The amount of n-butane required to give the needed RVP is calculated by:

$$M_t(RVP)_t = \sum_{i=1}^{n} M_i(RVP)_i$$

where

M_t = total moles blended product

$(RVP)_t$ = specification RVP for product, psi

M_i = moles of component i

$(RVP)_i$ = RVP of component i, psi

Example 11.1:

Base Stock	BPD	lb/hr	MW	mol/hr	mol %	RVP	PVP
LSR gasoline	4,000	39,320	86	457	21.0	11.1	2.32
Reformate	6,000	69,900	115	617	28.4	2.8	0.80
Alkylate	3,000	30,690	104	295	13.4	4.6	0.62
FCC gasoline	8,000	87,520	108	810	37.2	4.4	1.64
	21,000			2,179	100.0		

Blend for a 10 psi RVP
n-butane: MW = 58, RVP = 52

Butane requirement:

$$(2,179)(5.38) + M(52.0) = (2,179 + M)(10)$$
$$11,723 + 52.0M = 21,790 + 10.0M$$
$$42.0M = 10,067$$
$$M = 240 \text{ moles n-}C_4 \text{ required}$$

	BPD	lb/hr	MW	mol/hr
n-butane	1,640	13,920	58	240

Total 10 psi RVP gasoline = 21,000 + 1,640 = 22,640 BPD

Blending property data for many refinery streams are given in Table 11.1.

The theoretical method for blending to the desired Reid vapor pressure requires the average molecular weight of each of the streams be known. Although there are accepted ways of estimating the average molecular weight of a refinery stream from boiling point, gravity, and characterization factor, a more convenient way is to use the empirical method developed by Chevron Research Company. Vapor pressure blending indices (VPBI) have been compiled as a function of the RVP of the blending streams and are given in Table 11.2. The Reid vapor pressure of the blend is closely approximated by the sum of all the products of the volume fraction (v) times the VPBI for each component. In equation form:

$$RVP_{blend} = \Sigma v_i (VPBI)_i$$

In the case where the volume of the butane to be blended for a given RVP is desired:

$$A(VPBI)_a + B(BPBI)_b + \text{———} + W(VPBI)_w = (Y + W)(VPBI)_m$$

where

A = bbl of component a, etc.

W = bbl of n-butane (w)

Y = A + B + C + ———▶ (all components except n-butane)

$(VPBI)_m$ = VPBI corresponding to the desired RVP of the mixture

w = subscript indicating n-butane.

TABLE 11.1

Regular Blending Component Values

No.	Component	RVP	MON	MON 1.59 g*	MON 3.17 g*	RON	RON 1.59 g*	RON 3.17 g*	Gravity, °API
1	i-C_4	71.0	92.0	99.3	102.0	93.0	100.4	103.2	120.0
2	n-C_4	52.0	92.0	98.8	101.5	93.0	99.9	102.5	111.0
3	i-C_5	19.4	90.8	105.7	107.5	93.2	101.3	103.4	95.0
4	n-C_5	14.7	72.4	87.2	92.3	71.5	86.2	90.7	88.9
5	i-C_6	6.4	78.4	90.6	95.1	79.2	91.3	94.6	76.5
6	Hydrocrackate, C_5-C_6	15.5	85.5	96.7	100.8	89.2	96.8	99.3	86.4
7	Hydrocrackate, C_6-190°	3.9	73.7	86.3	91.4	75.5	88.4	93.4	
8	Hydrocrackate, 190-250°	1.7	75.6	84.6	87.9	79.0	88.3	92.2	55.5
9	Lt thermal gaso.	9.9	73.2	77.3	78.8	80.3	87.3	90.7	74.0
10	C_6^+ lt thermal gaso.	1.1	68.1	71.5	72.7	76.8	82.4	86.3	
11	Coker gaso.	3.6				67.2		82.5	57.2
12	C_5^+ TCC gaso.	4.0	76.6	80.6	82.1	85.5	90.7	93.0	
13	C_6^+ TCC gaso.	2.6	75.8	80.0	81.0	84.3	89.9	92.3	
14	FCC gaso., 200-300°	1.4	77.1	81.4	83.0	92.1	95.3	96.3	49.5
15	FCC C_5^+ gaso.	4.4	76.8	78.8	79.4	92.3	94.8	95.8	57.2
16	Hydrog C_5-200° FCC gaso.	14.1	81.7	88.9	91.7	91.2	98.4	100.3	
17	Hydrog C_5^+ FCC gaso.	13.1	80.7	90.0	91.7	91.0	97.7	100.0	
18	Hydrog 300-400° FCC gaso.	0.5	81.3	84.2	85.7	90.2	95.5	97.5	
19	Reformate, 94 RON	2.8	84.4	89.3	91.2	94.0	99.4	100.8	45.8
20	Reformate, 98 RON	2.2	86.5	91.1	92.5	98.0	101.0	102.2	43.1
21	Reformate, 100 RON	4.2	88.2	92.3	93.7	100.0	103.1	104.0	41.2
22	Aromatic concentrate	1.1	94.0	94.6	95.4	107.0	109.4	111.4	
23	Alkylate, $C_4^=$	4.6	95.9	101.9	103.4	97.3	102.0	104.0	70.3
24	Alkylate, $C_5^=$	1.0	88.8	99.2	103.0	89.7	99.2	103.1	
25	Polymer	8.7	84.0	85.4	86.5	96.9	99.0	99.7	59.5
26	Absorber gaso.	21.3	83.2	95.4	99.4	83.9	96.1	98.2	
27	LSR gaso. (C_5-180°)	11.1	61.6	73.7	80.6	66.4	77.3	83.5	78.6
28	LSR gaso. isomerized, once-through	13.5	81.1	98.1	100.6	83.0	92.3	98.0	80.4
29	Hvy SR gaso.	1.0	58.7	72.5	78.2	62.3	73.5	79.3	48.2
30	Lt hydrocrackate	12.9	82.4	93.7	97.9	82.8	94.8	98.4	79.0
31	Hvy hydrocrackate	1.1	67.3	79.2	83.5	67.6	80.8	85.6	49.0
32	Hydrog lt cat gaso., C_5^+	23.9	80.9	92.6	95.7	83.2	94.5	97.4	
33	Hydrog lt cat gaso., C_6^+	5.0	74.0	81.4	84.0	86.3	92.2	94.4	

*Grams Pb per gallon

TABLE 11.2

Reid Vapor Pressure Blending Index Numbers for Gasolines and Turbine Fuels

Vapor Pressure, psi	0.0	0.1	0.2	0.3	0.4	0.5	0.6	0.7	0.8	0.9
0	0.00	0.05	0.13	0.22	0.31	0.42	0.52	0.64	0.75	0.87
1	1.00	1.12	1.25	1.38	1.52	1.66	1.79	1.94	2.08	2.23
2	2.37	2.52	2.67	2.83	2.98	3.14	3.30	3.46	3.62	3.78
3	3.94	4.11	4.28	4.44	4.61	4.78	4.95	5.13	5.30	5.48
4	5.65	5.83	6.01	6.19	6.37	6.55	6.73	6.92	7.10	7.29
5	7.47	7.66	7.85	8.04	8.23	8.42	8.61	8.80	9.00	9.19
6	9.39	9.58	9.78	9.98	10.2	10.4	10.6	10.8	11.0	11.2
7	11.4	11.6	11.8	12.0	12.2	12.4	12.6	12.8	13.0	13.2
8	13.4	13.7	13.9	14.1	14.3	14.5	14.7	14.9	15.2	15.4
9	15.6	15.8	16.0	16.2	16.4	16.7	16.9	17.1	17.3	17.6
10	17.8	18.0	18.2	18.4	18.7	18.9	19.1	19.4	19.6	19.8
11	20.0	20.3	20.5	20.7	20.9	21.2	21.4	21.6	21.9	22.1
12	22.3	22.6	22.8	23.0	23.3	23.5	23.7	24.0	24.2	24.4
13	24.7	24.9	25.2	25.4	25.6	25.9	26.1	26.4	26.6	26.8
14	27.1	27.3	27.6	27.8	28.0	28.3	28.5	28.8	29.0	29.3
15	29.5	29.8	30.0	30.2	30.5	30.8	31.0	31.2	31.5	31.8
16	32.0	32.2	32.5	32.8	33.0	33.2	33.5	33.8	34.0	34.3
17	34.5	34.8	35.0	35.3	35.5	35.8	36.0	36.3	36.6	36.8
18	37.1	37.3	37.6	37.8	38.1	38.4	38.6	38.9	39.1	39.4
19	39.7	39.9	40.2	40.4	40.7	41.0	41.2	41.5	41.8	42.0
20	42.3	42.6	42.8	43.1	43.4	43.6	43.9	44.2	44.4	44.7
21	45.0	45.2	45.5	45.8	46.0	46.3	46.6	46.8	47.1	47.4
22	47.6	47.9	48.2	48.4	48.7	49.0	49.3	49.5	49.8	50.1
23	50.4	50.6	50.9	51.2	51.5	51.7	52.0	52.3	52.5	52.8
24	53.1	53.4	53.7	54.0	54.2	54.5	54.8	55.1	55.3	55.6
25	55.9	56.2	56.5	56.7	57.0	57.3	57.5	57.9	58.1	58.4
26	58.7	59.0	59.3	59.6	59.8	60.1	60.4	60.7	61.0	61.3
27	61.5	61.8	62.1	62.4	62.7	63.0	63.3	63.5	63.8	64.1
28	64.4	64.7	65.0	65.3	65.6	65.8	66.1	66.4	66.7	67.0
29	67.3	67.6	67.9	68.2	68.4	68.8	69.0	69.3	69.6	69.9
30	70.2									
40	101									
(nC_4) 51.6	138									
(iC_4) 72.2	210									
(C_3) 190	705									

Equation:

$$VPBI = VP^{1.25}$$

Example:

Calculate the vapor pressure of a gasoline blend as follows:

Component	Volume Fraction	Vapor Pressure, psi	Vapor Pressure Blending Index No.	Volume Fraction x VPBI
n-Butane	0.050	51.6	138	6.90
Light Straight Run	0.450	6.75	10.9	4.90
Heavy Refined	0.500	1.00	1.00	0.50
Total	1.000	7.4	12.3	12.30

From the brochure, "31.0°API Iranian Heavy Crude Oil," by arrangement with Chevron Research Company. Copyright © 1971 by Chevron Oil Trading Company.

Example 11.2:

Component	BPCD	RVP	VPBI	vol X VPBI
n-butane	W	51.6	138.0	138W
LSR gasoline	4,000	11.1	20.3	81,200
Reformate	6,000	2.8	3.62	21,720
Alkylate	3,000	4.6	6.73	20,190
FCC gasoline	8,000	4.4	6.37	50,960
	21,000 + W			174,070 + 138W

For 10 psi RVP, $(VPBI)_m = 17.8$

$$17.8(21,000 + W) = 174,070 + 138W$$
$$138 - 17.8W = 373,800 - 174,070$$
$$120.2W = 199,730$$
$$W = 1,660 \text{ bbl n-butane required.}$$

Total 10 psi RVP gasoline = 21,000 + 1,660 = 22,660 BPCD.

Although this differs slightly from the result obtained in Example 11.1, they agree well within the limits required for normal refinery operation.

11.2 OCTANE BLENDING

Octane numbers are blended on a volumetric basis using the blending octane numbers of the components. True octane numbers do not blend linearly and it is necessary to use blending octane numbers in making calculations. Blending octane numbers are based upon experience and are those numbers which, when added on a volumetric average basis, will give the true octane of the blend. True octane is defined as the octane number obtained using a CFR test engine.

The formula used for calculations is:

$$B_t(ON)_t = \sum_{i=1}^{n} B_i(ON)_i$$

where

B_t = total gasoline blended, bbl

$(ON)_t$ = desired octane of blend

B_i = bbl of component i

$(ON)_i$ = blending octane number of component i.

If tetraethyl lead is used to increase the octane of the blend it is necessary to calculate the amount that must be added to produce the desired leaded-product octane. The effectiveness of the TEL decreases with concentration and a special graph has been developed to calculate the quantity of TEL needed to produce the desired octane. It is necessary to know the clear and 3.17g Pb/gal octane numbers of each of the stocks being blended in order to calculate the TEL concentration needed to be added.

The same blending stocks used in Example 11.1 will be the basis for calculation of the amount of TEL required to produce a 94 RON gasoline.

Example 11.3:

Blend the components listed in Example 11.1 and calculate the amount of TEL needed in g Pb/gal for a 94 RON gasoline.

Component	BPD	A vol %	B Clear RON	A X B 100 ΣCl	C 3.17g RON	A X C 100 Σ3.17g
LSR gasoline	4,000	17.7	66.4	11.7	83.5	14.8
Reformate	6,000	26.5	94.0	24.9	100.8	26.7
Alkylate	3,000	13.2	97.3	12.8	104.0	13.7
FCC gasoline	8,000	35.3	92.3	32.5	95.8	33.8
n-butane	1,640	7.3	93.0	6.8	102.5	7.5
	22,640	100.0		88.7		96.5

Plot clear and +3.17g Pb/gal on chart (Fig. 11.1) and connect. At 94 octane number read that 1.23g Pb/gal TEL is required.

If TEL mixture cost is $0.128 per g Pb/gal per barrel of gasoline, the daily cost of TEL is:

$$(22,640 \text{ BPD})(1.23 \text{ g})(\$0.128) = \$3,564$$

Federal requirements with respect to lead contents of gasolines have caused the industry to substitute a number of other compounds to increase the octanes of gasoline blends. None are nearly as effective as lead and most have to be used at levels up to 10% by volume in order to have a significant effect on the octane improvement. Blending octane values for some of the more common additives are given in Table 11.6.

11.3 BLENDING FOR OTHER PROPERTIES

There are several methods of estimating the physical properties of a blend from the properties of the blending stocks. One of the most convenient methods of estimating properties that do not blend linearly is to substitute for the value of the inspection to be blended another value which has the property of blending approximately linear. Such values are called blending factors or blending index numbers. The Chevron Research Company has compiled factors or index numbers for viscosities, flash points, aniline points,

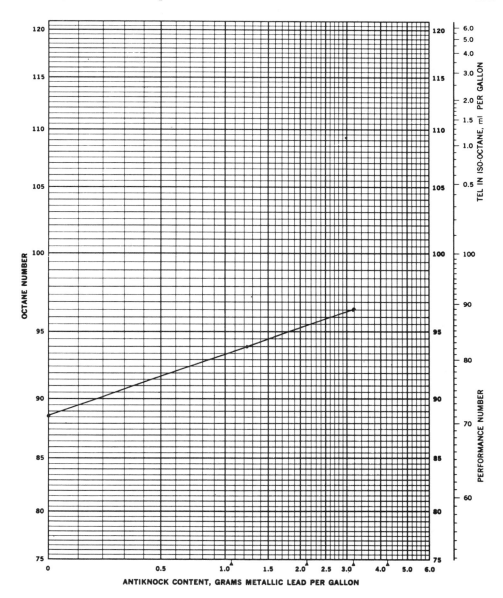

FIG. 11.1. Tetraethyllead requirement calculation. Chart reproduced with
permission of Ethyl Corporation.

and vapor pressures. These are tabulated in Tables 11.2, 11.3, 11.4, and
11.5, respectively. This material is copyrighted and reproduced with permission of the Chevron Research Company.

Examples are given in each table of the use of blending index numbers.
Since it is more complicated than the others, viscosity blending is more fully
discussed below.

TABLE 11.3

Viscosity Blending Index Numbers

FACTORS FOR VOLUME BLENDING OF VISCOSITIES AT CONSTANT TEMPERATURE CORRESPONDING TO VALUES OF KINEMATIC VISCOSITY

Centistokes	0.00	0.01	0.02	0.03	0.04	0.05	0.06	0.07	0.08	0.09
0.5	0.000	0.006	0.013	0.019	0.025	0.030	0.036	0.041	0.046	0.051
0.6	0.056	0.061	0.065	0.069	0.074	0.078	0.082	0.086	0.089	0.093
0.7	0.097	0.100	0.104	0.107	0.110	0.114	0.117	0.120	0.123	0.126
0.8	0.128	0.131	0.134	0.137	0.139	0.142	0.144	0.147	0.149	0.152
0.9	0.154	0.156	0.159	0.161	0.163	0.165	0.167	0.169	0.172	0.174

cSt	0.0	0.1	0.2	0.3	0.4	0.5	0.6	0.7	0.8	0.9
1	0.176	0.194	0.210	0.224	0.236	0.247	0.257	0.266	0.275	0.283
2	0.290	0.297	0.303	0.309	0.314	0.320	0.325	0.329	0.334	0.338
3	0.342	0.346	0.350	0.353	0.357	0.360	0.363	0.366	0.369	0.372
4	0.375	0.378	0.380	0.383	0.385	0.387	0.390	0.392	0.394	0.396
5	0.398	0.400	0.402	0.404	0.406	0.408	0.410	0.411	0.413	0.414
6	0.416	0.418	0.419	0.421	0.422	0.423	0.425	0.426	0.428	0.429
7	0.431	0.432	0.433	0.434	0.436	0.437	0.438	0.439	0.440	0.442
8	0.443	0.444	0.445	0.446	0.447	0.448	0.449	0.450	0.451	0.452
9	0.453	0.454	0.455	0.456	0.456	0.457	0.458	0.459	0.460	0.461

cSt	0	1	2	3	4	5	6	7	8	9
10	0.462	0.470	0.477	0.483	0.489	0.494	0.499	0.503	0.508	0.511
20	0.515	0.519	0.522	0.525	0.528	0.531	0.533	0.536	0.538	0.541
30	0.543	0.545	0.547	0.549	0.551	0.553	0.555	0.557	0.558	0.559
40	0.561	0.563	0.564	0.566	0.567	0.568	0.570	0.571	0.572	0.573
50	0.575	0.576	0.577	0.578	0.579	0.580	0.581	0.582	0.583	0.584
60	0.585	0.586	0.587	0.588	0.589	0.590	0.591	0.592	0.592	0.593
70	0.594	0.595	0.596	0.596	0.597	0.598	0.599	0.599	0.600	0.601
80	0.601	0.602	0.603	0.603	0.604	0.605	0.605	0.606	0.607	0.607
90	0.608	0.608	0.609	0.610	0.610	0.611	0.611	0.612	0.612	0.613

cSt	0	10	20	30	40	50	60	70	80	90
100	0.613	0.618	0.623	0.627	0.631	0.634	0.637	0.640	0.643	0.646
200	0.648	0.651	0.653	0.655	0.657	0.659	0.661	0.662	0.664	0.666
300	0.667	0.669	0.670	0.671	0.673	0.674	0.675	0.676	0.678	0.679
400	0.680	0.681	0.682	0.683	0.684	0.685	0.686	0.687	0.688	0.688
500	0.689	0.690	0.691	0.692	0.692	0.693	0.694	0.695	0.696	0.696
600	0.697	0.698	0.698	0.699	0.700	0.700	0.701	0.701	0.702	0.702
700	0.703	0.704	0.704	0.705	0.705	0.706	0.706	0.707	0.707	0.708
800	0.708	0.709	0.709	0.710	0.710	0.711	0.711	0.712	0.712	0.713
900	0.713	0.714	0.714	0.715	0.715	0.715	0.716	0.716	0.716	0.717

cSt	0	100	200	300	400	500	600	700	800	900
1,000	0.717	0.721	0.724	0.727	0.730	0.733	0.735	0.737	0.739	0.741
2,000	0.743	0.745	0.747	0.748	0.750	0.751	0.752	0.754	0.755	0.756
3,000	0.757	0.758	0.759	0.761	0.762	0.763	0.764	0.765	0.765	0.766
4,000	0.767	0.768	0.769	0.770	0.770	0.771	0.772	0.772	0.773	0.774
5,000	0.775	0.775	0.776	0.777	0.778	0.778	0.778	0.779	0.779	0.780
6,000	0.780	0.781	0.781	0.782	0.782	0.783	0.783	0.784	0.784	0.785
7,000	0.785	0.786	0.786	0.787	0.787	0.787	0.788	0.788	0.789	0.790
8,000	0.790	0.790	0.790	0.791	0.791	0.791	0.792	0.792	0.792	0.793
9,000	0.793	0.794	0.794	0.794	0.795	0.795	0.795	0.796	0.796	0.796

cSt	0	1000	2000	3000	4000	5000	6000	7000	8000	9000
10,000	0.796	0.799	0.802	0.804	0.806	0.808	0.810	0.812	0.814	0.815
20,000	0.817	0.818	0.820	0.821	0.822	0.823	0.824	0.825	0.826	0.827
30,000	0.828	0.829	0.830	0.831	0.832	0.833	0.833	0.834	0.835	0.836
40,000	0.836	0.837	0.838	0.838	0.839	0.839	0.840	0.841	0.841	0.842
50,000	0.842	0.843	0.843	0.844	0.844	0.845	0.845	0.846	0.846	0.847
60,000	0.847	0.848	0.848	0.848	0.849	0.849	0.850	0.850	0.850	0.851
70,000	0.851									
80,000	0.854									
90,000	0.858									

cSt	
100,000	0.860
200,000	0.877
300,000	0.887
400,000	0.894
500,000	0.899
600,000	0.903
700,000	0.906
800,000	0.909
900,000	0.912

cSt	
1,000,000	0.914
2,000,000	0.928
3,000,000	0.937
4,000,000	0.942
5,000,000	0.947
6,000,000	0.950
7,000,000	0.953
8,000,000	0.956
9,000,000	0.958

cSt	
10,000,000	0.960
20,000,000	0.973
30,000,000	0.980
40,000,000	0.985
50,000,000	0.989
60,000,000	0.992
70,000,000	0.995
80,000,000	0.997
90,000,000	0.999

Example: Calculate the viscosity of a three-component blend as follows:

Component	Vol. Fraction in Blend	Viscosity	Factor	Vol. Fraction x Factor
A	0.5	430 SFS at 122°F	0.700	0.350
B	0.3	82.5 SUS at 130°F	0.500	0.150
C	0.2	2.15 cSt at 130°F	0.300	0.060
Total	1.0	(39.5 cSt at 130°F) (or 183 SUS at 130°F) (or 26 SFS at 122°F)	(0.560)	0.560

FACTORS FOR VOLUME BLENDING OF VISCOSITIES AT CONSTANT TEMPERATURE CORRESPONDING TO VALUES OF SAYBOLT UNIVERSAL SECONDS

SUS	0.0	0.1	0.2	0.3	0.4	0.5	0.6	0.7	0.8	0.9
32	0.275	0.278	0.280	0.282	0.284	0.286	0.288	0.290	0.292	0.294
33	0.296	0.298	0.300	0.302	0.303	0.305	0.307	0.309	0.310	0.312
34	0.314	0.315	0.317	0.318	0.320	0.321	0.323	0.324	0.326	0.327
35	0.328	0.330	0.331	0.333	0.334	0.335	0.337	0.338	0.339	0.340
36	0.342	0.343	0.344	0.345	0.346	0.347	0.349	0.350	0.351	0.352
37	0.353	0.354	0.355	0.356	0.357	0.358	0.359	0.360	0.362	0.363
38	0.363	0.364	0.365	0.366	0.367	0.368	0.369	0.370	0.371	0.372
39	0.373	0.373	0.374	0.375	0.376	0.377	0.378	0.378	0.379	0.380

SUS	0	1	2	3	4	5	6	7	8	9
40	0.381	0.388	0.395	0.402	0.408	0.413	0.418	0.423	0.428	0.431
50	0.435	0.439	0.442	0.445	0.449	0.451	0.454	0.457	0.459	0.462
60	0.464	0.466	0.469	0.471	0.473	0.475	0.476	0.478	0.480	0.482
70	0.483	0.485	0.486	0.488	0.489	0.491	0.492	0.493	0.495	0.496
80	0.497	0.498	0.499	0.501	0.502	0.503	0.504	0.505	0.506	0.507
90	0.508	0.509	0.510	0.511	0.512	0.513	0.513	0.514	0.515	0.516

SUS	0	10	20	30	40	50	60	70	80	90
100	0.517	0.524	0.531	0.537	0.542	0.547	0.551	0.555	0.559	0.562
200	0.565	0.568	0.571	0.574	0.576	0.579	0.581	0.583	0.585	0.587
300	0.589	0.591	0.593	0.595	0.596	0.598	0.600	0.601	0.603	0.604
400	0.605	0.607	0.608	0.609	0.611	0.612	0.613	0.614	0.615	0.616
500	0.617	0.618	0.619	0.620	0.621	0.622	0.623	0.624	0.625	0.626
600	0.627	0.628	0.628	0.629	0.630	0.631	0.632	0.632	0.633	0.634
700	0.635	0.635	0.636	0.637	0.637	0.638	0.639	0.639	0.640	0.640
800	0.641	0.642	0.642	0.643	0.643	0.644	0.645	0.645	0.646	0.646
900	0.647	0.647	0.648	0.648	0.649	0.649	0.650	0.650	0.651	0.651

SUS	0	100	200	300	400	500	600	700	800	900
1,000	0.652	0.656	0.660	0.664	0.667	0.670	0.673	0.676	0.678	0.681
2,000	0.683	0.685	0.687	0.689	0.691	0.692	0.694	0.696	0.697	0.699
3,000	0.700	0.701	0.703	0.704	0.705	0.706	0.707	0.708	0.709	0.710
4,000	0.711	0.712	0.713	0.714	0.715	0.716	0.717	0.718	0.719	0.719
5,000	0.720	0.721	0.722	0.722	0.723	0.724	0.725	0.725	0.726	0.726
6,000	0.727	0.728	0.728	0.729	0.729	0.730	0.731	0.731	0.732	0.732
7,000	0.733	0.733	0.734	0.734	0.735	0.735	0.736	0.736	0.737	0.737
8,000	0.738	0.738	0.739	0.739	0.740	0.740	0.740	0.741	0.741	0.742
9,000	0.742	0.742	0.743	0.743	0.744	0.744	0.744	0.745	0.745	0.745

SUS	0	1000	2000	3000	4000	5000	6000	7000	8000	9000
10,000	0.746	0.749	0.752	0.755	0.758	0.760	0.762	0.764	0.766	0.768
20,000	0.770	0.771	0.773	0.774	0.776	0.777	0.778	0.779	0.781	0.782
30,000	0.783	0.784	0.785	0.786	0.787	0.788	0.789	0.790	0.790	0.791
40,000	0.792	0.793	0.793	0.794	0.795	0.795	0.796	0.797	0.797	0.798
50,000	0.799	0.799	0.800	0.800	0.801	0.802	0.802	0.803	0.803	0.804
60,000	0.804	0.805	0.805	0.806	0.806	0.807	0.807	0.807	0.808	0.808

SUS	
70,000	0.809
80,000	0.813
90,000	0.816

SUS	
100,000	0.819
200,000	0.838
300,000	0.849
400,000	0.856
500,000	0.862
600,000	0.867
700,000	0.870
800,000	0.874
900,000	0.877

SUS	
1,000,000	0.879
2,000,000	0.895
3,000,000	0.904
4,000,000	0.911
5,000,000	0.915
6,000,000	0.919
7,000,000	0.923
8,000,000	0.925
9,000,000	0.928

SUS	
10,000,000	0.930
20,000,000	0.944
30,000,000	0.952
40,000,000	0.957
50.000,000	0.961
60,000,000	0.965
70,000,000	0.968
80,000,000	0.970
90,000,000	0.972

FACTORS FOR VOLUME BLENDING OF VISCOSITIES AT 130°F CORRESPONDING TO VALUES OF SAYBOLT FUROL SECONDS AT 122°F

SFS at 122°F	0	1	2	3	4	5	6	7	8	9
20						0.558	0.561	0.563	0.566	0.568
30	0.570	0.572	0.574	0.576	0.578	0.580	0.582	0.584	0.585	0.587
40	0.588	0.590	0.591	0.593	0.594	0.595	0.597	0.598	0.599	0.600
50	0.601	0.602	0.604	0.605	0.606	0.607	0.608	0.609	0.610	0.610
60	0.611	0.612	0.613	0.614	0.615	0.616	0.616	0.617	0.618	0.619
70	0.619	0.620	0.621	0.622	0.622	0.623	0.624	0.624	0.625	0.626
80	0.626	0.627	0.627	0.628	0.629	0.629	0.630	0.630	0.631	0.632
90	0.632	0.633	0.633	0.634	0.634	0.635	0.635	0.636	0.636	0.637

SFS at 122°F	0	10	20	30	40	50	60	70	80	90
100	0.637	0.642	0.646	0.649	0.653	0.656	0.659	0.661	0.664	0.666
200	0.669	0.671	0.673	0.675	0.676	0.678	0.680	0.681	0.683	0.684
300	0.686	0.687	0.688	0.689	0.691	0.692	0.693	0.694	0.695	0.696
400	0.697	0.698	0.699	0.700	0.701	0.702	0.703	0.703	0.704	0.705
500	0.706	0.707	0.707	0.708	0.709	0.710	0.710	0.711	0.712	0.712
600	0.713	0.713	0.714	0.715	0.715	0.716	0.716	0.717	0.718	0.718
700	0.719	0.719	0.720	0.720	0.721	0.721	0.722	0.722	0.723	0.723
800	0.724	0.724	0.724	0.725	0.725	0.726	0.726	0.727	0.727	0.727
900	0.728	0.728	0.729	0.729	0.729	0.730	0.730	0.730	0.731	0.731

SFS at 122°F	0	100	200	300	400	500	600	700	800	900
1,000	0.732	0.735	0.738	0.741	0.743	0.746	0.748	0.750	0.752	0.754
2,000	0.755	0.757	0.759	0.760	0.761	0.763	0.764	0.764	0.766	0.767
3,000	0.769	0.770	0.771	0.772	0.773	0.773	0.774	0.775	0.776	0.777
4,000	0.778									
5,000	0.784									
6,000	0.790									
7,000	0.795									
8,000	0.798									
9,000	0.802									

Note: Values from this table are for 130°F, although the Saybolt Furol seconds are at 122°F. This table alone must not be used for any other temperatures. Values from this table may be used interchangeably with values for kinematic and Saybolt Universal viscosities if the latter are for 130°F.

For SFS at 210°F, assume SUS = 10 x SFS and use the Saybolt Universal table.

From the brochure, "31.0°API Iranian Heavy Crude Oil," by arrangement with Chevron Research Company. Copyright © 1971 by Chevron Oil Trading Company.

TABLE 11.4

Flash Point Blending Index Numbers [3]

May be used to blend flash temperatures determined in any apparatus but, preferably, not to blend closed cup with open cup determinations.

Flash Point, °F	0	1	2	3	4	5	6	7	8	9
0	168,000	157,000	147,000	137,000	128,000	120,000	112,000	105,000	98,600	92,400
10	86,600	81,200	76,100	71,400	67,000	62,900	59,000	55,400	52,100	49,000
20	46,000	43,300	40,700	38,300	36,100	34,000	32,000	30,100	28,400	26,800
30	25,200	23,800	22,400	21,200	20,000	18,900	17,800	16,800	15,900	15,000
40	14,200	13,500	12,700	12,000	11,400	10,800	10,200	9,680	9,170	8,690
50	8,240	7,810	7,410	7,030	6,670	6,330	6,010	5,700	5,420	5,150
60	4,890	4,650	4,420	4,200	4,000	3,800	3,620	3,441	3,280	3,120
70	2,970	2,830	2,700	2,570	2,450	2,330	2,230	2,120	2,020	1,930
80	1,840	1,760	1,680	1,600	1,530	1,460	1,400	1,340	1,280	1,220
90	1,170	1,120	1,070	1,020	978	935	896	857	821	786
100	753	722	692	662	635	609	584	560	537	515
110	495	475	456	438	420	404	388	372	358	344
120	331	318	305	294	283	272	261	252	242	233
130	224	216	208	200	193	186	179	172	166	160
140	154	149	144	138	134	129	124	120	116	112
150	108	104	101	97.1	93.8	90.6	87.5	84.6	81.7	79.0
160	76.3	73.8	71.4	69.0	66.7	64.5	62.4	60.4	58.4	56.5
170	54.7	52.9	51.3	49.6	48.0	46.5	45.1	43.6	42.3	40.9
180	39.7	38.4	37.3	36.1	35.0	33.9	32.9	31.9	30.9	30.0
190	29.1	28.2	27.4	26.6	25.8	25.0	24.3	23.6	22.9	22.2
200	21.6	20.9	20.3	19.7	19.2	18.6	18.1	17.6	17.1	16.6
210	16.1	15.7	15.2	14.8	14.4	14.0	13.6	13.3	12.9	12.5
220	12.2	11.9	11.6	11.2	10.9	10.6	10.4	10.1	9.82	9.56
230	9.31	9.07	8.83	8.60	8.37	8.16	7.95	7.74	7.55	·7.35
240	7.16	6.98	6.80	6.63	6.47	6.30	6.15	5.99	5.84	5.70
250	5.56	5.42	5.29	5.16	5.03	4.91	4.79	4.68	4.56	4.45
260	4.35	4.24	4.14	4.04	3.95	3.86	3.76	3.68	3.59	3.51
270	3.43	3.35	3.27	3.19	3.12	3.05	2.98	2.91	2.85	2.78
280	2.72	2.66	2.60	2.54	2.48	2.43	2.37	2.32	2.27	2.22
290	2.17	2.12	2.08	2.03	1.99	1.95	1.90	1.86	1.82	1.79

Flash Point, °F	0	10	20	30	40	50	60	70	80	90
300	1.75	1.41	1.15	0.943	0.777	0.643	0.535	0.448	0.376	0.317
400	0.269	0.229	0.196	0.168	0.145	0.125	0.108	0.094	0.082	0.072
500	0.063	0.056	0.049	0.044	0.039	0.035	0.031	0.028	0.025	0.022

Example:	Component	Volume	Flash Point, °F	Blending Index	Volume X Blending Index
	A	0.30	100	753	226
	B	0.10	90	1,170	117
	C	0.60	130	224	134
	Total	1.00	111	477	477

In blending some products, viscosity is one of the specifications that must be met. Viscosity is not an additive property and it is necessary to use special techniques to estimate the viscosity of a mixture from the viscosities of its components. The method most commonly accepted is the use of special charts developed by and obtainable from ASTM.

Blending of viscosities may be calculated conveniently by using viscosity factors from Table 11.3. It is usually true to a satisfactory approximation that the viscosity factor of the blend will be the sum of all the products of the volume fraction times the viscosity factor (VF) for each component. In equation form:

TABLE 11.5

Aniline Point Blending Index Numbers

Aniline Point, °F	0	-1	-2	-3	-4	-5	-6	-7	-8	-9
-10	20.0	17.4	14.9	12.6	10.3	8.10	6.06	4.17	2.46	1.00
0	49.1	46.0	42.8	39.8	36.8	33.8	30.9	28.1	25.3	22.6

Aniline Point, °F	0	1	2	3	4	5	6	7	8	9
0	49.1	52.4	55.6	58.9	62.3	65.7	69.1	72.6	76.1	79.6
10	83.2	86.8	90.5	94.2	97.9	102	105	109	113	117
20	121	125	129	133	137	141	145	149	153	157
30	162	166	170	174	179	183	187	192	196	200
40	205	209	214	218	223	227	232	237	241	246
50	250	255	260	264	269	274	279	283	288	293
60	298	303	308	312	317	322	327	332	337	342
70	347	352	357	362	367	372	377	382	388	393
80	398	403	408	414	419	424	429	435	440	445
90	451	456	461	467	472	477	483	488	494	491
100	505	510	516	521	527	532	538	543	549	554
110	560	566	571	577	582	588	594	599	605	611
120	617	622	628	634	640	645	651	657	663	669
130	674	680	686	692	698	704	710	716	722	727
140	733	739	745	751	757	763	769	775	781	788
150	794	800	806	812	818	824	830	836	842	849
160	855	861	867	873	880	886	892	898	904	911
170	917	923	930	936	942	948	955	961	967	974
180	980	986	993	999	1,006	1,012	1,019	1,025	1,031	1,038
190	1,044	1,050	1,057	1,064	1,070	1,077	1,083	1,090	1,096	1,103
200	1,110	1,116	1,122	1,129	1,136	1,142	1,149	1,156	1,162	1,169
210	1,176	1,182	1,189	1,196	1,202	1,209	1,216	1,222	1,229	1,236
220	1,242	1,249	1,256	1,262	1,269	1,276	1,283	1,290	1,330	1,337
230	1,310	1,317	1,324	1,331	1,337	1,344	1,351	1,358	1,365	1,372
240	1,379	1,386	1,392	1,400	1,406	1,413	1,420	1,427	1,434	1,441

Mixed Aniline Point, °F	0	1	2	3	4	5	6	7	8	9
0	-736	-730	-723	-716	-709	-703	-696	-689	-682	-675
10	-668	-660	-653	-646	-639	-631	-623	-616	-608	-600
20	-593	-584	-577	-569	-561	-552	-544	-536	-528	-519
30	-511	-503	-494	-486	-477	-468	-460	-451	-442	-433
40	-425	-416	-407	-398	-389	-380	-371	-361	-352	-343
50	-334	-324	-315	-306	-296	-287	-277	-267	-258	-248
60	-239	-229	-219	-210	-200	-190	-180	-170	-160	-150
70	-140	-130	-120	-110	-100	-89.6	-79.4	-69.2	-58.9	-48.6
80	-38.3	-27.9	-17.5	-7.06	3.39	13.9	24.4	35.0	45.5	56.1
90	66.8	77.4	88.1	98.8	110	120	131	142	153	164
100	175	186	197	208	219	230	241	252	263	274
110	285	297	308	319	330	342	353	364	376	387
120	399	410	422	433	445	456	468	479	491	503
130	514	526	538	550	561	573	585	597	609	620
140	632	644	656	668	680	692	704	716	728	741

Example:	Component	Volume	Aniline Point, °F	Index	Volume X Index
	A	0.8	70	347	278
	B	0.2	40 (Mixed)	-425	-85
	Total	1.0	37 (Or 102 Mixed)	193	193

From the brochure, "31.0°API Iranian Heavy Crude Oil,"by arrangement with Chevron Research Company. Copyright © 1971 by Chevron Oil Trading Company.

TABLE 11.6

Blending Values of Octane Improvers [1, 2]

Compound	RVP, psi	Blending Octane		
		RON	MON	(R + M)/2
Methanol	40	135	105	120
Ethanol	11	132	106	119
tert-Butanol (TBA)	6	106	89	98
MTBE	9	118	101	110
Diisobutylene (DIB)	2	120	95	108
tertiary Butyl Formate (TBF)	6	116	96	106
TEL	—	10,000	13,000	

$$VF_{blend} = \Sigma v_i \times VF_i$$

Table 11.3 shows an example calculation.

Blending of kinematic viscosities (centistokes) may be done at any temperature, but the viscosities of all components of the blend must be expressed at the same temperature. Blending of Saybolt Universal viscosities also may be done at any temperature and interchangeably with kinematic viscosities at the same temperature. Therefore, Table 11.3 may be used to convert viscosities expressed in centistokes to Saybolt Universal seconds (SUS) and vice versa.

Viscosity factors also are given in Table 11.3 for viscosities expressed in Saybolt Furol seconds (SFS). It is important that Saybolt Furol viscosities be blended only at 122°F. If SFS viscosities are at any other temperature, they must be converted to centistokes or SUS before blending.

Viscosity factors for SFS at 122°F may be used interchangeably with viscosity factors for SUS at 130°F and with centistokes at 130°F. Thus, Table 11.3 may be used also to convert viscosities in SFS at 122°F to either kinematic or Saybolt Universal viscosities at 130°F.

A similar method has been developed by Reid and Allen of Chevron Research Company for the estimation of "wax" pour points of distillate blends [4]. Pour point indices for distillate stocks are given in Table 11.7. The pour point index of the blend is the sum of the products of the volume fraction times the pour point blending index (PPBI) for each component, or

$$PPBI_{blend} = \Sigma v_i (PPBI)_i$$

The viscosity of a blend can also be estimated by API Procedure 11A4.3, given on pages 11-35 of the API Technical Data Book—Petroleum Refining.

TABLE 11.7

Pour Point Blending Indices for Distillate Stocks

Pour Point	ASTM 50% Temp															
	300	350	375	400	425	450	475	500	525	550	575	600	625	650	675	700
70	133	131	129	128	127	125	123	120	118	115	113	110	108	105	103	100
65	114	111	109	107	105	103	101	98	96	94	91	88	85	82	79	76
60	99	94	92	90	87	85	82	80	77	74	72	69	67	64	62	60
55	88	79	77	75	73	71	68	66	63	61	58	56	53	50	48	46
50	72	68	66	63	61	59	56	54	52	49	47	44	42	39	37	35
45	60	56	54	52	50	48	46	44	42	40	38	35	33	31	29	27
40	52	48	46	44	42	40	38	36	34	32	30	28	26	24	22	21
35	44	41	39	37	35	33	32	30	28	26	24	23	21	19	18	16
30	37	34	32	31	29	27	26	24	23	21	19	18	16	15	14	13
25	32	29	27	26	24	23	21	20	18	17	15	14	13	12	11	10
20	27	24	23	21	20	19	17	16	15	14	12	11	10	9.1	8.3	7.5
15	23	20	19	18	17	16	14	13	12	11	10	9.0	8.1	7.2	6.4	5.8
10	20	17	16	15	14	13	12	11	9.8	8.8	8.0	7.1	6.3	5.6	5.0	4.5
5	17	15	14	13	12	11	9.7	8.8	7.9	7.1	6.3	5.6	5.0	4.4	3.8	3.5
0	14	12	11	10	9.6	8.7	7.9	7.1	6.3	5.6	5.0	4.4	3.8	3.4	3.0	2.7
-5	12	10	9.5	8.7	8.0	7.2	6.5	5.8	5.1	4.5	3.9	3.4	3.0	2.7	2.4	2.1
-10	10	8.8	8.0	7.3	6.6	5.9	5.3	4.7	4.1	3.6	3.2	2.8	2.5	2.2	1.9	1.6
-15	8.8	7.4	6.8	6.1	5.5	4.9	4.4	3.9	3.4	3.0	2.6	2.2	1.9	1.7	1.4	1.2
-20	7.5	6.3	5.7	5.1	4.6	4.1	3.6	3.2	2.8	2.4	2.1	1.8	1.5	1.3	1.1	0.94
-25	6.4	5.3	4.7	4.2	3.7	3.3	2.9	2.5	2.2	1.9	1.7	1.4	1.2	1.0	0.90	0.72
-30	5.5	4.5	4.0	3.6	3.2	2.8	2.4	2.1	1.8	1.5	1.3	1.1	0.96	0.80	0.67	0.56
-35	4.6	3.7	3.3	2.9	2.6	2.3	2.0	1.7	1.4	1.2	1.0	0.90	0.75	0.62	0.51	0.43
-40	4.0	3.2	2.8	2.5	2.2	1.9	1.6	1.4	1.2	1.0	0.86	0.73	0.62	0.51	0.41	0.33
-45	3.3	2.7	2.4	2.1	1.8	1.5	1.3	1.1	0.98	0.82	0.68	0.58	0.48	0.38	0.31	0.25
-50	2.8	2.3	2.0	1.7	1.5	1.3	1.1	0.93	0.78	0.66	0.56	0.47	0.38	0.31	0.25	0.20
-55	2.5	1.9	1.7	1.4	1.2	1.1	0.90	0.77	0.65	0.55	0.46	0.37	0.30	0.24	0.19	0.15
-60	2.1	1.6	1.4	1.2	1.0	0.87	0.74	0.62	0.52	0.43	0.36	0.30	0.24	0.19	0.14	0.10
-65	1.8	1.4	1.2	1.0	0.85	0.72	0.60	0.50	0.41	0.34	0.28	0.23	0.18	0.14	0.10	0.07
-70	1.5	1.1	0.99	0.84	0.71	0.60	0.50	0.42	0.36	0.30	0.25	0.20	0.15	0.11	0.08	0.05

By arrangement with Chevron Research Company.

11.4 CASE-STUDY PROBLEM: GASOLINE BLENDING

The gasoline blending streams available from the various units are:

Base stock	BPCD	Clear RON
LSR gasoline	5,600	66.4
Lt hydrocrackate	3,640	82.8
Alkylate	7,460	97.3
Hvy hydrocrackate	12,530	67.6
FCC C_5^+ gasoline	16,470	92.3
Reformate	15,610	94.0
Total:	61,310	

The requirements are to produce a 50/50 split of premium and regular gasolines having 100 and 94 research octane numbers, respectively, and Reid vapor pressures of 10 psi.

As the premium gasoline requirements are more severe, it is customary to select the higher octane stocks to blend into premium gasoline. This is a trial-and-error process at this stage. After selecting the stocks, the quantity of n-butane required to give the desired vapor pressure is calculated first because the n-butane contributes significantly to the octane of the finished product.

For the first round of calculations, the blending stocks selected for the premium gasoline should total approximately the volume fraction of premium gasoline times the total blending stocks. The following stocks are used:

Component	Volume	RVP	VPBI	Vol X VPBI
n-butane	W	51.6	138.0	138W
Lt hydrocrackate	3,640	12.9	24.4	88,816
Reformate	10,500	2.8	3.62	38,010
FCC C_5^+ gasoline	9,200	4.4	6.37	58,604
Alkylate	7,460	4.6	6.73	50,206
	30,800 + W			

Case-Study Problem: Gasoline Blending

$$17.8(30,800 + W) = 235,636 + 138W$$
$$548,240 + 17.8W = 235,636 + 138W$$
$$120.2W = 312,604$$
$$W = 2,600 \text{ bbl } C_4$$

Total volume of 10 psi RVP premium gasoline = 33,400 BPCD.

Octane Calculations for Premium Gasoline

Research Octane Number:

Component	BPCD	Vol frac	Clear RON	Σcl	3.17g Pb RON	Σ3.17g
n-butane	2,600	0.078	93.0	7.25	102.5	8.00
Lt hydrocrackate	3,640	0.109	82.8	9.03	98.4	10.73
Reformate	10,500	0.314	94.00	29.52	100.8	31.65
FCC C_5^+ gasoline	9,200	0.275	92.3	25.38	95.8	26.35
Alkylate	7,460	0.224	97.3	21.80	104.0	23.30
	33,400	1.000		92.98		100.03

Motor Octane Number:

Component	BPCD	Vol frac	Clear RON	Σcl	3.17g Pb MON	Σ3.17g
n-butane	2,600	0.078	92.0	7.12	101.5	7.92
Lt hydrocrackate	3,640	0.109	82.4	8.98	97.9	10.67
Reformate	10,500	0.314	84.4	26.50	91.2	28.64
FCC C_5^+ gasoline	9,200	0.275	76.8	21.12	92.3	25.38
Alkylate	7,460	0.224	95.9	21.48	103.4	23.16
	33,400	1.000		85.20		95.77

To meet specifications, 3 ml tetraethyl lead (TEL) mix (3.17 grams Pb) per gallon must be blended into the premium gasoline to give a 100.0 RON and 95.8 MON with 10 psi RVP.

Butane Required for Regular Gasoline

Component	Volume	RVP	VPBI	Vol X VPBI
n-butane	W	51.6	138	138W
LSR gasoline	5,600	11.1	20.3	113,680
Reformate	5,110	2.8	3.62	18,498
Hvy hydrocrackate	12,530	1.1	1.12	14,034
FCC C_5^+ gasoline	7,270	4.4	6.37	46,310
	30,510 + W			192,522 + 138W

$$192,522 + 138W = 17.8(30,510 + W)$$
$$120.2W = 543,078 - 192,522 = 350,556$$
$$W = 2,920 \text{ bbl n-butane required to give 10 psi RVP}$$

Octane Calculations for Regular Gasoline

Research Octane Number:

Component	BPCD	Vol frac	Clear RON	Σcl	3.17g Pb RON	Σ3.17g Pb
n-butane	2,920	.087	93.0	8.09	102.5	8.92
LSR gasoline	5,600	.168	66.4	11.16	83.5	14.03
Reformate	5,110	.153	94.0	14.35	100.8	15.42
Hvy hydrocrackate	12,530	.375	67.6	25.35	85.6	32.10
FCC C_5^+ gasoline	7,270	.217	92.3	20.03	95.8	20.79
	33,430	1.000		79.01		91.26

This is not acceptable, as the requirement for regular gasoline is 94 RON with 3.17g Pb/gal.

There are several ways of correcting this. Among the possibilities are:

1. Increase severity of reforming to produce a 98 or 100 RON clear reformate.
   a. This does not appear to be attractive because the leaded RON would change very little.

2. Install an isomerization unit to increase the clear RON of the LSR gasoline from about 67 to 83 and the leaded RON, with 3.17g Pb as TEL/gal to about 98.
   a. It is questionable whether a once-through process would be sufficient. Substituting 98 RON leaded LSR gasoline into the blend gives a calculated RON of 93.7 for the regular gasoline. This does not give sufficient leeway to allow for inaccuracies in estimating properties.
   b. By fractionating and recycling both the C_5 and C_6 paraffin portions of the reactor effluent the leaded octane number of the LSR gasoline fraction can be increased to approximately 102 RON and that of the blend to 94.4. Little flexibility is afforded here.

3. Reforming about one-half of the heavy hydrocrackate to a 94 RON clear.
   a. Increasing the capacity of the reforming unity by 6,000 BPCD will require a capital investment less than half that of installing a 5,600 BPCD isomerization unit.
   b. Operating costs per barrel will be lower than those of an isomerization unit and there will be greater flexibility of operation.

Recalculating regular gasoline RVP and RON (3.17g Pb) after reforming 55% of the heavy hydrocrackate to 94 RON clear, gives the following:

Component	Volume	RVP	VPBI	Vol X VPBI
n-butane	W	51.6	138.0	138W
LSR gasoline	5,600	11.1	20.3	113,680
Reformate	11,070	2.8	3.62	40,073

Hvy hydrocrackate	5,640	1.1	1.12	6,317
FCC C_5^+ gasoline	7,270	4.4	6.37	46,310
	29,580 + W			206,380 + 138W

$$206,380 + 138W = 17.8(29,580 + W)$$
$$120.2W = 526,524 - 206,380 = 320,144$$
$$W = 2,660 \text{ bbl}$$

Total regular 10 psi RVP gasoline = 32,240 BPCD.

Component	BPCD	Vol frac	Clear RON	Σcl	3.17g Pb RON	Σ3.17g Pb
n-butane	2,660	.083	93.0	7.72	102.5	8.51
LSR gasoline	5,600	.174	66.4	11.55	83.5	14.53
Reformate	11,070	.343	94.0	32.24	100.8	34.57
Hvy hydrocrackate	5,640	.175	67.6	11.83	85.6	14.98
FCC C_5^+ gasoline	7,270	.225	92.3	20.77	95.8	21.56
	33,240	1.000		84.11		94.15

This blend will require slightly less than 3.17g Pb/gal to produce a regular gasoline with a 94 RON and 10 psi RVP.

The motor octane number of the regular gasoline is 89.1 with 3.17g Pb/gal.

PROBLEMS

1. Using values from Table 11.1, calculate the number of barrels of n-butane that have to be added to a mixture of 1,250 barrels of HSR gasoline, 750 barrels of LSR gasoline, and 620 barrels of C_5^+ TCC gasoline to produce a 9.0 psi Reid vapor pressure. What are the research and motor octane numbers of the blend?

2. For the blend of components in problem 1, what would be the sensitivity of the 9.0 psi RVP gasoline if 2.8g Pb/gal were added in the form of TEL motor gasoline mixture?

3. Calculate the amount of butane needed to produce a 12.5 psi RVP for a mixture of 2,730 barrels of LSR gasoline, 2,490 barrels of 94 RON-clear reformate, 6,100 barrels of heavy hydrocrackate, and 3,600 barrels of C_5^+ FCC gasoline. How much TEL must be added to produce a 90 RON product?

4. What is the flash point of a mixture of 2,500 barrels of oil with a flash point of 120°F, 3,750 barrels with a flash point of 35°F, and 5,000 barrels with a 150°F flash point?

5. Calculate the pour point of the following mixture:

Component	Barrels	ASTM 50% temp °F	Pour point °F
A	5,200	575	10
B	3,000	425	50
C	6,500	500	65
D	3,250	550	45

6. What is the viscosity of a blend of 2,000 barrels of oil with a viscosity of 75.5 SUS at 130°F, 3,000 barrels with 225 SUS at 130°F, and 5,000 barrels with 6,500 SUS at 130°F.

7. Calculate the clear octane numbers of the final blend and amount of butane needed for producing a 9.5 psi RVP gasoline from 5,100 BPSD of LSR gasoline, 3,000 BPSD light hydrocrackate, 4,250 BPSD alkylate, 10,280 BPSD heavy hydrocrackate, 14,500 BPSD FCC C_5^+ gasoline, 14,200 BPSD of 96 RON reformate and 2,500 BPSD of polymer gasoline.

8. Recommend the best method for increasing the clear posted octane number of the pool gasoline in Prob. 11.7 by 3 numbers. Estimate the cost involved. Assume any necessary processing units are available and have the necessary capacity.

9. Calculate the number of barrels of n-butane that have to be added to a mixture of 1,000 barrels of light thermal gasoline, 1,000 barrels of polymer gasoline, and 1,000 barrels of C_4^- alkylate to produce a gasoline product having 10.0 psi Reid vapor pressure.

10. What is the clear posted octane number of the gasoline product of problem 3?

11. Calculate the clear octane numbers (RON and MON) and the amount of butane needed for a 12.0 psi RVP gasoline produced from the following:

	BPSD
LSR naphtha	4,200
Lt. hydrocrackate	1,800
C_5^+ alkylate	4,500
Hvy. hydrocrackate	9,150
Reformate (94 RON)	11,500
C_5^+ FCC gasoline	15,600

12. Recommend the best method (lowest capital cost) for increasing the clear posted octane number of the pool gasoline in Prob. 11.11 by 5.5 octane numbers. Estimate the size of the unit and its 1983 construction cost.

NOTES

1. H. L. Hoffman, *Hydro Proc.* 59(2), 57-59, 1980.
2. J. L. Keller, *Hydro Proc.* 58(5), 127-138, 1979.
3. R. O. Wickey and D. H. Chittenden, Flash Points of Blends Correlated, Petrol. Refiner 42(6), 157-158 (1963).
4. E. B. Reid and H. I. Allen, Petrol Refiner 30(5), 93-95 (1951).

Chapter 12
SUPPORTING PROCESSES

There are a number of processes which are not directly involved in the production of hydrocarbon fuels but serve in a supporting role. These include the hydrogen unit, to produce hydrogen for hydrocracking and hydrotreating; the gas processing unit, which separates the low-boiling hydrocarbons; the acid-gas treating unit, which removes hydrogen sulfide and other acid gases from the hydrocarbon gas stream; the sulfur recovery unit; and effluent water-treating systems.

12.1 HYDROGEN MANUFACTURE

Many refineries produce sufficient quantities of hydrogen for hydrotreating from their naphtha-fed platinum catalyst reforming operations. Some of the more modern plants with extensive hydrotreating and hydrocracking operations, however, require more hydrogen than is produced by their catalytic reforming units. This supplemental hydrogen requirement can be provided by one of two processes: partial oxidation of heavy hydrocarbons such as fuel oil, or steam reforming of light ends such as methane (natural gas), ethane, or propane [1, 2]. The steam reforming process employs catalysts, but it should not be confused with the catalytic reforming of naphtha for octane improvement.

Relative hydrogen production costs by the two processes are primarily a function of feedstock cost. Steam reforming of methane usually produces hydrogen at a lower cost than partial oxidation of fuel oil if the cost of methane is less than about 65% of fuel oil cost on a BTU basis. For this reason steam reforming is more widely used in North America than partial oxidation for hydrogen production.

Steam reforming for hydrogen production is accomplished in four steps:

1. *Reforming.* This involves the catalytic reaction of methane with steam at temperatures in the range of 1,400 to 1,500°F, according to the following equation:

$$CH_4 + H_2O \rightarrow CO + 3H_2$$

This reaction is endothermic and is carried out by passing the gas through catalyst-filled tubes in a furnace. The catalyst usually is in the form of hollow cylindrical rings ranging up to 3/4 inch in diameter. It consists of 25 to 40% nickel oxide deposited on a low-silica refractory base.

2. *Shift conversion.* More steam is added to convert the CO from step 1 to an equivalent amount of hydrogen by the following reaction:

$$CO + H_2O \rightarrow CO_2 + H_2$$

This is an exothermic reaction and is conducted in a fixed-bed catalytic reactor at about 650°F. Multiple catalyst beds in one reactor with external cooling between beds are commonly employed to prevent the temperature from getting too high, as this would adversely affect the equilibrium conversion. The catalyst used is a mixture of chromium and iron oxide.

3. *Gas purification.* The third step is removal of carbon dioxide by absorption in a circulating amine or hot potassium carbonate solution. Several other treating solutions are in use. The treating solution contacts the hydrogen and carbon dioxide gas in an absorber containing about 24 trays, or the equivalent amount of packing. Carbon dioxide is absorbed in the solution, which is then sent to a still for regeneration.

4. *Methanation.* In this step, the remaining small quantities of carbon monoxide and carbon dioxide are converted to methane by the following reactions:

$$CO + 3H_2 \rightarrow CH_4 + H_2O$$

$$CO_2 + 4H_2 \rightarrow CH_4 + 2H_2O$$

This step is also conducted in a fixed-bed catalytic reactor at temperatures of about 700 to 800°F. Both reactions are exothermic and, if the feed concentration of CO and CO_2 is more than 3%, it is necessary to recycle some of the cooled exit gas to dissipate the heat of reaction. The catalyst contains 10 to 20% nickel on a refractory base.

The preceding description is somewhat idealized, since the actual reactions are more complicated than those shown. The actual process conditions of temperature, pressure, and steam/carbon ratios are a compromise of several factors.

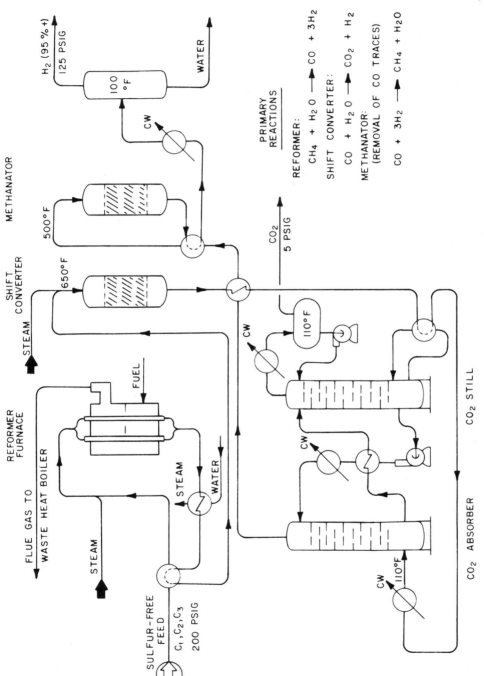

FIG. 12.1. Hydrogen production by steam reforming.

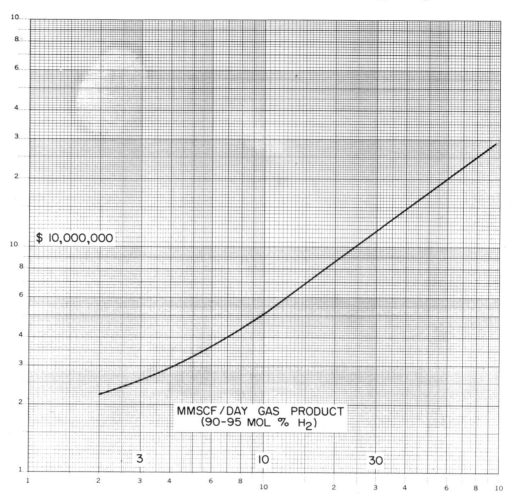

FIG. 12.2*. Hydrogen production by steam-methane reforming investment
           cost—1982 U.S. Gulf Coast.
*See Table 12.1, page 207.

Figure 12.1 is a simplified flow diagram for steam reforming production
of hydrogen. Investment and operating costs can be estimated from Figure
12.2 and Table 12.1.

Partial oxidation of fuel oils is accomplished by burning the fuel at high
pressures (800 to 1300 psig) with an amount of pure oxygen which is limited
to that required to convert the fuel oil to carbon monoxide and hydrogen.
Enough water (steam) is added to shift the carbon monoxide to hydrogen in a
catalytic shift conversion step. The resulting carbon dioxide is removed by
absorption in hot potassium carbonate or other solvents.

TABLE 12.1*

Hydrogen Production [20, 21]

Costs Included:
1. Feed gas desulfurization.
2. Reformer, shift converter, methanator, waste heat boiler, amine (MEA) unit.
3. Hydrogen delivery to battery limits at 250 psig, 100° F, 90 to 95% pure.
4. Initial catalyst charges.

Costs Not Included:
1. Boiler feed water treating.
2. Cooling water.
3. Dehydration of hydrogen product.
4. Waste water treating and disposal.
5. Power supply.

Utility Data:

Power, kWh/lb H_2	0.15
Cooling water,[a] gal crclt/lb H_2	65
Fuel,[b] MBtu/lb H_2	45
Treated boiler feed water,[c] gal/lb H_2	1.0
Feed gas,[d] mol/lb H_2	0.13
Catalysts and chemicals, ¢/lb H_2	0.50

---

*See Fig. 12.2, page 206.
[a]30° F rise.
[b]LHV basis, heater efficiency taken into account.
[c]For waste heater boiler.
[d]Does not include fuel.

Ideally the partial oxidation reactions are as follows:

$$2C_nH_m + nO_2 \longrightarrow 2nCO + m\,H_2$$

$$2nCO + 2nH_2O \longrightarrow 2nCO_2 + 2nH_2$$

A good summary of partial oxidation processes has been published [5].

In some refineries a significant amount of hydrogen is present in gas streams vented from hydrocracking or hydrotreating operations. Recovery of this hydrogen should be considered whenever it is necessary to supplement the catalytic reformer hydrogen to produce an overall refinery hydrogen balance. Three industrial processes are used for recovery of a concentrated hydrogen stream from a dilute gas where the other major components are methane and other hydrocarbon vapors. The three processes are:

1. Cryogenic phase separation
2. Adsorption
3. Diffusion

In the cryogenic phase separation method the gas is cooled to about $-200°F$ to $-250°F$ at pressures ranging from 200 to 500 psig. The resulting vapor phase is typically 90 mol % hydrogen and the liquid phase contains most of the methane and other hydrocarbons. The liquid phase is expanded to about 50 psig and used to cool the feed gas. This revaporizes the liquid phase. The cold vapor phase is also used to cool the feed gas. Carbon dioxide, hydrogen sulfide, and water vapor must be removed from the feed gas prior to chilling. Conditions required to achieve various hydrogen purities are described in the literature [6].

In the adsorption process the hydrocarbons are adsorbed from the gas on a solid adsorbent (usually a molecular sieve) and the hydrogen leaves the adsorber at the desired purity. Several adsorbent vessels are used and the feed gas flow is periodically switched from one vessel to another so that the adsorbent can be regenerated. The adsorbed methane and other impurities are released from the adsorbent by simple pressure reduction and purging [7]. Hence, this process is called Pressure Swing Adsorption.

The diffusion process separates hydrogen from methane and other gases by allowing the hydrogen to permeate through a membrane composed of small synthetic hollow fibers [8]. The driving force for this process is the difference between the hydrogen partial pressures on each side of the membrane. Thus a substantial pressure drop must be taken on the hydrogen product to achieve high recoveries.

The most economical method for hydrogen recovery from refinery gas streams depends on the volume of the gas to be processed, the desired hydrogen recovery and purity, and the type of components to be separated. For relatively small streams (less than 2 to 3 MMSCFD) the diffusion process should be considered. For bulk removal of hydrocarbons from relatively large streams (greater than 20 MMSCFD) the cryogenic process should be considered. The adsorption process usually has advantages when the hydrogen must be recovered at purities over 95 mol %.

12.2 GAS PROCESSING UNIT

The main functions of refinery gas processing units are:

1. Recovery of valuable C_3, C_4, C_5, and C_6 components from the various gas streams generated by processing units such as crude distillation units, cokers, cat crackers, reformers and hydrocrackers.
2. Production of a desulfurized dry gas consisting mostly of methane and ethane which is suitable for use as a fuel gas or as feedstock for hydrogen production [8].

In the typical gas processing unit shown in Fig. 12.3, low pressure (0 to 20 psig) gases are collected and compressed to approximately 200 psig and fed to an absorber-deethanizer. This column usually contains 20 to 24 trays

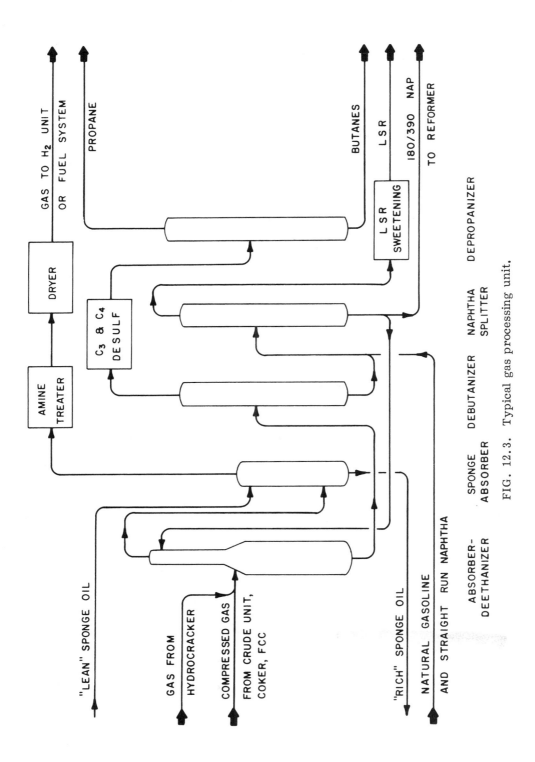

FIG. 12.3. Typical gas processing unit.

in the absorption section (top) and 16 to 20 trays in the stripping section (bottom). Lean absorption oil is fed to the top tray in sufficient quantity to absorb 85 to 90% of the C_3's and almost all of the C_4's and heavier components from the feed gas and from vapor rising from the stripping section. The lean absorption oil is usually a dehexanized naphtha with an end point (ASTM) of 350 to 380°F.

Due to the vapor-liquid equilibrium conditions on the top tray a significant amount of the lighter hydrocarbons (such as C_7) are vaporized from the lean oil and leave the top of the column with the residue gas. This material is recovered in the sponge absorber. The sponge absorber usually contains 8 to 12 trays. A heavy molecuar-weight, relatively nonvolatile material, such as kerosine or No. 2 fuel oil, is used as sponge oil. This sponge oil is derived as a side cut from the coker fractionator or cat cracker fractionator. The rich sponge oil is returned as a side feed to the column from which it was derived for stripping of the recovered lean oil light engs.

Sufficient reboil heat is added to the bottom of the stripping section of the absorber-deethanizer to eliminate any absorbed ethane and methane from the bottom liquid product. This deethanized rich oil then flows to a debutanizer column where essentially all the recovered propane, propylene, butanes, and butylenes are fractionated and taken off as overhead product. This type of debutanizer usually operates in the range of 125 to 150 psig and contains 26 to 30 trays. The bottom product from the debutanizer contains pentanes and heavier hydrocarbons recovered from the gas feed to the absorber-deethanizer plus the lean oil. This material is fed to a naphtha splitter. Natural gasoline and/or straight-run naphtha are sometimes fed to this same column. The naphtha splitter produces a C_5, C_6 light straight-run cut overhead and suitable lean absorption oil from the bottom. Bottoms product in excess of the lean oil requirement can be fed to a hydrotreater and reformer. The light straight-run product is desulfurized (or sweetened) and used directly as gasoline blend stock or else isomerized and then used as gasoline blend stock.

The overhead C_3, C_4 product from the debutanizer is condensed, desulfurized, and fed to a depropanizer for separation into propane and butane. The desulfurization is usually accomplished by molecular sieve treating which simultaneously dehydrates the stream.

Overhead gas from the sponge oil absorber is treated by contact with an aqueous solution of diethanolamine (DEA) or other solvents for removal of carbon dioxide and hydrogen sulfide. Extracted hydrogen sulfide is converted to elemental sulfur in a separate unit. See Figure 12.2 and Photo 12, Appendix F.

See Tables 12.2 and 12.3 and Figures 12.4 and 12.5 for gas processing unit cost data.

12.3 ACID GAS REMOVAL

Gases from various operations in a refinery processing sour crudes contain hydrogen sulfide and occasionally carbonyl sulfide. Some hydrogen sulfide occurs in minor amounts in solution in crude oil. Most of the hydrogen sulfide in refinery gases is formed as a result of conversion of sulfur compounds in

processes such as hydrotreating, cracking, and coking. Until the period of
about 1965 to 1970, it was common practice simply to burn this hydrogen sulfide along with other light gases as refinery fuel, since its removal from the
gases and conversion to elemental sulfur was not economical. Recent air pollution regulations, however, require that most of the hydrogen sulfide be
removed from refinery fuel gas and converted to elemental sulfur.

In addition to hydrogen sulfide many crudes contain some dissolved
carbon dioxide which through distillation finds its way into the refinery fuel
gas. These components—hydrogen sulfide and carbon dioxide—are generally
termed acid gases. They are removed simultaneously from the fuel gas by a
number of different processes, some of which are:

a. Chemical solvent processes:
   1. Monoethanolamine (MEA)
   2. Diethanolamine (DEA)
   3. Triethanolamine (TEA)
   4. Diglycolamine (DGA)
   5. Hot potassium carbonate
b. Physical solvent processes:
   1. Selexol
   2. Propylene carbonate
   3. Sulfinol
   4. Rectisol
c. Dry adsorbents:
   1. Molecular sieve
   2. Activated charcoal
   3. Iron sponge
   4. Zinc oxide

An excellent bibliography on this subject is given in the NGPSA data book
[9]. The selection of available processes for a given application involves
many considerations. In general, the diethanolamine process has been the
most widely used for refinery gas treating [9]. This process uses an aqueous
solution of diethanolamine with concentrations of the DEA in the range of 15
to 30 wt %. The solution is pumped to the top of an absorber containing about
20 to 24 trays or an equivalent amount of packing such as Pall rings. Hydrogen sulfide and carbon dioxide are removed from the gas by absorption into
the solution. Rich solution from the absorber flows into a flash tank which is
operated at a lower pressure than the absorber and permits any dissolved or
entrained methane and ethane to be vented from the system. The rich solution
is then preheated and flows to a regenerator or still where the acid gases are
stripped from the solution by steam generated in a reboiler. The still also
contains about 20 to 24 trays or an equivalent amount of packing. The acid
gases are taken from the top of the still through a condenser where most of
the steam is condensed. This condensate is separated from the acid gases
and returned to the top of the still as reflux. The acid gases are then sent
to a sulfur recovery unit where the hydrogen sulfide is converted to elemental
sulfur. The lean solution is cooled and returned to the top of the absorber
(Fig. 12.6).

Operating conditions are usually such that the treated gas will contain less than 0.25 grain of sulfur per 100 scf and less than 1.5 vol % carbon dioxide. This is controlled by the amount of solution circulated and the amount of steam generated in the reboiler. The solution rate is set at a value which is in the range of 0.15 to 0.40 gallon per scf of acid gases absorbed. The heat input to the reboiler is typically about 600 to 1,200 Btu per gallon of circulating solution. It is generally considered to be advisable to have high solution rates to minimize corrosion. However, high solution rates result in increased investment and utility costs and therefore a compromise is made in the design and operating conditions.

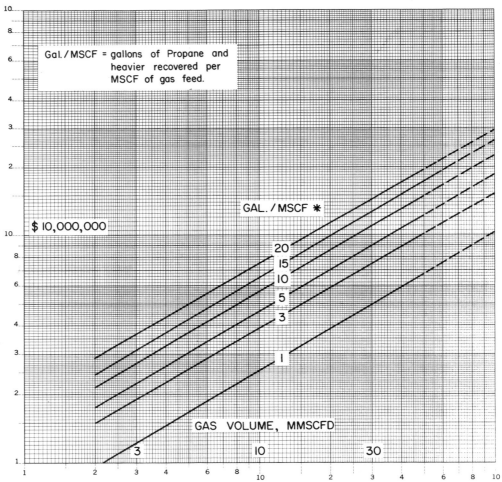

FIG. 12.4†. Refinery gas processing units investment cost—1982 U.S. Gulf Coast.

*GAL/MSCF = gallons of propane and heavier recovered per MSCF of gas feed
†See Table 12.2, page 213.

TABLE 12.2*

Refinery Gas Processing Units

Costs Included:
1. Compressors to raise gas pressure from 5 psig to approximately 200 psig.
2. Absorber deethanizer utilizing naphtha as absorption oil.
3. Sponge oil absorber.
4. Debutanizer.
5. Naphtha splitter.[a]
6. C_3/C_4 mol sieve desulfurizer.
7. Depropanizer.
8. All related heat exchangers, pumps, scrubbers, accumulators, etc.
9. Initial charge of mol sieves.

Costs Not Included:
1. Gas sweetening.[b]
2. Butane splitter.
3. Hot oil or steam supply for reboilers.
4. Liquid product storage.
5. Cooling water supply.
6. Sponge oil distillation facilities. (This is usually accomplished in cat cracker fractionator, coker fractionator, or crude tower.)
7. LSR sweetening.[c]

Utility Data:

Process fuel, Btu/gal total liquid products (LHV)[d]	14,000
Compressor bhp/MMscf/day gas[e]	150
kWh/gal total liquid products[f]	0.06
Cooling water,[g] gal CW/gal total liquid products	100

---

*See Fig. 12.4, page 212.

[a]When "outside" natural gasoline or straight-run naphtha is fed to naphtha splitter, increase investment obtained from Figure 12.4 by $85 for each BPD of such material. This material is not to be included when determining the gal/Mscf parameter for Figure 12.4.

[b]See figure 12.5 and Table 12.3.

[c]Add $20 for each BPD of LSR naphtha to be sweetened by caustic washing or "doctor treating."

[d]Process fuel requirement can frequently be obtained from heat exchange with hot oil from other refinery units. If sufficient detail is not available to ascertain that heat input can be obtained by exchange as stated above, assume that fired heater must be provided.

[e]Gas compressor will generally be motor driven when gas volume is less than 5 MMscfd. In this case it is necessary to add the compressor horsepower to the power requirement shown. For gas volumes in excess of 5 MMscfd compressor may be driven by electric motors, gas turbines, reciprocating gas engines or steam turbines. Appropriate adjustments should be made to the utility requirements to provide for operation of gas compressors.

[f]Power shown is for operation of process pumps only. All cooling is done with cooling water. No power included for aerial coolers.

[g]30°F rise.

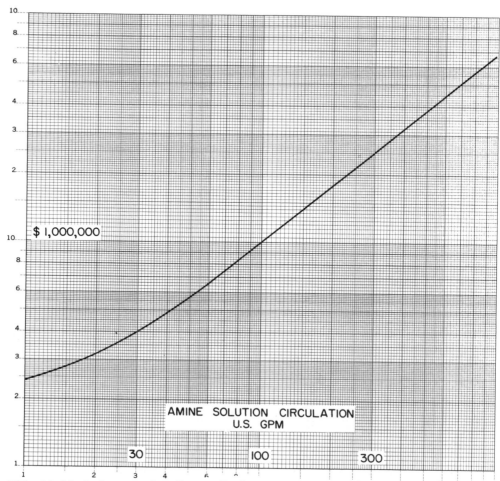

FIG. 12.5*. Amine-gas treating units investment cost--1982 U.S. Gulf Coast.
*See Table 12.3, page 215.

12.4 SULFUR RECOVERY PROCESSES

Until 1970, the major reason to recover sulfur from refinery gases was an economic one. The hydrogen sulfide was commonly used with other gases as a refinery fuel and the sulfur dioxide concentrations in the flue gases were within acceptable limits. Even in those refineries with sulfur recovery units the efficiency of sulfur recovery was in the range of 90 to 93% of that contained in the hydrogen sulfide stream. The methods used were of the dry-bed catalytic conversion type such as the modified Claus and Direct Oxidation processes. On a once-through operation these are limited to 90 to 93% sulfur recovery.

After the implementation of the Environmental Pollution Act it became necessary to recover greater than 95% of the sulfur in order to meet pollution standards. This generally requires a two-stage process with a modified Claus or Direct Oxidation unit for the first stage followed by a second stage such as the Stretford or Sulfreen process [10]. Sulfur recoveries of 98% and above can be achieved by these combinations [11].

TABLE 12.3*

Amine Gas Treating Units

Costs Included:
1. Conventional, single flow, MEA or DEA treating system.
2. Electric-motor driven pumps.
3. Steam-heated reboiler.
4. Water-cooled reflux condenser and solution cooler.

Costs Not Included:
1. Acid gas disposal.
2. Cooling water supply.
3. Steam (or hot oil) supply for regenerator reboiler.

Utility Data:

Power, kWh/gal solution circulated[a]	0.01
Fuel, Btu/gal solution circulated[b]	1000
Cooling water, gal CW/gal solution[c]	4.4
Amine consumption averages about 2.5 lb per MMscf gas to contactor[d]	

*See Fig. 12.5, page 214.

[a]Assumes amine pumps driven by electric motors and cooling done with water.

[b]Reboiler heat usually supplied as 60 psig steam.

[c]30°F rise.

[d]In actual practice, amine solution circulation varies in the range of 0.15 to 0.40 gallon per scf of acid gas (H_2S plus CO_2) removed. For preliminary estimates, a value of 0.30 gallon of solution circulation per scf of acid gas can be assumed.

Claus Processes

The Claus processes work well for gas streams containing greater than 20% (by volume) hydrogen sulfide and less than 5% hydrocarbons. The original process was reported by Chance and Claus in 1885 [12]. Today, variations of the process are used depending upon the hydrogen sulfide and hydrocarbon concentrations. The limitations of thes processes are (mol %):

1. Partial combustion process:
   a. Hydrogen sulfide concentrations: ≥50%
   b. Hydrocarbon concentration: <2%
2. Split-stream process:
   a. Hydrogen sulfide concentration: 20 to 50%
   b. Hydrocarbon concentration: <5%

In the partial combustion (once-through—process, the hydrogen sulfide-rich gas stream is burned with one-third the stoichiometric quantity of air and the hot gases are passed over a bauxite catalyst to react sulfur dioxide with unburned hydrogen sulfide to produce additional free sulfur. The reactions are:

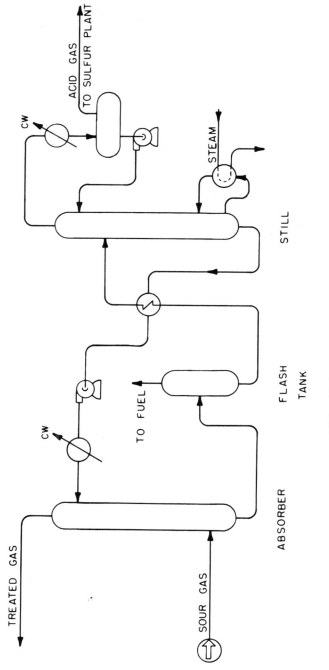

FIG. 12.6. Amine treating unit.

Burner: $2H_2S + 2O_2 \rightarrow SO_2 + S + 2H_2O$

Reactor: $2H_2S + SO_2 \rightarrow 3S + 2H_2O$

If the hydrogen sulfide content of the gas stream is less than 25% and the hydrocarbon content is 2% or greater (and less than 5%), the split stream process is used [13]. Here one-third of the gas stream is burned with the stoichiometric quantity of air to give

$$H_2S + 3/2O_2 \rightarrow SO_2 + H_2O$$

The remaining two-thirds of the gas stream is mixed with the combustion products and enters the reactor containing a bauxite catalyst. The reaction is

$$2H_2S + SO_2 \rightarrow 3S + 2H_2O$$

Flow schemes of these processes are shown in Figures 12.7 and 12.8. Installed costs and utility and chemical requirements are given in Figure 12.9 and Table 12.4.

Direct Oxidation Process

The direct oxidation process is a variation of the Claus process developed by Pan American Petroleum Corporation to handle hydrogen sulfide rich gas streams containing relatively low concentrations of hydrogen sulfide and large amounts of hydrocarbons. The normal limits of operation are:

1. Hydrogen sulfide concentrations: 2 to 15%
2. Hydrocarbon concentrations: >5%

This process requires the hydrogen sulfide rich gas stream be preheated, mixed with air, and fed to a catalytic reactor containing bauxite followed by a condenser and another reactor. Sulfur recoveries of 80 to 85% are usual with this process [11, 14].

Partial Combustion (Once-Through) Claus Process Details

Present refinery practice generally provides for removal of hydrogen sulfide from refinery gas streams by solvent absorption as discussed in the previous section. The acid gas stream recovered from these treating processes will contain some carbon dioxide and minor amounts of hydrocarbons, but in most cases, the hydrogen sulfide content will be over 50%. Therefore, the once-through Claus process is used in most sour crude refineries to convert the hydrogen sulfide to elemental sulfur.

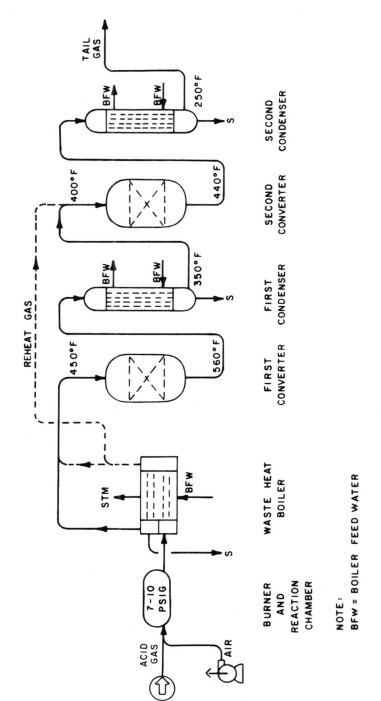

FIG. 12.7. Once-through Claus sulfur process.

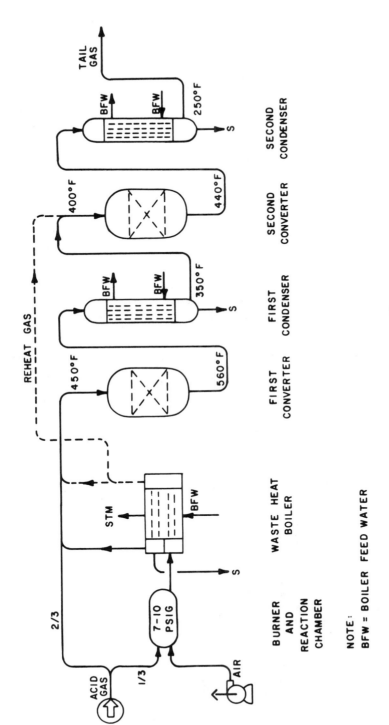

FIG. 12.8. Split-flow Claus sulfur process.

In the once-through process, sufficient air is introduced to the burner to combust one-third of the hydrogen sulfide to sulfur dioxide. This burner is located in a reaction chamber which may be either a separate vessel or a part of the waste heat boiler. The purpose of the reaction chamber is to allow sufficient time for the combustion reaction to be completed before the gas temperature is reduced in the waste heat boiler.

Ammonia frequently is present in the Claus unit feed streams and must be completely destroyed in the reaction furnace to avoid plugging of equipment with ammonium salts. Specially designed burners have been developed for this purpose which provide very complete mixing of the acid gas and air prior to combustion.

The waste heat boiler removes most of the exothermic reaction heat from gases by steam generation. Many types of waste heat boilers are in use. Usually they are arranged so that the gas flows through several tube passes in series with chambers or "channels" where a portion of the gases may be withdrawn at elevated temperatures to use for reheating the main gas flow stream prior to the catalytic converters [15]. Some elemental sulfur is often condensed and removed from the gas in the waste heat boiler. In some plants a separate condenser is used after the waste heat boiler. The gas temperature entering the first catalytic converter is controlled at about 425 to 475°F which is necessary to maintain the catalyst bed above sulfur dewpoint in order to avoid saturating the catalyst with sulfur and therby deactivating the catalyst. The reaction between hydrogen sulfide and sulfur dioxide in the converter is also exothermic. Gases from the converter are cooled in the following condenser for removal of most of the elemental sulfur as liquid. Condenser outlet temperatures are set to reduce sulfur vapor partial pressures to allow further conversion in the following converter. The condenser outlet temperatures must be maintained above about 250°F to avoid solidifying the

TABLE 12.4*

Sulfur Recovery Plant

Costs Included:
1. Packaged, skid-mounted type unit designed for 90 to 93% recovery.
2. Two converters (reactors) with initial charge of catalyst.
3. Incinerator and 50-ft tall stack.
4. Sulfur receiving tank and loading pump.
5. Waste heat boiler.

Costs Not Included:
1. Boiler feed water treating.
2. Boiler blowdown disposal.
3. Solid sulfur storage or reclaiming.
4. Sulfur loading facilities (except for loading pump).
5. Supply of power and water.

Utility Data:

Power, kWh/long ton of product sulfur	100
Boiler feed water, gal/long ton of product sulfur	820
Waste heat steam production @ 250 psia, lb stm/long ton of product sulfur	6,500
Fuel	None
Cooling water	None

*See Fig. 12.9, page 221.

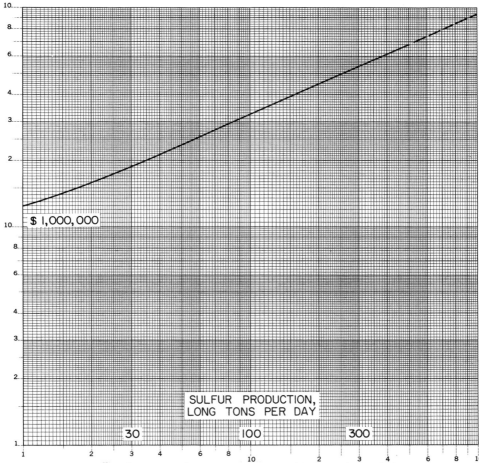

FIG. 12.9*. Claus sulfur plant investment cost—1982 U.S. Gulf Coast.
*See Fig. 12.4, page 220.

sulfur. Two converters and condensers in series are generally provided, but some of the newer plants have three converters. Overall recovery as previously stated is usually not over 95% and is limited by thermodynamic considerations as described in the literature [16, 17, 18, 19]. Modifications of the once-through process include various reheat methods for the converter feed temperature control such as heat exchange with converter outlet gases, in-line burners, and fired reheaters.

Carbon-Sulfur Compounds

Carbonyl sulfide (COS) and carbon disulfide (CS_2) have presented problems in many Claus plant operations due to the fact that they cannot readily be converted to elemental sulfur and carbon dioxide. These compounds are formed in the combustion step by reaction of hydrocarbons and carbon dioxide as shown below:

$$CH_4 + SO_2 \rightarrow COS + H_2O + H_2$$

$$CO_2 + H_2S \rightarrow COS + H_2O$$

$$CH_4 + 2S_2 \rightarrow CS_2 + 2H_2S$$

Many more complex reactions are possible. These compounds, if unconverted, represent a loss of recoverable sulfur and an increase in sulfur emission to the atmosphere. Recent studies indicate that a special alumina catalyst is significantly more effective in converting both COS and CS_2 to elemental sulfur than the conventional bauxite catalyst [20].

Stretford Process

The North Western Gas Board and Clayton Aniline Co., Ltd., of England developed an absorption process (Fig. 12.10) which can reduce the hydrogen sulfide content of exit gases to less than one part per million. The sulfur produced is removed from the liquid by filtration, centrifugation, or froth flota-

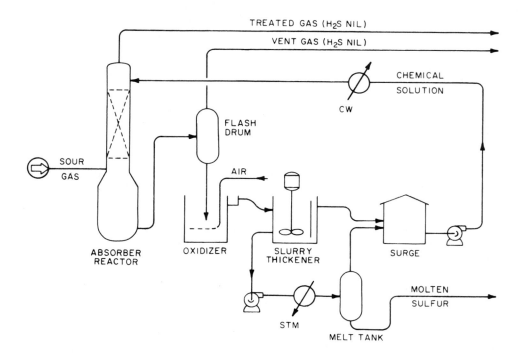

FIG. 12.10. Stretford process.

tion. This process is generally only economical on gas streams containing less than 2% hydrogen sulfide.

The hydrogen sulfide is absorbed in a quinone solution which may contain sodium metavanadate to accelerate the oxidation reactions [20, 21]. The dissolved hydrogen sulfide is then oxidized to form free sulfur and hydro-quinone. Air or oxygen is injected into the solution to oxidize the hydro-quinone back to the original quinone for reuse. The solution is filtered or centrifuged to remove the sulfur.

Gas containing small amounts of sulfur dioxide, carbonyl sulfide, and carbon disulfide in addition to hydrogen sulfide can be processed for essentially complete removal of sulfur by first converting all the sulfur compounds to hydrogen sulfide and then treating the gas in a Stretford unit. The Beavon process is one example of this combination of steps. Conversion of various sulfur compounds to hydrogen sulfide is done at 500°F to 700°F over a cobalt-molybdate catalyst. Some fuel gas is partially combusted and mixed with the sulfur containing gas upstream of the reactor. This provides the necessary preheat and the hydrogen required for the reaction. A typical use of the Beavon process is final sulfur removal from the tail gas of Claus units. Another process used for this purpose is the SCOT process [22].

Since the costs of these tail gas clean-up processes are so sensitive to the specific conditions at the plant, it is suggested for preliminary evaluations an allowance be made for the capital investment for tail gas clean-up in the range of 75 to 100% of the cost of a Claus unit. Typical utility and chemical requirements are shown in Table 12.5.

TABLE 12.5

Stretford Utility Data

Power, kWh/lb sulfur recovered	1
Cooling water, gal/lb sulfur recovered	6
Chemicals, $/long ton of sulfur recovered	2.65

12.5 ECOLOGICAL CONSIDERATIONS IN PETROLEUM REFINING

Since the end of World War II, petroleum refineries have made special efforts to minimize discharge of wastes into the surrounding environment. This voluntary control of emissions was done on the basis of safety, fuel economy, and the economic advantages of good maintenance. The rules and regulations adapted by local, state, and federal governments in the last three or four years have, however, generally required added investment and operating costs for the refiner. These extra costs vary widely but for a new refinery the costs are frequently considered to be in the range of 10 to 15% of the total investment.

The major potential types of pollution which must be carefully controlled include discharge of liquid hydrocarbons into streams, rivers, lakes, and oceans, and relief of hydrocarbon vapors into the atmosphere. In addition to these sources, waste water and combustion products must meet strict specifications relative to toxic impurities. Noise levels must also be controlled.

Complex facilities are necessary to meet these requirements during both normal operations and periods of abnormal operations.

The more common methods of control of these various potential pollution sources are discussed in the following paragraphs.

Waste Water Treatment

Typical sources of waste water in refineries are:

1. Runoff surface drainage from leaks, open drains, and spills carried away by rain.
2. Crude and product storage tank water drains.
3. Desalter water.
4. Water drains from atmospheric still reflux drums.
5. Water drains from barometric sumps or accumulators on vacuum tower ejectors.
6. Water from hydraulic decoking of coke drums.
7. Condensed steam from coke-drum purging operations.
8. Product fractionator reflux drums on units such as catalytic crackers, hydrotreaters, alkylation units, light ends recovery and others.
9. Cooling tower and boiler water blowdown.

Surface water is collected in open trenches and sewer systems and water from process vessels is collected in pipe drain systems. Practically all vessels, tanks, pumps, and low spots in piping are connected to a closed drain system. Any water which may be contaminated with oil is skimmed in large concrete sumps called API separators. The skimmed oil is pumped to slop tanks and then reprocessed. Water from the API separators is further purified by co-agulation of impurities in flotation tanks. In this step a mixture of ferric hydroxide and aluminum hydroxide is used to cause the impurities to form a froth or slurry which floats to the top of the water. The froth is withdrawn and thickened or settled. The resulting sludge is then incinerated.

Water from the flotation tanks is oxygenated under pressure and then fed to digestion tanks which may also receive sanitary sewage from the re-finery. A controlled flock of bacteria is maintained in the digestion tanks to consume any remaining oils or phenolic compounds. A certain amount of the bacteria is continuously withdrawn from the digestion tanks and incinerated.

Water from the digestion tanks may be given a final "polish" in sand filters and reused in the refinery or aerated to increase the oxygen content and subsequently discharged to natural drainage.

Oil-free water drains from the cooling towers and boilers are neutralized and either evaporated in solar ponds, injected into disposal wells, or diluted with other treated waste water to lower the dissolved solids content and then aerated and discharged to natural drainage.

Acid sludge from sources such as alkylation units is collected in a separate system and neutralized before going to the API separators.

Sour water drained from vessels such as the atmospheric still reflux drums is stripped in a bubble tower with either gas or steam to eliminate dissolved hydrogen sulfide and other organic sulfur compounds before feeding this water to the API separators. The stripped vapors are processed for sulfur recovery.

As a precaution against spills all storage tanks are normally surrounded by earthen dikes of sufficient size to retain the entire volume of oil which the tank can hold.

12.6 CONTROL OF ATMOSPHERIC POLLUTION FROM REFINERIES

The major sources of potential atmospheric pollution include combustion gases exhausted from boilers and process furnaces and hydrocarbon vapors vented from process equipment and storage tanks.

The sulfur dioxide content of combustion gases is controlled to local regulations by limiting the sulfur content of the fuel. Tail gases from sulfur recovery units such as Claus plants are sometimes reprocessed in a unit specially designed to convert low concentrations of hydrogen sulfide and sulfur dioxide into elemental sulfur and thus achieve over 99% recovery of the inlet sulfur. The final tail gas is then incinerated and vented through tall stacks, often 200 feet or more in height, and at sufficient velocity so that resulting ground level sulfur dioxide concentrations are well within safe values.

Hydrocarbon vapors from process equipment and storage tanks are collected in closed piping systems and used for refinery fuel or, in the event of high venting rates during the process upset, the vapors are incinerated in a flare or burn pit, with special provisions to prevent visible smoke and to ensure complete combustion.

Fluid catalytic crackers are provided with two- to three-stage cyclones to minimize loss of catalyst dust to the atmosphere. In some cases, electrostatic precipitators are employed along with waste heat boilers to eliminate essentially all visible dust from catalytic cracker regenerator flue gases.

Crushing and screening of coke from delayed coking units is generally done in the wet condition to prevent dust losses to the air. The final coke product is often stored in buildings to prevent wind from carrying fine particles into the atmosphere.

12.7 CONTROL OF NOISE LEVEL IN REFINERIES

Noise in refineries is caused primarily from rotating machinery such as turbines, compressors, engines, and motors. High velocity flow of fluids through valves, nozzles, and piping also contributes to the general noise level. To control these noises, equipment causing the noise is enclosed or insulated. Proper intake and exhaust silencers are provided on blowers, combustion engines, and turbines. In newer refineries the land area used is sufficient so that, combined with the above noise control measures, essentially no noise is heard outside the refinery boundaries.

12.8 CASE-STUDY PROBLEM: GAS RECOVERY UNIT, AMINE UNIT, AND SULFUR RECOVERY UNIT

In order to select the size of the auxiliary units, it is necessary to summarize the outputs of all the major process units to find the quantities of the light components and sulfur available and the quantity of hydrogen consumed.

TABLE 12.6

Light Ends Summary

Unit	H_2 (lb/hr)	C_2 & ltr (lb/hr)	C_3 (lb/hr)	i-C_4 (BPCD)	n-C_4 (BPCD)	S in H_2S (lb/hr)
Crude units		550	2,970	200	700	
Coker	270	20,790	5,520	110	270	2,380
Reformer*	4,810	2,850	5,130	500	700	540
FCC		28,290	6,680	1,710	780	1,990
Hydrocracker	(-10,440)	1,450	5,400	1,370	680	3,910
Alkylation			3,730	(-6,230)	370	
Blending					(-5,260)	
Net	(-5,630)	53,930	29,430	(-2,340)	(-1,760)	8,820

Note: Parentheses indicate material consumed or needed.
*Includes hydrotreater.

TABLE 12.7

Hydrotreater and Catalytic Reformer Material Balance (Revised)

Basis: 100,000 BPCD   North Slope, Alaska, Crude Oil
Severity: 94 RON clear K_W = 11.7

Component	vol %	BPCD	°API	(lb/hr)/BPD	lb/hr	wt % S	S lb/hr
Feed:							
190-380° HSR	50.1	12,500	48.5	11.47	143,380	0.1	140
Coker gasoline	22.2	5,540	54.6	11.09	61,420	0.65	400
180-400° hyd crk	22.7	6,890	46.0	11.63	80,130		
	100.0	24,930			284,930		540
Products:							
H_2, wt % total	1.7				4,844		
$C_1 + C_2$, wt %	1.0				2,850		
C_3, wt %	1.8		146.5	7.42	5,130		
i-C_4	2.0	500	119.4	8.22	4,110		
n-C_4	2.8	700	110.8	8.51	5,960		
C_5^+ reformate	86.5	21,570	38.4	12.12	261,496		
Hydrogen:							
H_2S					574		540
H_2 net					4,810		

$$H_2S = \frac{540}{32.06} = 16.84 \text{ lb-mol/hr}$$

$$H_2 \text{ in } H_2S = (16.84)(2) = 34 \text{ lb/hr}$$

TABLE 12.8

Hydrotreater and Catalytic Reformer
Chemical and Utility Requirements
(per bbl feed)

	Treater	Reformer	Total required
BPCD feed	18,040	24,930	
Steam, lb	6	30	35,670 lb/hr
Power, kWh	2	2	110,870 kWh/day
CW, gal	300	600	14,150 gpm
Fuel, MMBtu	0.1	0.3	9,280 MMBtu/day
Catalyst, lb	0.002	—	36 lb/day
Catalyst, lb		0.004	100 lb/day

First, however, the catalytic reformer must be recalculated because the gasoline blending calculations disclosed that 55% of the heavy hydrocrackate (6,890 BPCD) must be reformed to 94 RON clear to produce the quality of gasoline needed.

The light ends summary using the revised reformer balance is given in Table 12.6.

From this summary, it is apparent that hydrogen and sulfur units must be constructed to produce 25,600,000 scf/day of hydrogen and 95 long tons per day of sulfur.

In addition, 2,340 BPCD of isobutane and 2,460 BPCD of normal butane must be purchased to meet alkylation unit and gasoline blending requirements.

The revised catalytic reformer balance is shown in Tables 12.7 and 12.8.

12.9 CASE-STUDY PROBLEM: HYDROGEN UNIT

For simplicity of calculation, it will be assumed the feed gas to the hydrogen unit will be purchased natural gas. Using a 95% pure hydrogen product stream as a basis, the size and requirements of the hydrogen unit are summarized in Table 12.9.

TABLE 12.9

Hydrogen Unit Utility Requirements

Unit capacity: 25,600,000 scf/day or 135,120 lb H_2/day

	Requirements/lb H_2	Total requirements
Power	0.15 kWh	20,270 kWh/day
Cooling water	65 gal	6,100 gpm
Fuel	45 MBtu	6,080 MMBtu/day
Boiler feed water	1 gal	94 gpm
Feed gas	0.13 mol	6,660 Mscf/day
Catalysts and chemicals	0.10 ¢	$135/day

TABLE 12.10

Gas Processing Unit Feed Balance

Feed to unit	lb/hr	Mol wt	mol/hr	MMscfd	gal liq /mscf gas feed
C_2 and ltr	53,930	23		21.3	
C_3	29,430	44	669	6.1	4.6
n-C_4	23,740	58		3.7	3.2
i-C_4	31,890	58	550	5.0	4.5
				36.2	12.3

Total liquid products = (36.2)(12.3 × 10^3) = 445,260 gal/day

12.10 REFINERY GAS PROCESSING UNIT

The refinery gas processing unit will handle all C_4 and lighter gases. In this refinery, propane will be burned for refinery fuel and the n-butane product in the units will be liquified and used for blending into gasoline. The iso-butane is also recovered and fed to the alkylation unit. The unit feed components are tabulated in Table 12.10 and unit utility requirements in Table 12.11.

12.11 AMINE GAS TREATING UNIT

The amine gas treating unit removes the acid gases (hydrogen sulfide and carbon dioxide) from the gaseous streams in the refinery gas plant. No information is available concerning the concentration of carbon dioxide but a reasonable estimate is 1% by volume of the gases not liquefied. Therefore

$$\text{Amount of CO}_2 = \frac{(21,300,000)(0.01)}{(24)(60)} = 148 \text{ scfm CO}_2$$

The volume of hydrogen sulfide in the gases is:

$$\frac{8,820 \text{ lb}}{\text{hr}} \frac{\text{lb-mol}}{32.07 \text{ lb}} \frac{\text{hr}}{60 \text{ min}} \frac{379 \text{ ft}^3}{\text{lb-mol}} = 1,737 \text{ scfm}$$

Total acid gases = 1,737 + 148 = 1,885 scfm

TABLE 12.11

Gas Processing Unit Utility Requirements

Process fuel	6,234	MMBtu/day
Compressor (5,430 bhp)	97.2	MMWh/day
Power, excluding compressor	26.7	MMWh/day
Cooling water	30,920	gpm

TABLE 12.12

Amine Unit Utility Requirements

Power		8.2 MkWh/day
Fuel	815	MMBtu/day
Cooling water	2,490	gpm
Amine consumption	54	lb/day

$$\text{Amine solution circulation rate} = \frac{1885 \text{ scf}}{\text{min}} \frac{1 \text{ gal}}{3.33 \text{ scf}} = 566 \text{ gpm}$$

The amine unit utility and chemical requirements are summarized in Table 12.12.

12.12 SULFUR RECOVERY PLANT

To comply with environmental factors, at least 98% of the sulfur must be recovered from the hydrogen sulfide-rich gases. A Claus sulfur unit followed by a Stretford absorption unit will reduce the hydrogen sulfide content of the exit gases to <5 ppm.

To provide a conservatively sized Stretford unit, a sulfur recovery of 90% will be assumed for the Claus unit.

The material balance and utilities for the Claus unit are given in Tables 12.13 and 12.14 and those for the Stretford unit in Tables 12.15 and 12.16.

Total sulfur production is 96.2 long tons per day.

TABLE 12.13

Claus Sulfur Recovery Unit

Basis: 100,000 BPCD   North Slope, Alaska, Crude Oil
        90% Sulfur recovery   86.6 LT/day sulfur

Component	vol % dry basis	scfm	lb/hr	S (lb/hr)
Feed:				
H_2S	92.1	1,737	9,370	8,820
CO_2	7.9	148	1,030	
N_2		3,124	13,850	
O_2		781	3,960	
	100.0	5,790	28,210	8,820
Products:				
H_2S	5.0	174	940	880
CO_2	4.3	148	1,030	
N_2	90.7	3,124	13,850	
H_2O		1,562	4,450	
Sulfur			7,940	7,940
	100.0	3,446	28,210	8,820

TABLE 12.14

Claus Sulfur Unit Utility Requirements

Power	3,507 kWh/day
Boiler feed water	48 gpm
Steam production, 250 psia	23,040 lb/hr

TABLE 12.15

Stretford Absorption Unit

Basis: 100,000 BPCD   North Slope, Alaska, Crude Oil

Component	vol %	scfm	lb/hr	S (lb/hr)
Feed:				
H_2S	5.0·	174	940	880
CO_2	4.3	148	1,030	
N_2	90.7	3,124	13,850	
	100.0	3,446	15,820	
Products:				
CO_2	4.5	148	1,030	
N_2	95.5	3,124	13,850	
H_2 in H_2O			60	
			880	880
	100.0	3,272	15,820	

Sulfur production = 9.6 LT/day

TABLE 12.16

Stretford Absorption Unit Utility Requirements

Power	21,100 kWh/day
Cooling water	90 gpm
Chemical costs	$25.44/day

PROBLEMS

1. Determine the utility and feed gas requirements for a 10 million standard cubic feet per day hydrogen unit.

2. Estimate the 1985 construction and direct operating costs for a 15 MMscfd hydrogen unit using as feed natural gas at $2.75 per Mscf. Labor costs are $25.00/hr.

3. Make an overall material and utility balance and estimate the 1985 construction cost for a gas processing unit handling the following gas stream:

	lb/hr
H_2S	12,760
CO_2	1,540
C_1	39,000
C_2	39,100
C_3	42,580
$n-C_4$	34,350
$i-C_4$	27,540

4. Calculate the amine circulation rate for an amine gas treating unit processing the gas stream listed in problem 3.

5. Calculate the annual sulfur production in long tons for the refinery having the gas stream of problem 3, assuming 99.5% recovery of the acid gases by the amine unit and 93.5% efficiency of the Claus sulfur unit.

6. If a Stretford absorption unit is used after the Claus sulfur unit to reduce the hydrogen sulfide content of the exit gases to 5 ppm, what will be the annual production of sulfur in long tons?

7. What will the total 1985 construction cost be for the complete gas plant including sulfur conversion units for the gas stream given in problem 3? Assume the figures given are the average pounds per hour on an annual basis.

8. Recommend a process scheme for handling the gas stream in Problem 3, meeting all current environmental restrictions.

9. Estimate the present construction cost of the processing units of Prob. 12.8.

10. Using the heating value of the gas stream of Prob. 3 as its raw material value, calculate the net cost, per Mscf, of operating the gas plant of Prob. 12.8 assuming market value ("Chemical Marketing Reporter") of the products can be realized.

NOTES

1. H. S. Pyland, Oil Gas J. *59*(4), 94 (1961).
2. W. D. Rosen, Oil Gas J. *65*(12), 75-77 (1968).
3. J. Voogel, Hydrocarbon Process. *46*(9), 115 (1967).
4. W. L. Nelson, Oil Gas J. *61*(22), 70 (1964).
5. S. Strelzoff, Hydrocarbon Process. *53*(12), 79-87 (1974).
6. L. Lehman and E. Van Baush, Chem. Eng. Prog. *72*(1), 44-49 (1976).
7. J. Heck, Oil Gas J. Fev. 11, 1980, 122-130.
8. W. Schell and C. Houston, Hydrocarbon Process. *61*(9), 249-252 (1982).
9. Hydrocarbon Process. *49*(9), 257 (1970).
10. L. C. Cameron, Proc. 53rd GPA Annual Convention, Mar. 25-27 (Natural Gas Proc. Assn., Tulsa, Okla., 1974).
11. B. G. Goar, Hydrocarbon Process. *47*(9), 248-252 (1968).
12. C. L. Thomas, *Catalytic Processes and Proven Catalysts* (Academic Press, New York, 1970), p. 184.
13. C. L. Thomas, *Catalytic Processes and Proven Catalysts* (Academic Press, New Yor, 1970), pp. 101-107.
14. H. Greckel, L. V. Kunkel, and R. McGalliard, Chem. Eng. Prog. *61*(9), 70-73 (1965).
15. A. R. Valdes, Hydrocarbon Process. *44*(5), 223-229 (1965).
16. B. W. Gamson and R. H. Elkins, Chem. Eng. Progr. *49*(4), 203-215 (1953).
17. F. G. Sawyer and R. N. Hader, Ind. Eng. Chem. *42*(10), 1938-1950 (1950).
18. A. R. Valdes, Hydrocarbon Process. *43*(3), 104-108 (1964).
19. Ibid, *43*(4), 122-124 (1964).
20. M. J. Pearson, Gas Proc. Canada *65*(5), 22-36 (1973).
21. P. Ellwood, Chem. Eng. *71*(15), 128-130 (1964).
22. Oil Gas J. *69*(41), 68-69 (1971).
23. A. Kohl and F. Riesenfeld, Gas Purification, 3rd Edition, p. 679-686, Gulf Publishing Co., 1979.

ADDITIONAL READING

Hydrocarbon Process. *49*(9), 269-270 (1970).
R. N. Maddox and M. D. Burns, Oil Gas J. *64*(38), 112-121 (1967).
Nautral Gas Proc. Suppliers Assoc. Eng. Data Book, 9th ed. (Tulsa, Oklahoma, 1972), pp. 15-25.

Chapter 13
LUBRICATING OILS

The large number of natural lubricating and specialty oils sold today are produced by blending a small number of lubricating oil base stocks and additives. The lube oil base stocks are prepared from selected crude oils by distillation and special processing to meet the desired qualifications. The additives are chemicals used to give the base stocks desirable characteristics which they lack or to enhance and improve existing properties. The properties considered important are:

1. Viscosity
2. Viscosity change with temperature (VI)
3. Pour point
4. Oxidation resistance
5. Flash point
6. Boiling temperature
7. Acidity (neutralization number)

Viscosity. The viscosity of a fluid is a measure of its internal resistance to flow. The higher the viscosity the thicker the oil and the thicker the film of the oil that clings to a surface. Depending upon the service for which it is used, the oil needs to be very thin and free-flowing or thick with a high resistance to flow. From a given crude oil, the higher the boiling point range of a fraction the greater the viscosity of the fraction. Therefore the viscosity of a blending stock can be selected by the distillation boiling range of the cut.

Viscosity index. The rate of change of viscosity with temperature is expressed by the viscosity index (VI) of the oil. The higher the VI, the smaller its change in viscosity for a given change in temperature. The VI's

233

of natural oils range from negative values for oils from naphthenic crudes to about 100 for paraffinic crudes. Specially processed oils and chemical additives can have VI's of 130 and higher. Additives, such as polyisobutylenes and polymethacrylic acid esters, are frequently mixed with lube blending stocks to improve the viscosity-temperature properties of the finished oils. Motor oils must be thin enough at low temperatures to permit easy starting and viscous enough at engine operating temperatures (80 to 120°C, 180 to 250°F) to reduce friction and wear by providing a continuous liquid film between metal surfaces.

Pour point. The lowest temperature at which an oil will flow under standardized test conditions is reported in 5°F or 3°C increments as the pour point of the oil. For motor oils, a low pour point is very important to obtain ease of starting and proper start-up lubrication on cold days.

There are two types of pour points, a viscosity pour point and a wax pour point. The viscosity pour point is approached gradually as the temperature is lowered and the viscosity of the oil increases until it will not flow under the standardized test conditions. The wax pour point occurs abruptly as the paraffin wax crystals precipitate from solution and the oil solidifies. Additives that affect wax crystal properties can be used to lower the pour point of a paraffin base oil.

A related test is the CLOUD POINT which reports the temperature at which wax or other solid materials begin to separate from solution. For paraffinic oils, this is the starting temperature of crystallization of paraffin waxes.

Oxidation resistance. The high temperatures encountered in internal combustion engine operation promote the rapid oxidation of motor oils. This is especially true for the oil coming in contact with the piston heads where temperatures can range from 260 to 400°C (500 to 750°F). Oxidation causes the formation of coke and varnish-like asphaltic materials from paraffin-base oils and sludge from naphthenic-base oils [1]. Antioxidant additives, such as phenolic compounds and zinc dithiophosphates, are added to the oil blends to suppress oxidation and its effects.

Flash point. The flash point of an oil has little significance with respect to engine performance and serves mainly to give an indication of the source of the oils in the blend. For example, whether it is a blend of high and low viscosity oils to give an intermediate viscosity or is comprised of a blend of center cut oils.

Boiling temperature. The higher the boiling temperature range of a fraction, the higher the molecular weights of the components and, for a given crude oil, the greater the viscosity. The boiling ranges and viscosities of the fractions are the major factors in selecting the cut points for the lube oil blending stocks on the vacuum distillation unit.

Acidity. The corrosion of bearing metals is largely due to acid attack on the oxides of the bearing metals [2]. These organic acids are formed by the oxidation of lube oil hydrocarbons under engine operating conditions and by acids produced as byproducts of the combustion process which are introduced into the crankcase by piston blow-by. Motor oils contain buffering materials to neutralize these corrosive acids. Usually the dispersant and detergent additives are formulated to include alkaline materials which serve to neutralize the acid contaminants. Lube oil blending stocks from paraffinic crude oils have excellent thermal and oxidation stability and exhibit lower acidities than do oils from naphthenic crude oils. The neutralization number is used as the measure of the organic acidity of an oil; the higher the number the greater the acidity.

13.1 LUBE OIL PROCESSING

The first step in the processing of lubricating oils is the separation on the crude oil distillation units of the individual fractions according to viscosity and boiling range specifications. The heavier lube oil raw stocks are included in the vacuum fractionating tower bottoms with the asphaltenes, resins, and other undesirable materials.

The raw lube oil fractions from most crude oils contain components which have undesirable characteristics for finished lubricating oils. These must be removed or reconstituted by processes such as liquid-liquid extraction, crystallization, selective hydrocracking, and/or hydrogenation. The undesirable characteristics include high pour points, large viscosity changes with temperature (low VI), poor oxygen stability, poor color, high cloud points, high organic acidity, and high carbon- and sludge-forming tendencies.

The processes used to change these characteristics are:

1. Reduce carbon- and sludge-forming tendencies
   Solvent deasphalting
2. Improve viscosity index
   Solvent extraction
   Hydrocracking
3. Lower cloud and pour points
   Solvent dewaxing
   Selective hydrocracking
4. Improve color and oxygen stability
   Hydrotreating
5. Lower organic acidity
   Hydrotreating

Although the main effects of the processes are as shown above, there are also secondary effects which are not shown. For example, although the main result of solvent dewaxing is the lowering of the cloud and pour points of the oil, solvent dewaxing also slightly reduces the VI of the oil.

For economic reasons as well as process ones, the process sequence is usually in the order of deasphalting, solvent extraction, dewaxing, and finishing. In general, the processes increase in cost and complexity in this same order.

13.2 PROPANE DEASPHALTING

The lighter distillate feedstocks for producing lubricating oil base stocks can
be sent directly to the solvent extraction units but the atmospheric and
vacuum still bottoms require deasphalting to remove the asphaltenes and
resins before undergoing solvent extraction. In some cases the highest
boiling distillate stream may also contain sufficient asphaltenes and resins
to justify deasphalting.

Propane is usually used as the solvent in deasphalting but it may also
be used with ethane or butane in order to obtain the desired solvent properties. Propane has unusual solvent properties in that from 40 to 60°C (100
to 140°F) paraffins are very soluble in propane but the solubility decreases
with an increase in temperature until at the critical temperature of propane
(96.8°C or 206°F) all hydrocarbons become insoluble. In the range of 40
to 96.8°C (100 to 206°F) the high molecular weight asphaltenes and resins
are largely insoluble in propane. Separation by distillation is generally by
molecular weight of the components and solvent extraction is by type of
molecule. Propane deasphalting falls in between these catagories because
separation is a function of both molecular weight and type of molecular
structure.

The feedstock is contacted with 4 to 8 volumes of liquid propane at the
desired operating temperature. The extract phase contains from 15 to 20% by
weight of oil with the remainder solvent. The heavier the feed stock, the
higher the ratio of propane to oil required.

The raffinate phase contains from 30 to 50% propane by volume and is
not a true solution but an emulsion of precipitated asphaltic material in
propane [3].

As in most other refinery processes, the basic extraction section of the
process is relatively simple, consisting of a cylindrical tower with angle iron
baffles arranged in staggered horizontal rows [4] or containing perforated
baffles [3] using counter-current flow of oil and solvent. Some units use the
rotating disc contactor (RDC) for this purpose.

A typical propane deasphalting unit (Figure 13.1) injects propane into
the bottom of the treater tower and the vacuum tower bottoms feed enters
near the top of the tower. As the propane rises through the tower, it
dissolves the oil from the residuum and carries it out of the top of the tower.
Between the residuum feed point and the top of the tower, heating coils
increase the temperature of the propane-oil extract phase thus reducing the
solubility of the oil in the propane. This causes some of the oil to be ex-
pelled from the extract phase creating a reflux stream. The reflux flows
down the tower and increases the sharpness of separation between the oil
portion of the residuum and the asphaltene and resin portion. The asphaltene
and resin phase leaving the bottom of the tower is the raffinate and the propane-oil mixture leaving the top is the extract.

The solvent recovery system of the propane deasphalting process as in all
solvent extraction processes, is much more complicated and costly to operate
than the treating section. Two-stage flash systems are used to recover the
propane from the raffinate and extract phases. The first stages are operated
at pressures high enough to condense the propane vapors with cooling water
as the heat exchange medium. In the high pressure raffinate flash tower
foaming and asphalt entrainment can be a major problem. To minimize this,
the flash tower is operated at about 290°C (550°F) to keep the asphalt viscosity at a reasonably low level.

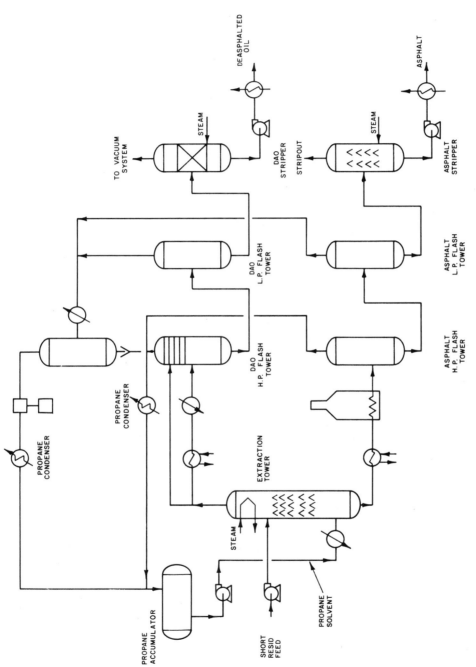

FIG. 13.1. Typical Propane Deasphalter.

The second stages strip the remaining propane from the raffinate and extract at near atmospheric pressure. This propane is compresed and condensed before being returned to the propane accumulator drum.

The propane deasphalting tower is operated at a pressure sufficiently high to maintain the solvent in the liquid phase. This is usually about 34 bars (500 psig).

The asphalt recovered from the raffinate can be blended with other asphalts, into heavy fuels, or used as a feed to the coking unit.

The heavy oil product from vacuum residuum is called brightstock. It is a high viscosity blending stock that, after further processing, is used in the formulation of heavy-duty lubricants for truck, automobile, and aircraft services.

13.3 SOLVENT EXTRACTION

There are three solvents used for the extraction of aromatics from lube oil feed stocks and the solvent recovery portions of the systems are different for each. The solvents are furfural, phenol, and N-methyl-2-pyrrolidone (NMP). The purpose of solvent extraction is to improve the viscosity index (VI), oxidation resistance, and color of the lube oil base stock and to reduce the carbon- and sludge-forming tendencies of the lubricants by separating the aromatic portion from the naphthenic and paraffinic portion of the feed stock.

Furfural extraction. The process flow through the furfural extraction unit is similar to that of the propane deasphalting unit except for the solvent recovery section which is more complex. The oil feedstock is introduced into a continuous counter-current extractor at a temperature which is a function of the viscosity of the feed; the greater the viscosity the higher the temperature used. The extraction unit is usually a raschig ring packed tower or a rotating disc contactor (RDC) with a temperature gradient from top to bottom of 30 to 50°C (60 to 90°F). The temperature at the top of the tower is a function of the miscibility temperature of the furfural and oil. It is usually in the range 105 to 150°C (220 to 300°F).

The oil phase is the continuous phase and the furfural dispersed phase passes downward through the oil. Extract is recycled at a ratio of 0.5:1 to improve the extraction efficiency.

Furfural-to-oil ratios range from 2:1 for light stocks to 4.5:1 for heavy stocks. Solvent losses are normally less than 0.02% (wt) of raffinate and extract flow rates. Furfural is easily oxidized and inert gas blankets are maintained on the system to reduce oxidation and polymerization. Sometimes deaeration towers are used to remove dissolved oxygen from the feed. Furfural is subject to thermal decomposition and skin temperatures of heat exchange equipment used to transfer heat to furfural-containing streams must be carefully controlled to prevent polymerization of the furfural and fouling of the heat exchange surfaces.

The furfural is removed from the raffinate and extract streams by flashing and steam stripping. Furfural forms an azeotrope with water and this results in a unique furfural recovery system. Furfural is purified in the furfural tower by distilling overhead the water-furfural azeotrope vapor which upon condensing separates into water-rich and furfural-rich layers. The furfural-rich layer is recycled to the furfural tower as reflux and the furfural in the water-rich layer is separated from the water by steam-stripping. The overhead vapors, consisting of the azeotrope, are condensed and returned to the furfural-water separator. The bottoms product from the furfural tower is the pure furfural stream which is sent to the furfural solvent drum.

The most important operating variables for the furfural extraction unit are the furfural-to-oil ratio (F/O ratio), extraction temperature, and extract recycle ratio. The F/O ratio has the greatest effect on the quality and yield of the raffinate while the temperature is selected as a function of the viscosity of the oil and the miscribility temperature. The extract recycle ratio determines to some extent the rejection point for the oil and the sharpness of separation between the aromatics and naphthenes and paraffins.

Phenol extraction. The process flow for the phenol extraction unit is somewhat similar to that of the furfural extraction unit but differs markedly in the solvent recovery section because phenol is easier to recover than is furfural.

The distillate or deasphalted oil feed is introduced near the bottom of the extraction tower and phenol enters near the top. The oil-rich phase rises through the tower and the phenol-rich phase descends the tower. Trays or packing are used to provide intimate contact between the two phases. Some of the newer phenol extraction units use either rotating disc contactors (RDC) or centrifugal extractors to contact the two phases. Both the RDC and the centrifugal extractors offer the advantage that much smaller volumes are needed for the separations. As all lube oil refineries operate on a blocked operation basis (that is, charging one feed stock at a time), the lower inventories make it possible to change from one feed stock to another with a minimum loss of time and small loss of off-specification product.

The extraction tower and RDC are operated with a temperature gradient which improves separation by creating an internal reflux. The phenol is introduced into the tower at a higher temperature than the oil. The temperature of the phenol-rich phase decreases as it proceeds down the column and the solubility of the oil in this phase decreases. The oil coming out of the phenol-rich phase reverses direction and rises to the top as reflux. The tower top temperature is kept below the miscible temperature of the mixture and the tower bottom temperature is usually maintained about 10°C (20°F) lower than the top.

Phenol will dissolve some of the paraffins and naphthenes as well as the aromatics. Water acts as an antisolvent to increase the selectivity of the phenol and typically from 3 to 8% (vol) water is added to the phenol. A decrease in reaction temperature has a similar effect. Raffinate yield is increased by increasing water content and/or decreasing temperature.

The important extraction tower operating variables are:

Phenol-to-oil ratio (treat rate)

Extraction temperature

Percent water in phenol

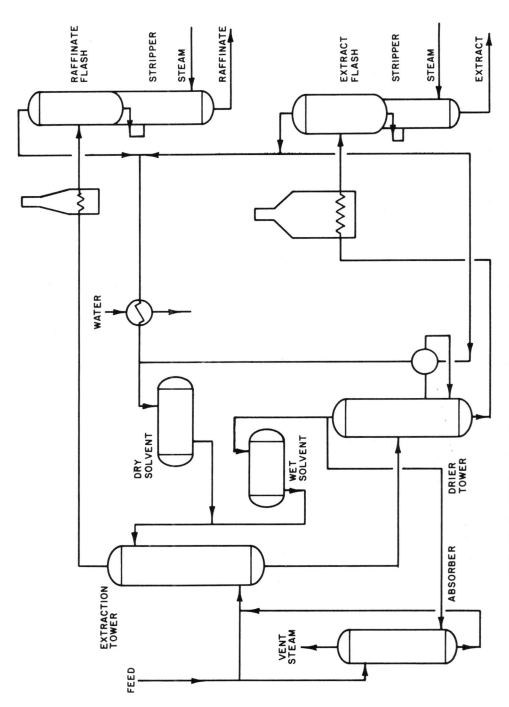

FIG. 13.2. NMP extraction unit using steam stripping for solvent recovery.

Treat rates vary from 1:1 to 2.5:1 depending upon the quality and viscosity of the feed and the quality of the product desired. Increasing the
treat rate for a given stock improves the VI of the product and decreases
the yield.

Phenol is recovered from the extract and raffinate streams by distillation and gas or steam stripping. Phenol losses average from 0.02 to 0.04%
of circulation rate.

NMP extraction. The NMP extraction process uses N-methyl-pyrrolidone
as the solvent to remove the condensed ring aromatics and polar components
from the lubricating oil distillates and brightstocks. This process was developed as a replacement for phenol extraction because of the safety, health,
and environmental problems associated with the use of phenol. Several differences between the characteristics of NMP and phenol make it necessary to
modify the phenol plant design. These differences include a 22°C (40°F)
higher boiling point for NMP, a 64°C (115°F) lower melting point, complete
miscibility of NMP with water, no azeotrope formation of NMP with water, and
a 69% lower viscosity than phenol at 50°C (122°F). A simplified process flow
diagram in which the solvent is recovered by flashing and stripping with
steam at pressures above atmospheric is shown in Figure 13.2 [1].

A portion of the distillate or deasphalted oil feed is used as the lean oil
in an absorption tower to remove the NMP from the exiting stripping steam.
The rich oil from the absorption tower is combined with the remainder of the
feed which is heated to the desired temperature before being introduced near
the bottom of the treater tower. The hot solvent enters near the top of the
tower. Specially designed cascade weir trays are used to mix and re-mix
the NMP-rich and oil-rich phases as they pass through the tower.

The solvent is stripped from the raffinate and extract by distillation
and steam stripping. Recovery of NMP is better than that for phenol and NMP
losses are only 25 to 50% those of phenol.

The lower viscosity of NMP gives greater through-put for a given size
tower. This results in lower construction costs for a grass-root plant and
up to a 25% increase in through-put for converted phenol plants.

Solvent-to-oil ratios for a given feedstock and quality product are the
same for NMP and phenol extraction but raffinate oil yields average 3 to 5%
higher for the NMP extraction.

13.4 DEWAXING

All lube stocks, except those from a relatively few highly naphthenic crude
oils, must be dewaxed or they will not flow properly at ambient temperatures.
Dewaxing is one of the most important and most difficult processes in lubricating oil manufacturing. There are two types of processes in use today.
One uses refrigeration to crystallize the wax and solvent to dilute the oil
portion sufficiently to permit rapid filtration to separate the wax from the
oil. The other uses a selective hydrocracking process to crack the wax
molecules to light hydrocarbons.

Solvent dewaxing. There are two principal solvents used in the United
States in solvent dewaxing processes, propane and ketones. Dichloroethane-
methylene is also used in some other countries. The ketone-processes use
either a mixture of methyl ethyl ketone (MEK) with methyl isobutyl ketone

(MIBK) or MEK with toluene. The solvents act as a diluent for the high molecular weight oil fractions to reduce the viscosity of the mixture and provide sufficient liquid volume to permit pumping and filtering. The process operations for both solvent processes are similar but differ in the equipment used in the chilling and solvent recovery portions of the process. About 80% of the dewaxing installations use ketones as the solvent and the other 20% use propane. The comparative advantages and disadvantages of the processes are [3]:

Propane:

1. Readily available, less expensive and easier to recover
2. Direct chilling can be accomplished by vaporization of the solvent thus reducing the capital and maintenance costs of scraped-surface chillers
3. High filtration rates can be obtained because of its low viscosity at very low temperatures
4. Rejects asphaltenes and resins in the feed
5. Large differences between filtration temperatures and pour point of finished oils (15 to 25°C, 25 to 45°F)
6. Requires use of a dewaxing aid

Ketones:

1. Small difference between filtration temperature and pour point of dewaxed oil (5 to 10°C, 9 to 18°F)
   a. Lower pour point capability
   b. Greater recovery of heat by heat exchange
   c. Lower refrigeration requirements
2. Fast chilling rate. Shock chilling can be used to improve process operations
3. Good filtration rates but lower than for propane.

Ketone dewaxing. The most widely used processes use mixtures of MEK-toluene and MEK-MIBK for the solvent. MEK-benzene was used originally but health hazards associated with the handling of benzene as well as its cost resulted in the use of MEK-toluene in its place. A simplified process flow diagram is shown in Figure 13.3.

Solvent is added to the solvent-extracted oil feed and the mixture is chilled in a series of scraper-surface exchangers and chillers. Additional solvent is added to the feed to maintain sufficient liquid for easy handling as the temperature is decreased and the wax crystallizes from solution. Cold filtrate from the rotary filters is used as the heat exchange medium in the exchangers and a refrigerant, usually propane, in the chillers. The cold slurry is fed to the rotary filters for removal of the crystalline wax. The filter cake is washed with cold solvent and an inert gas blanket, at a slight positive pressure, is maintained on the filters and also used to "blow back" the wax cake from the filter cloth and release it before it reaches the filter drum scraper knife.

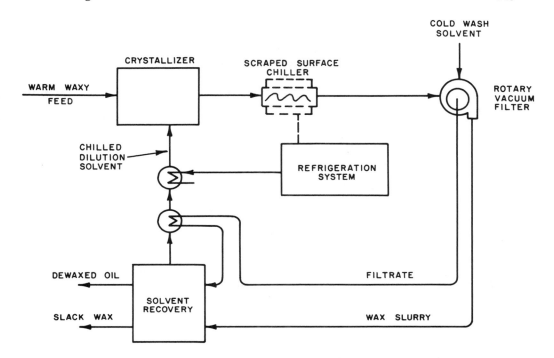

FIG. 13.3. Exxon DILCHILL dewaxing unit [7].

The solvent is recovered from the filtrate and the wax by heating and two-stage flashing followed by steam stripping.

The water introduced into the system by the steam stripping is removed continuously by a series of phase separations and distillations. The overheads from the steam strippers are combined and condensed. The condensate separates into two phases in the decanter drum with the aqueous phase containing about 3.5 mol % ketones and the ketone-rich phase about 3 mol % water. The ketone phase is fed into the top of the solvent dehydration tower as reflux and the water-rich phase is fed into the ketone stripper. The ketone is stripped from the water-rich phase with steam and water is discharged from the bottom of the tower into the sewer.

The products from the dewaxing unit are a dewaxed oil and a slack wax. The dewaxed oil next must go through a finishing step to improve its color and color stability. The slack wax is used either for catalytic cracker feed or undergoes a de-oiling operation before being sold as industrial wax.

Dilchill dewaxing. The Dilchill Dewaxing Process was developed by Exxon. It is a modification of the ketone dewaxing process which uses shock chilling of the waxy oil feed by direct injection with very cold solvent in a highly agitated mixer. The wax crystals formed are larger and more dense than the crystals formed in scraped-surface chillers. Higher filtration rates and better oil removal are achieved, resulting in lower capital and operating costs as well as higher dewaxed oil yields. The scraped-surface coolers are eliminated but the chillers are still necessary.

Propane dewaxing. Propane dewaxing differs from ketone dewaxing in that the propane is used both as a diluent and a refrigerant for direct chilling. As a diluent, it dilutes the oil phase so it is fluid and acts as the carrying medium after the wax crystallizes. As a refrigerant, it chills the oil by being evaporated batch-wise from the propane-waxy oil solution. The cooling rate usually averages about 1.5°C (3°F) per minute but ranges from almost 4°C (7°F) per minute at the beginning of the chilling cycle to about 1°C (2°F) at the end. In propane dewaxing, the chilling rate is limited to avoid shock chilling because it results in poor wax crystal formation and growth. The amount of dilution required is a function of the viscosity of the oil and ranges from a 1.5:1 ratio of propane to fresh feed for light stocks to 3:1 for heavy stocks [4].

Waxy oil feed is mixed with propane, cooled to about 27°C (80°F) and charged into the warm solution drum under sufficient pressure to prevent vaporization of the propane. From the solution drum, it is charged into one of the two batch chillers where it is cooled at a controlled rate by evaporation of the propane. It usually takes about 30 minutes to chill the mixture to the desired temperature. The slurry is then discharged into the filter feed drum. The cycle is so arranged that while one chiller is being used to chill a batch, the other chiller is first discharged into the filter feed drum then refilled from the warm solution drum.

The wax is separated from the oil on rotary filters. The wax cake is washed with cold propane while on the filter to remove and recover as much oil as practical. Cole propane vapors are used to release the filter cake.

Most of the propane in the dewaxed oil and wax stream is recovered by heating each of the streams to about 160°C (320°F) and flashing at a pressure sufficiently high that the propane can be condensed with refinery cooling water. The remaining propane is recovered by low-pressure flashing and a final steam stripping of each of the streams. The propane from the low-pressure flash is compressed, condensed and returned to the propane storage drum.

A dewaxing aid such as Paraflow⁰ (a naphthalene-chlorinated wax condensation product) and/or Acryloid⁰ (a methacrylate polymer) is frequently added to the waxy oil feed at a level of 0.05 to 0.2 liquid volume percent on feed to modify the wax crystal structure. This results in higher filtration rates and denser wax cakes from which it is easier to wash the oil.

Selective hydrocracking. There are two types of selective hydrocracking processes for dewaxing oil, one uses a single catalyst for pour point reduction only and the other uses two catalysts to reduce the pour point and improve the oxygen stability of the product [5, 6].

For the pour point reduction operation, both processes use synthetic shape-selective zeolite catalysts which selectively crack the n-paraffins and slightly branched paraffins. Zeolites with openings about 6A in diameter provide rapid cracking rates for n-paraffins with the rate decreasing rapidly as the amount of branching increases.

The Mobil Lube Dewaxing Process uses a fixed-bed reactor packed with two catalysts and the process flow is similar to that of hydrotreating unit. Severity of operation is controlled by the furnace outlet temperature (reactor) temperature). Essentially no methane or ethane is formed in the reaction. A

⁰Trademark of Exxon Chemical Americas

similar British Petroleum (BP) process produces propane, butane, and pentane in the ratio 2:4:3 by weight. Reaction conditions for the Mobil and BP processes appear to be similar with the following considered typical ranges:

Reactor temp, °C (°F)	290-370	(560-700)
Reactor pres, bars (psig)	20-135	(300-2000)
H_2 partial pres, bars (psig)	17-100	(250-1500)
Gas rate, N cu m/t (scf/bbl)	100-1000	(500-5000)
H_2 consumption, N cu m/t (scf/bbl)	20-40	(100-200)

Yields of dewaxed oil of similar pour point from the same feed stocks vary from 0 to 15% greater than from solvent dewaxing (SDW) with the increase reflecting the difficulty of separating the oil from the wax in the SDW processes.

The feed to the selective hydrocracking unit is solvent-extracted oil from the aromatic extraction units. The advantages claimed over conventional solvent dewaxing units include:

1. Production of very low pour and cloud point oils from paraffinic stocks
2. Lower capital investment
3. Improved lube oils base stock yields
4. A separate hydrofinishing operation is not necessary.

13.5 HYDROFINISHING

Hydrotreating of dewaxed lube oil stocks is needed to remove chemically active compounds that affect the color and color stability of lube oils. Most hydrotreating operations use cobalt-molybdate catalysts and are operated at a severity set by the color improvement needed. Organic nitrogen compounds seriously affect the color and color stability of oils and their removal is a major requirement of the operation [3].

The process flow is the same as that for a typical hydrotreating unit. Representative operating conditions are:

Reactor temp, °C (°F)	200-340	(400-650)
Reactor pres, bars (psig)	34-55	(500-800)
LHSV, v/h/v	0.5-2.0	
H_2 as rate, N cu m/t (scf/bbl)	100	(500)

Usually finished oil yields are approximately 98% of dewaxed oil feed.

NOTES

1. J. D. Bushnell and R. J. Fiocco, Hydrocarbon Process. *59*(5), 119-123 (1980).
2. P. A. Asseff, *Petroleum Additivies and their Functions*, April, 1966. Lubrizol Corp., Cleveland, Ohio.
3. M. Soudel, Hydrocarbon Process. *53*(12), 59-66 (1974).
4. W. L. Nelson, *Petroleum Refining Engineering*, 4th Ed., p. 62-70 (McGraw-Hill Book Co., New York, 1958).
5. J. D. Hargrove, G. J. Elkes, and A. H. Richardson, Oil and Gas J., 77(2), p. 103-105 (15 Jan. 1979).
6. K. W. Smith, W. C. Starr, and N. Y. Chen, Oil and Gas J. *78*(21), 75-84 (26 May 1980).

ADDITIONAL READING

J. D. Bushnell and J. F. Eagen, *Commercial Experience with Dilchill Dewaxing*, Presented at NPRA meeting, Sept. 11-12, 1975.

Chapter 14
PETROCHEMICAL FEEDSTOCKS

The preparation of hydrocarbon feed stocks for petrochemical manufacture can be a significant operation in today's petroleum refineries. There are three major classes of these feedstocks according to use and method of preparation. These are aromatics, unsaturates (olefins and diolefins), and saturates (paraffins and cycloparaffins). Aromatics are produced using the same catalytic reforming units used to upgrade the octanes of heavy straight-run naphtha gasoline blending stocks. For petrochemical use the aromatic fraction is concentrated by solvent extraction techniques. Some of the unsaturates are produced by the fluid catalytic cracking unit but most have to be produced from refinery feedstocks by steam cracking or low-molecular-polymerization of low molecular weight components.

To provide better continuity from regular refinery operations, the petrochemical feedstock preparation will be covered in the order (1) aromatics, (2) unsaturates, and (3) saturates.

14.1 AROMATICS PRODUCTION

For aromatics production for petrochemical use, the catalytic reformer can be operated at higher severity levels than for motor gasoline production. Highly naphthenic feedstocks also aid in high aromatics yields because the dehydration of naphthenes is the most efficient reaction taking place and provides the highest yield of aromatics. Table 14.1 illustrates the increase in aromatics yields as the reformer operating severity is increased (the clear research octane number (RON) is the measure of operating severity).

The C_6-C_8 aromatics (benzene, toulene, xylene, and ethylbenzene) are the large volume aromatics used by the petrochemical industry with benzene having the greatest demand. The product from the catalytic reformer contains

247

TABLE 14.1

Aromatics Yield as a Function of Severity

Product RON	% Aromatics
90	54
95	60
100	67
103	74

all of these aromatics and it is separated into its pure components by a combination of solvent extraction, distillation, and crystallization. In addition, because of the much greater demand for benzene, the excess of the toluene and xylenes over market needs can be converted to benzene by hydrodealkylation.

Solvent extraction of aromatics. Present separation methods for recovery of aromatics from hydrocarbon streams use liquid-liquid solvent extraction to separate the aromatic fraction from the other hydrocarbons. Most of the processes used by U.S. refineries use either polyglycols or sulfolane as the extracting solvent. The polyglycol processes are the Udex process developed by Dow Chemical Company and licensed by UOP and the Tetra process licensed by the Linde Division of Union Carbide. The solvents used are tetraethylene glycol for the Tetra process and usually diethylene glycol for the Udex process although dipropylene glycol and triethylene glycol can also be used. The Suifolane process was originally developed by the Royal Dutch/Shell group and is licensed worldwide by UOP. It uses sulfolane (tetrahydrothiophene 1-1 dioxide) as the solvent and a simplified process flow diagram is shown in Figure 14.1.

The important requirements for a solvent are [1]:

1. High selectivity for aromatics versus non-aromatics
2. High capacity (solubility of aromatics)
3. Capability to form two phases at reasonable temperatures
4. Capability of rapid phase separation
5. Good thermal stability
6. Non-corrosive and non-reactive.

The Udex process, using polyglycols, was the first of the solvent extraction processes to find wide-spread usage. Many of these units are still in operation in U.S. refineries but since 1963 most new units have been constructed using the Sulfolane or Tetra processes. At a given selectivity, the solubility of aromatics in sulfolane is about double that in triethylene glycol. The higher solubility permits lower solvent circulation rates and lower operating costs.

Both the polyglycol and sulfolane solvent processes use a combination of liquid-liquid extraction and extractive stripping to separate the aromatics from the other hydrocarbons [1, 2, 3] because of the characteristics of polar solvents. As the concentration of aromatics in the solvent increases the

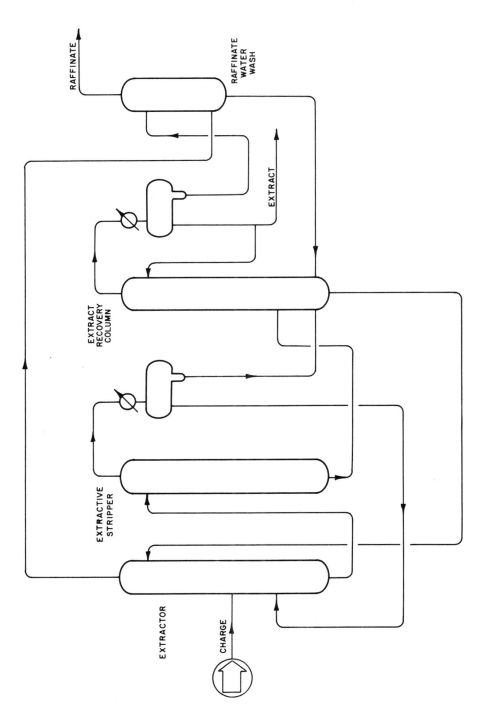

FIG. 14.1a. Sulfolane aromatics extraction unit.

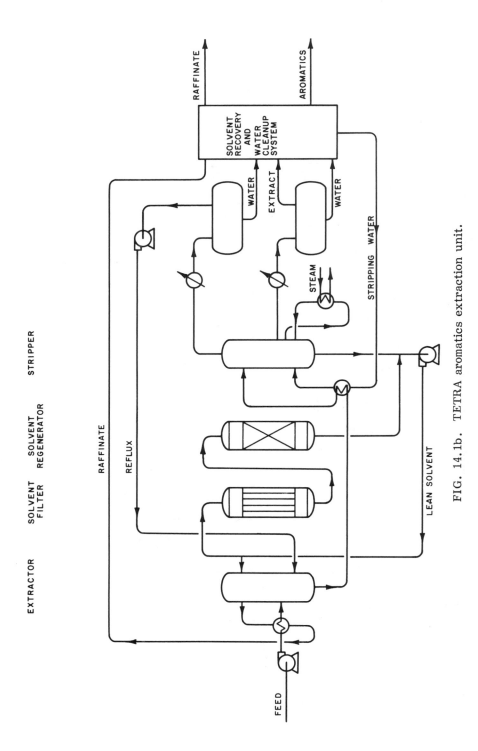

FIG. 14.1b. TETRA aromatics extraction unit.

TABLE 14.2

Boiling and Melting Points of Aromatics

	Boiling Points		Melting Points	
	°F	C	°F	C
Benzene	176.2	80.1	42.0	5.5
Toluene	231.2	110.6	−139.0	−95.0
Ethyl benzene	277.1	136.2	−139.0	−95.0
p-Xylene	281.0	138.4	+55.9	13.3
m-Xylene	282.4	139.1	−54.2	−47.9
o-Xylene	291.9	144.4	−13.3	−25.2

solubility of the non-aromatic hydrocarbons in the extract phase also increases. This results in a decrease in the selectivity of the solvent and a carry-over of some of the non-aromatic hydrocarbons with the extract phase to an extractive stripper. In the extractive stripper an extractive distillation occurs and the non-aromatic hydrocarbons are stripped from the aromatic-solvent mixture and returned as reflux to the extraction column. The solvent is then recovered from the non-aromatic hydrocarbon-free extract stream leaving an extract which contains less than 1000 ppm (<0.1%) aliphatics.

Aromatics recoveries are typically equal to or better than 99.9, 99.0, and 97.0% for benzene, toluene, and xylenes, respectively.

Because of the high affinity of the polar solvents for water and the low solubilities in the non-aromatic hydrocarbon raffinate phase, the solvent is recovered from the raffinate phase by washing with water. The water is returned to either the extractive stripper (Tetra[1]) or extract recovery column (Sulfolane[2]) for recovery of the solvent.

The water content of the solvents is carefully controlled and used to increase the selectivity of the solvents. The water content of the polyglycols is kept in the range of 2 to 10 wt % and that of sulfolane is maintained at about 1.5 wt %.

The solvent quality is maintained in the Sulfolane process by withdrawing a slipstream from the main solvent recirculation stream and processing it in a solvent reclaiming tower to remove any high-boiling contaminants that accumulated. The Tetra process solvent quality is maintained by using a combination of filtration and adsorption to remove accumulated impurities.

Sulfolane and the polyglycols have thermal stability problems and the skin temperatures of heat exchange equipment must be limited. For sulfolane, skin temperatures must be kept below 232°C (450°F).

Aromatics separation. Benzene and toluene can be recovered from the extract product stream of the extraction unit by distillation. The boiling points of the C_8 aromatics are so close together (Table 14.2) that separation by distillation becomes more difficult and a combination of distillation and crystallization is used. The ethylbenzene is first separated from the mixed xylenes in a three unit fractionation tower with 120 trays per unit for a total of 360 trays. Each unit is about 200 feet in height and the units are connected

[1]Tetra is a Union Carbide licensed process.
[2]Sulfolane is a UOP licensed process.

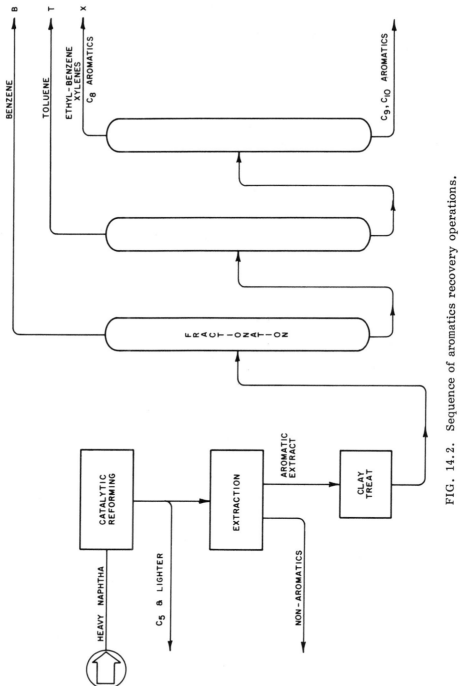

FIG. 14.2. Sequence of aromatics recovery operations.

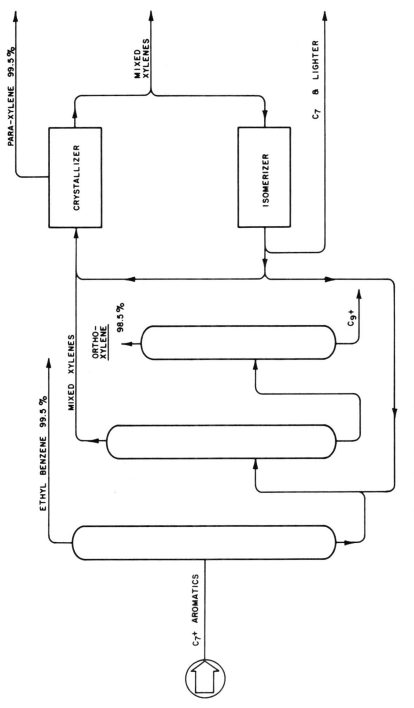

FIG. 14.3. Processing sequence to produce C$_8$ aromatics.

so they operate as a single fractionation tower of 360 trays. High reflux
ratios are needed to provide the desired separation efficiency because
the difference between the boiling points of ethylbenzene and p-xylene is
only about 2°C (3.9°F). This is a very energy intensive operation and,
with today's energy costs, it is usually less costly to make ethylbenzene by
alkylating benzene with ethylene.

A typical processing sequence for the separation of C_8 aromatics is
shown in Figure 14.2. After removal of the ethylbenzene by super-fractiona-
tion, the o-xylene along with the higher boiling C_9 aromatics is separated
from the p- and m-xylenes by fractionation. The boiling point of o-xylene
is more than 5°C (9.6°F) greater than that of its closest boiling isomer,
m-xylene, and it can be separated economically in a 160 plate distillation
column. The overhead stream from this column is a mixture of m- and p-
xylenes and the bottoms stream contains o-xylene and the C_9 aromatics.
The bottoms stream is processed further in a 50-tray rerun column to
separate 99+ % purity o-xylene from the C_9 aromatics.

The mixed m- and p-xylene overhead stream from the fractionator is
fed to the crystallization unit to separate the m- and p-xylenes. The solidi-
fication point of p-xylene is +13.3°C (55.9°F) while that of m-xylene is
−47.9°C (−54.2°F). A simplified flow diagram of the solvent extraction
process is shown in Figure 14.3 [1].

The xylenes feed is dried, chilled to about −40°C (−40°F) and charged
to staged crystallizers. Ethylene is used as the refrigerant in indirect
chilled surface-scraped heat exchangers and the mixture is chilled to −50°C
(−60°F) in the first crystallizer stage and to −68°C (−90°F) in the second
stage. The second stage crystallizer effluent is charged to a combination
solid and screen bowl centrifuge to separate the p-xylene crystals from the
mother liquor. The operation of this centrifuge is critical to the cost effectiveness of this process because it controls the subsequent purification
costs and the overall refrigeration requirements of this separation [1].

The p-xylene crystals from the centrifuge are purified in a series of
partial remelting, recrystallization, and centrifugation stages. The temperatures are increased for each remelting and recrystallization stage and
the filtrate from each stage is recycled to the preceeding stage. This
counter-current operation gives a high-purity p-xylene product from the
last stage. The m-xylene rich filtrate stream is used as feedstock for an
xylene isomerization unit or sent to gasoline blending.

The m-xylene rich stream from the crystallizer and the o-xylene stream
from the rerun column can be converted to p-xylene by a xylene isomerization
process if demand for p-xylene warrants it. The process flow diagram for
this process is similar to the isomerization process in Chapter 6. The
catalyst used is a zeolite-non-noble metal catalyst developed by Mobil.

The aromatic-rich feedstock is mixed with hydrogen, heated to 375-425°C
(700-800°F) and charged to a fixed-bed reactor containing a non-noble
metal catalyst at a pressure of about 150 psig. The mixture is isomerized to
a near-equilibrium mixture of o-, m-, and p-xylenes [1, 4]. After cooling,
the xylenes are separated from the hydrogen-rich gas in a flash separator
and from other hydrocarbons in distillation columns. The xylene-rich stream
is returned to the p-xylene crystallization unit or the Parex[1] unit as feed.

Other isomerixation processes are used to maximize p-xylene production
by both isomerizing o- and m-xylenes and by converting ethylbenzene
into xylenes [1]. A two-stage process is generally used for conversion

---

[1]UOP design.

of aromatic hydrocarbons with two or more carbons in the side-chains
with the first stage for partial hydrogenation and the second for dehydrogenation. A single-stage process can also be used. The catalysts are
either silica-alumina containing a small amount of platinum or molecular
sieve silica-alumina catalysts. The equilibrium concentration of p-xylene
obtained in the effluent is about 24%.

Benzene. The primary source of benzene is from the refinery catalytic
reforming unit but substantial amounts of benzene are also produced by the
hydrodealkylation of toluene. The process flow for a hydrodealkylation unit
(HDA) is also similar to that of an isomerization unit. The feed can be
toluene or a mixture of toluene and xylenes. It is heated to 630°C (1175°F)
at 600 psig and charged to an open non-catalytic reactor (L/D > 20) where
thermal dealkylation of toluene and xylenes takes place during the residence
time of 25-30 seconds. The hydrogenation step in the dealkylation reaction
is highly exothermic (22,000 Btu/lb-mole of H_2 consumed) and temperature
is controlled by injection of quench hydrogen at several points along the
reactor. The hydrodealkylation reaction results in the conversion of about
90% of the aromatics in the feed and selectively favors about 95% of the
conversion to benzene. A small amount of polymer is also formed, primarily
diphenyl. A small amount of hydrogen sulfide or carbon disulfide is added to
the feed to prevent catastrophic corrosion of the furnace tubes and a small
amount of the polymer is recycled to minimize polymer formation [1].

14.2 UNSATURATE PRODUCTION

Although a substantial amount of C_3 and C_4 olefins are produced by the fluid
catalytic cracking unit in the refinery and some C_2 and C_3 olefins by the cat
cracker and coker, the steam cracking of gas oils and naphthas is the most
important process for producing a wide range of unsaturated hydrocarbons
for petrochemical use. Ethylene is also produced by thermal cracking
ethane and propane.

Steam cracking is the thermal cracking and reforming of hydrocarbons
with steam at low pressure, high temperature and very short residence
times (generally <1 second).

The hydrocarbon is mixed with steam in the steam/hydrocarbon weight
ratio of 0.2 to 0.8 and fed into a steam cracking furnace. Residence times
in the cracking zone range from 0.3 to 0.8 seconds with a coil outlet temperature between 760 and 840°C (1400-1550°F), a coil outlet pressure in the
range 10 to 20 psig. The coil outlet stream is quickly quenched to about
320°C (600°F) to stop the cracking and polymerization reactions. Recent
designs of steam crackers utilize heat transfer in a steam generator or
transfer-line heat exchanger with a low pressure drop to lower the coil outlet
temperature quickly with a maximum amount of heat recovery. The quenched
furnace outlet stream is sent to a primary fractionator where it is separated
into a gas stream and liquid product streams according to boiling point range.
The gases are separated into individual components as desired by compression
and high pressure fractionation. Streams involved in low temperature
rectification must be dessicant dried to remove water and prevent ice and hy-
drate formation.

Paraffins obtained from the dewaxing of lubricating oil base stocks are
frequently steam cracked to produce a wide range of linear olefins, usually
with from 11 to 16 carbon atoms per molecule. Mild steam cracking conditions

TABLE 14.3

Steam Cracking Conditions

	Naphtha		Wax	
Coil outlet:				
Temp, °F (C)	1100-1650	(730-900)	1100-1200	(595-650)
Pres, psig (bars)	45	(3)	15-30	(1-2)
Steam/HC (wt)	0.3-0.9		0.12-0.15	
C_3-conversion, wt %	60-80		8-10	

are employed because the object is to minimize the formation of high molecular
weight olefins. As with catalytic cracking, the reactions favor the breaking
of the paraffinic chain near its center with the formation of one paraffinic
molecule and one olefinic molecule. Typical operating conditions are given in
Table 14.3 [5].

The purity of the feed stock has a major effect on the product quality.
The petrolatum (high molecular weight microcrystalline waxes) obtained
from the dewaxing of lubricating oil base stocks contains from 85 to 90%
straight and branched chain paraffins with the remainder naphthenes and
aromatics. Aromatics are especially undesirable and the wax stream must be
deoiled before being used as a steam cracker feed stock. By choosing a
feedstock with an initial boiling point greater than the end point of the desired
product, it not only increases the yield of the desired olefins but also makes
it possible to separate the product olefins from the feed paraffins by distillation. A typical product stream contains above 80% mono-olefins with only a
small percentage less than 11 carbons or more than 16 carbons.

Linear mono-olefins are also produced by several catalytic processes.
The overall process consists of a vapor-phase catalytic dehydrogenation unit
followed by an extraction unit to extract the linear olefins from the paraffin
hydrocarbons by adsorption on a bed of solid adsorbent material. The olefins
are desorbed with a hydrocarbon boiling at a lower temperature than the olefin product to make it easy to separate the olefin product from the desorbant.
The desorbant hydrocarbon is recycled internally in the extraction section.
The olefin product contains about 96 wt % linear olefins of which about 98% are
mono-olefins. Approximately 93 wt % of the linear paraffins in the feed are
converted to linear olefins [1, 6].

14.3 SATURATED PARAFFINS

Normal paraffins. Normal paraffins are recovered from petroleum fractions by vapor-phase adsorption on molecular sieves having an average pore
diameter of 5 Angstroms. A simplified process flow diagram for the Exxon
Ensorb process is shown in Figure 14.5. The adsorption takes place in the
vapor phase at pressures of 5 to 10 psig and temperature from 300 to 350°C
(575-650°F). Ammonia is used to desorb the normal paraffins. A semi-batch
operation using two beds of 5A molecular sieves permits continuous operation
with one bed being desorbed while the other bed is on-stream. The sieves
gradually lose capacity because of contamination with polymerization products
and are regenerated by controlled burn-off of the high molecular weight

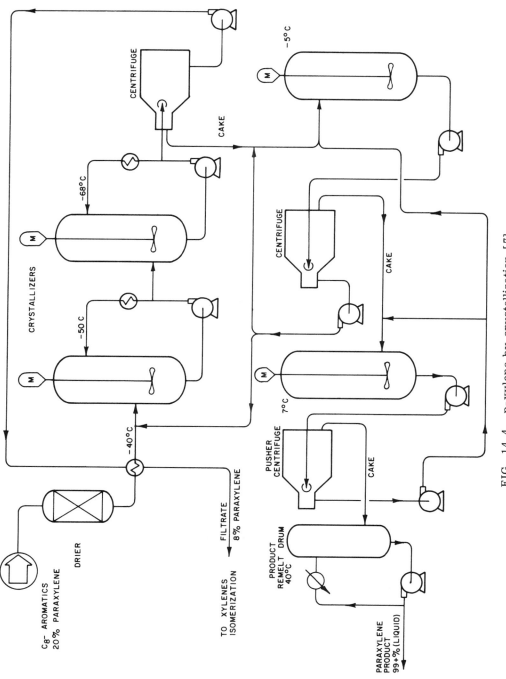

FIG. 14.4. p-xylene by crystallization [7].

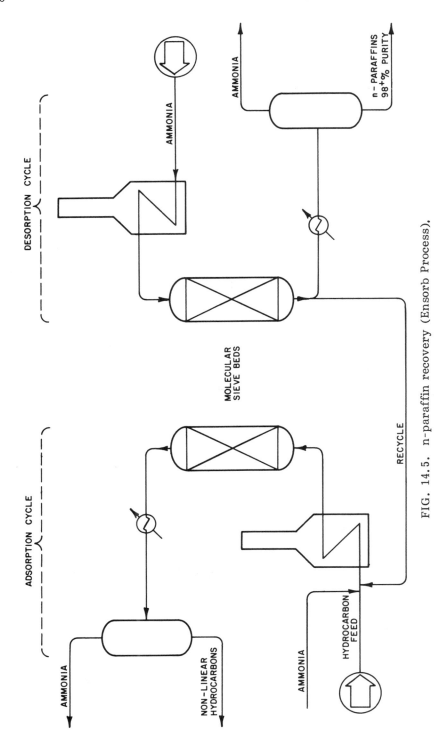

FIG. 14.5. n-paraffin recovery (Ensorb Process).

hydrocarbons. Run times of about 12 months between regenerations are common.

The greatest demand for n-paraffins is for use in the manufacture of biogradable detergents using straight chain paraffins with from 11 to 14 carbons per molecule.

Cycloparaffins. Cycloparaffins are usually prepared by the hydrogenation of the corresponding aromatic compound. Cyclohexane is prepared by the hydrogenation of benzene. There are a number of processes licensed for this purpose [1].

The hydrogenation of benzene is carried out over platinum or Raney nickel supported on alumina or silica alumina. The hydrogenation reaction is highly exothermic with the release of about 1,150 Btu/lb of benzene converted to cyclohexane. Reactor temperature is controlled by recycling and injecting a portion of the cyclohexane produced into the reactors [1].

Tetralin and decalin are prepared by the hydrogenation of naphthalene using the same hydrogenation catalysts as used for producing cyclohexane. Here, benzothiophene is a common impurity in naphthalene and causes the rapid deactivation of the nickel and platinum catalysts. It is necessary under these conditions to use a cobalt-molybdenum catalyst which has been presulfided to activate it [1].

NOTES

1. C. L. Thomas, *Catalytic Processes and Proven Catalysts* (Academic Press, New York, 1970).
2. D. B. Broughton, and Asselin, G. F., The Proceedings of the Seventh World Petroleum Congress, Vol. 4, 65-73 (1967).
3. Hydrocarbon Process. *59*(9), 203-4 (1980).
4. J. H. Prescott, Chem. Eng., 138-140 (June 1969).
5. H. C. Schutt and S. B. Zdonik, Oil Gas J. *54*(7), 98 (1956).
6. Hydrocarbon Process. *59*(11), 185 (1980).
7. Personal Communication, M. K. Lambert, 1983.

ADDITIONAL READING

Hydrocarbon Process. *48*(6), 185-7 (1969).
A. M. Brownstein, *Trends in Petrochemical Technology* (Petroleum Publishing Co., Tulsa, 1976, p. 17-51.

Chapter 15
COST ESTIMATION

Although detailed discussions of various capital cost estimating methods are not part of the intended scope of this work, some comments are pertinent.

All capital cost estimates of industrial process plants can be classified as one of four types:

1. "rule-of-thumb" estimates,
2. cost-curve estimates,
3. major equipment factor estimates,
4. definitive estimates.

The capital cost data presented in this work are of the second type—cost-curve estimates.

15.1 RULE-OF-THUMB ESTIMATES

The rule-of-thumb estimates are, in most cases, only an approximation of the order of magnitude of cost. These estimates are simply a fixed cost per unit of feed or product. Some examples are:

Complete coal-fired electric power plant	$ 1,000/kW
Complete synthetic ammonia plant	$75,000/TPD
Complete petroleum refinery	$10,000/BPD

These rule-of-thumb factors are useful for quick "ball-park" costs. Many assumptions are implicit in these values and the average deviation from actual practice can often be more than 50%.

15.2 COST-CURVE ESTIMATES

The cost-curve method of estimating corrects for the major deficiency illustrated above by reflecting the significant effect of size or capacity on cost. These curves indicate that costs of similar process units or plants are related to capacity by an equation of the following form:

$$\frac{\text{Plant A cost}}{\text{Plant B cost}} = \left(\frac{\text{Plant A capacity}}{\text{Plant B capacity}}\right)^{x}$$

This relationship was reported by Lang [1], who suggested an average value of 0.6 for the exponent (x). Other authors have further described this function [2, 3].

Cost curves of this type have been presented for petroleum refinery costs in the past [4, 5].

The curves presented herein have been adjusted to eliminate certain costs such as utilities, storage, offsite facilities and location cost differentials. Separate data are given for estimating the costs of these items. The facilities included have been defined in an attempt to improve accuracy.

It is important to note that most of the cost plots have an exponent which differs somewhat from the 0.6 value. Some of the plots actually show a curvature in the log-log slope which indicates that the cost exponent for these process units varies with capacity. Variations in the log-log slope (cost exponent) range from about 0.5 for small capacity units up to almost 1.0 for very large units. This curvature which is not indicated in the previously published cost curves is due to paralleling equipment in large units and to disproportionately higher costs of very large equipment such as vessels, valves, pumps, etc. The curvature in the log-log slope of cost plots has been described by Chase [6].

The cost-curve method of estimating, if carefully used and properly adjusted for local construction conditions, can predict actual costs within 20%. Except in unusual circumstances, errors will probably not exceed 30%.

15.3 MAJOR EQUIPMENT FACTOR ESTIMATES

Major equipment factor estimates are made by applying multipliers to the costs of all major equipment required for the plant or process facility. Different factors are applicable to different types of equipment such as pumps, heat exchangers, pressure vessels, etc [7]. Equipment size also has an effect on the factors.

It is obvious that prices of major equipment must first be developed to use this method. This requires that heat and material balances be completed in order to develop the size and basic specifications for the major equipment.

This method of estimating, if carefully followed, can predict actual costs within 10%.

A shortcut modification of this method uses a single factor for all equipment. A commonly used factor for petroleum refining facilities is 3.0. The accuracy of this shortcut is of course less than when using individual factors.

15.4 DEFINITIVE ESTIMATES

Definitive cost estimates are the most time-consuming and difficult to prepare but they are also the most accurate. These estimates require preparation of plot plans, detailed flow sheets and preliminary construction drawings. Scale models are sometimes used. All material and equipment is listed and priced. The number of man-hours for each construction activity is estimated. Indirect field costs such as crane rentals, costs of tools, supervision, etc. are also estimated.

This type of estimate usually results in an accuracy of ±5%.

15.5 SUMMARY FORM FOR COST ESTIMATES

The items to be considered when estimating investment from cost curves are:

    Process units
    Storage facilities
    Steam systems
    Cooling water systems
       Subtotal A
    Offsites
       Subtotal B
    Special costs
       Subtotal C
    Location factor
       Subtotal D
    Contingency
       Total

15.6 STORAGE FACILITIES

Storage facilities represent a significant item of investment costs in most refineries. Storage capacity for crude oil and products varies widely at different refineries. In order to properly determine storage requirements the following must be considered: the number and type of products, method of marketing, source of crude oil, and location and size of refinery.

Installed costs for "tank-farms" vary from $25 to $35 per barrel of storage capacity. This includes tanks, piping, transfer pumps, dikes, fire protection equipment, and tank-car or truck loading facilities. The value is applicable to low vapor pressure products such as gasoline and heavier liquids. Installed costs for butane storage ranges from $50 to $75 per barrel, depending on size. Costs for propane storage range from $60 to $90 per barrel.

15.7 LAND AND STORAGE REQUIREMENTS

Each refinery has its own land and storage requrements depending on location
with respect to markets and crude supply, methods of transportation of the
crude and products, and number and size of processing units. Availability
of storage tanks for short-term leasing is also a factor as the maximum
amount of storage required is usually based on shutdown of processing units
for turnaround at 18- to 24-month intervals rather than on day-to-day processing requirements. If sufficient rental tankage is available to cover
turnaround periods, the total storage and land requirements can be materially
reduced, as the land area required for storage tanks is a major portion of
refinery land requirements.

Three types of tankage are required: crude, intermediate, and product. For a typical refinery which receives the majority of its crude by pipe
line and distributes its products in the same manner, about 13 days of crude
storage and 25 days of product storage should be provided. The 25 days
of product storage is based on a three-week shutdown of a major process
unit. This generally occurs only every 18 months or two years but sufficient
storage is necessary to provide products to customers over this extended
period. A rule-of-thumb figure for total tankage, including intermediate
storage, is approximately 50 barrels of storage per BPD crude oil processed.

The trend to producing no-lead gasolines will make more intermediate
storage necessary to hold the components required for blending the final
gasoline product. This will increase total storage requirements by about 10%.

W. L. Nelson indicates that in 1973 the average refinery had 69 days of
total storage capacity [8]. The range was from 23 to 120 days of total storage
with refineries receiving crude by pipeline averaging 54 days and those
receiving crude by tanker averaging 78 days.

The land requirements are frequently dictated by considerations other
than process or storage because of the desire to provide increased security,
to be isolated from neighboring buildings, etc. If operational matters are
the prime consideration, the land necessary for operational and storage facilities is about four acres per 1,000 BPCD crude capacity.

Nelson has summarized land used for 32 refineries from 1948 to 1971 [9].
The land in use when the refineries were built ranged as a function of refinery
complexity from 0.8 acres to 5.7 acres per 1,000 BPD crude capacity. Land
actually purchased, though, was much more than this, and varied from 8 to
30 acres per 1,000 BPD. This additional land provided a buffer zone between
the refinery and adjacent property and allowed for expanding the capacity
and complexity of the refinery. In a summary article on land and storage
costs [10], Nelson suggests that five acres of land per 1,000 BPCD crude
capacity be used for planning purposes.

Land provided for growth and expansion to other processes, such as
petrochemicals, should not be included in the investment costs against which
the return on investment is calculated. Land purchased for future use should
be charged against the operation for which it is intended or against overall
company operations.

15.8 STEAM SYSTEMS

An investment cost of $50.00 per lb/hr of total steam generation capacity is
recommended for preliminary estimates. This represents the total installed

costs for gas or oil fired, forced draft boilers, operating at 250 to 300 psig and all appurtenant items such as water treating, deaerating, feed pumps, yard piping for steam, and condensate.

Total fuel requirements for steam generation can be assumed to be 1,200 Btu (LHV) per pound of steam.

A contingency of 25% should be applied to preliminary estimates of steam requirements.

Water makeup to the boilers is usually 5 to 10% of the steam produced.

15.9 COOLING WATER SYSTEMS

An investment cost of $75.00 per gpm of total water circulation is recommended for preliminary estimates. This represents the total installed costs for a conventional induced-draft cooling tower, water pumps, water treating equipment, and water piping. Special costs for water supply and blowdown disposal are not included.

The daily power requirements (kWh/day) for cooling water pumps and fans is estimated by multiplying the circulation rate in gpm by 0.6. This power requirement is usually a significant item in total plant power load and should not be ignored.

The cooling tower makeup water is about 5% of the circulation. This is also a significant item and should not be overlooked.

An "omission factor" or contingency of 15% should be applied to the cooling water circulation requirements.

15.10 OTHER UTILITY SYSTEMS

Other utility systems required in a refinery are electric power distribution, instrument air, drinking water, fire water, sewers, waste collection, etc. Since these are difficult to estimate without detailed drawings, the cost is normally included in the offsite facilities.

Offsites

"Offsites" are the facilities required in a refinery which are not included in the costs of major facilities. A typical list of offsites is shown below. Obviously, the offsite requirements vary widely between different refineries.

The values shown below can be considered as typical for grassroots refineries when estimated as outlined in this text.

Crude oil feed (BPSD)	Offsite costs, % of total major facilities costs*
Less than 30,000	20
30,000-100,000	15
More than 100,000	10

*Major facilities as defined herein include process units, storage facilities, cooling water systems, and steam systems.

Offsite costs for the addition of individual process units in an existing refinery can be assumed to be about 15 to 20% of the process unit costs.
Items considered to be offsites in this work are as follows:

Electric power distribution
Fuel oil and fuel gas facilities
Water supply and disposal
Plant air systems
Fire protection systems
Flare, drain and waste disposal systems
Plant communication systems
Roads and walks
Railroads
Fence
Buildings
Vehicles
Ethyl blending plant

Special Costs

Special costs include the following: land, spare parts, inspection,
project management, chemicals, miscellaneous supplies, office and laboratory
furniture. For preliminary estimates these costs can be estimated as 4% of
the cost of the process units, storage, steam systems, cooling water systems
and offsites.

Engineering costs and contractor's fees are included in the various
individual cost items.

Contingencies

Most professional cost estimators recommend that a contingency of at
least 15% be applied to the final total cost determined by cost curve estimates
of the type presented herein.

The term contingencies covers many loopholes in cost estimates of
process plants. The major loopholes include cost data inaccuracies when
applied to specific cases and lack of complete definition of facilities required.

Escalation

All cost data presented in this book are based on U.S. Gulf Coast
construction averages for the year 1982. This statement applies to the process unit cost curves, as well as values given for items such as cooling water
systems, steam plant systems, storage facilities and catalyst costs. Therefore, in any attempt to use the data for current estimates some form of escalation or inflation factor must be applied. Many cost index numbers are
available from the federal government and from other published sources. Of
these, the Chemical Engineering Plant Cost Index and the Nelson Refinery
Inflation Index are the most readily available and probably the most commonly
used by estimators and engineers in the U.S. refining industry.

The use of these indices is subject to errors inherent in any generalized estimating procedure, but some such factor must obviously be incorporated in projecting costs from a previous time basis to a current period. It should be noted that the contingencies discussed in the previous section are not intended to cover escalation.

Escalation or inflation of refinery investment costs is influenced by items which tend to increase costs as well as by items which tend to decrease costs. Items which increase costs include obvious major factors, such as:

1. Increased cost of steel, concrete, and other basic materials on a per ton basis
2. Increased cost of construction labor and engineering on a per hour basis
3. Increased costs for higher safety standards and better pollution control
4. An increase in the number of reports and amount of data necessary to obtain local, state, and federal construction permits

Items which tend to decrease costs are basically all related to technological improvements. These include:

1. Process improvements developed by the engineers in research, design, and operation
2. More efficient use of engineering and construction manpower

Examples of such process improvements include improvement of fractionator tray capacities, improved catalysts which allow smaller reactors, and improved instrumentation allowing for consistently higher plant feed rates.

Plant Location

Plant location has a significant influence on plant costs. The main factors contributing to these variations are: climate and its effect on design requirements and construction conditions; local rules, regulations, codes, taxes, etc; and availability and productivity of construction labor.

Relative hydrocarbon process plant costs on a 1982 basis at various locations are given below:

Location	Relative cost
U.S. Gulf Coast	1.0
Los Angeles	1.2
Portland, Seattle	1.2
Chicago	1.3
St. Louis	1.3
Detroit	1.3

(continued)

New York	1.3
Philadelphia	1.3
Alaska, North Slope	2.0-2.5
Alaska, Anchorage	1.5-2.0

15.11 APPLICATION OF COST ESTIMATION TECHNIQUES

Although economic evaluation will not be discussed until the next chapter, an example problem illustrating the methods to estimate capital and operating costs and return on investment is included here to aid in clarifying the principles discussed in this chapter. The illustrative problem is relatively simple but the same techniques and procedures can be applied to the most complex refinery economic evaluation.

15.12 STATEMENT OF PROBLEM

For the following example of a simplified refinery, calculate:

1. The products available for sale
2. Investment
3. Operating costs
4. Simple rate of return on investment
5. True rate of return on investment

Also prepare a basic block flow diagram (Fig. 15.1). The following data are available:

1.	Crude charge rate	30,000 BPSD
2.	Crude oil sulfur content	1.0 wt %
3.	Full-range naphtha in crude	4,000 BPSD
		240°F MBP
		56°API grav
		11.8 K_w
4.	Light gas oil in crude	4,000 BPSD
5.	Heavy gas oil in crude	4,000 BPSD
6.	Vacuum gas oil in crude	6,000 BPSD
7.	Vacuum residual in crude	12,000 BPSD
8.	On-stream factor	93.15%
9.	Cost of makeup water	20¢/1,000 gal
10.	Cost of power	3¢/kWh
11.	LHV of heavy gas oil	5.5 MMBtu/bbl
12.	Replacement cost for desulfurizer catalyst is $1/lb	
13.	Replacement cost for reformer catalyst is $5/lb	
14.	Insurance annual cost is 0.5% of plant investment	

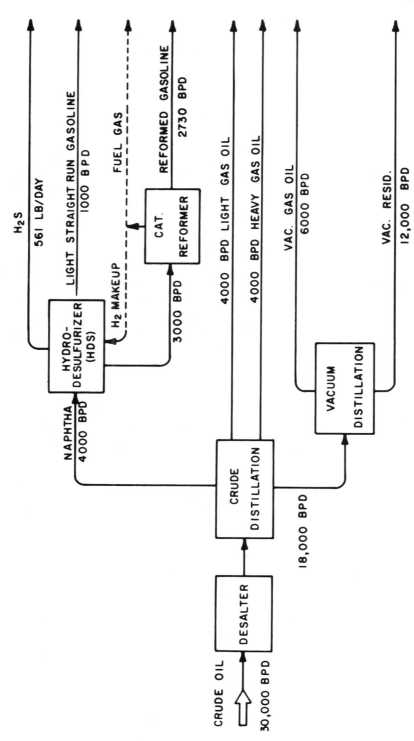

FIG. 15.1. Block flow diagram for sample problem.

15. Local taxes annual cost is 1.0%
    of plant investment
16. Maintenance annual cost is 5.5%
    of plant investment
17. Misc supplies annual cost is
    0.15% of plant investment
18. Average annual salary plus
    payroll burden for plant staff
    and operators is $35,000
19. Value of crude oil and products
    at refinery is:

	$/bbl
Crude	29.50
Gasoline	39.00
Light gas oil	38.00
Heavy gas oil	36.00
Vacuum gas oil	35.00
Vacuum residual	28.00

20. Depreciation allowance: 15 year, straight-line
21. Corporate income tax: 50% of taxable income
22. Location: St. Louis, Missouri
23. Construction period: 1984
24. Escalation rate (applicable to
    construction costs only) is 15%
    per year.

Process Requirements

The crude oil is to be desalted and fractionated to produce full-range naphtha, light gas oil, heavy gas oil, and "atmospheric bottoms." The latter cut is fed to a vacuum unit for fractionation into vacuum gas oil and vacuum residuals. The full-range naphtha is to be hydrodesulfurized. After desulfurization, the light straight-run (LSR) portion (i.e., the material boiling below 180°F) of the full-range naphtha is separated for blending into the gasoline product. The balance of the naphtha is fed to a catalytic reformer which is operated to produce a reformate having a research octane number (clear) of 93. This reformate plus the LSR are mixed to make the final gasoline product. Propane and lighter hydrocarbons, including the hydrogen, which are produced in the catalytic reformer are consumed as fuel. The necessary hydrogen makeup for the hydrodesulfurizer is, of course, taken from these gases before they are burned as fuel. The balance of the fuel requirement is derived from light gas oil. The hydrogen sulfide produced in the hydrodesulfurizer is a relatively small amount and is burned in an incinerator. No other product treating is required. It can be assumed that sufficient tankage for approximately 12 days storage of all products is required. The total storage requirement will thus be approximately 360,000 barrels.

The above information, in addition to that contained in this book, is sufficient for solution of the problem.

Catalytic Reformer

Calculate properties of feed to reformer. Given total naphtha stream properties as follows:

Mid-boiling point: 240°F

API gravity: 56°API

K_w: 11.8

The material boiling below 180°F (LSR) is not fed to the reformer. After desulfurizing the total naphtha, the LSR is fractionated out. The problem is to estimate the volume and weight of the LSR assuming a distillation curve is not available. This LSR is then deducted from the total naphtha to find the net reformer feed. This is done as shown in the following steps:

1. Assume that butane and lighter hydrocarbons in the naphtha are negligible. Thus, the lightest material would be isopentane (i-C_5) with a boiling point of 82°F. Hence, the mid-boiling point of the LSR would be approximately (82 + 180) (0.5) or 131°F.

2. Assume that the LSR has the same K_w as the total naphtha (i.e., 11.8).

3. From general charts [see ref. 4, p. 16] relating K_w, mean average boiling point, and gravity, find the gravity of the LSR. This is 76.5°API.

4. The gravity of the naphtha fraction boiling above 180°F is next determined by a similar procedure. The mid-boiling point for the total naphtha is given above as 240°F, and the initial boiling point was estimated in Step 1 as 82°F. Therefore, the naphtha end point can be estimated as shown:

(240) + (240 − 82) = 398°F.

Now the approximate mid-boiling point of the reformer feed is estimated as

(180 + 398)(0.5) = 289°F.

Using a K_w of 11.8, the reformer feed gravity is found (as in Step 3) to be 52.5°API.

5. With the above estimates of gravity of both the LSR and reformer feed, it is now possible to estimate the relative amounts of each cut which will exist in the total naphtha stream. This is done by weight and volume balances as shown below. Use simplified nomenclature:

V_{LSR} = gallons LSR
V_{RF} = gallons reformer feed
V_N = gallons total naphtha
W_{LSR} = pounds LSR

W_{RF} = pounds reformer feed
W_N = pounds total naphtha
lb/gal LSR = 5.93 (67.5°API)
lb/gal RF   = 6.41 (52.5°API)
lb/gal V_N = 6.29 (56°API)

Volume balance (hourly basis)

$$V_N = 4,000 \times \frac{42}{24} = 7,000 \text{ gal/hr}$$

$$(V_{LSR}) + (V_{RF}) = 7,000 \text{ gal/hr}$$

Weight balance (hourly basis)

$$V_N \times 6.29 = 7,000 \times 6.29 = 44,030 \text{ lb/hr}$$

$$(V_{LSR})(5.93) + (V_{RF})(6.41) = 44,030 \text{ lb/hr}$$

Simultaneous solution of the two equations for (V_{LSR}) and (V_{RF}) gives:

$$V_{LSR} = 1,750 \text{ gal/hr} = 1,000 \text{ BPD}$$

$$V_{RF} = 5,250 \text{ gal/hr} = 3,000 \text{ BPD}$$

6. The above information should then be tabulated as shown below, to ascertain that all items balance.

Stream	°API	lb/gal	gal/hr	lb/hr	BPD
LSR	67.5	5.93	1,750	10,378	1,000
Rfmr feed	52.5	6.41	5,250	33,652	3,000
Total naphtha			7,000	44,030	4,000

7. Before proceeding, it should be emphasized that the above methods for approximating the naphtha split into LSR and reformer feed are satisfactory for preliminary cost and yield computations such as this example, but not for final design calculations.

8. The reformer feed properties can now be used with the yield curves in the text. The following yields are based on production of 93 RON (clear) reformate. From the yield curves:

vol % C_5^+ 86.0

vol % C_4's   5.0 (i-C_4/n-C_4 = 41.5/58.5)

wt % C_1, C_2 1.1

wt % C_3 1.92

wt % H_2 1.75

With the above data, complete the following table:

Component	gal/hr	lb/gal	lb/hr	BPD	Mscf/day
H_2			589		2,682
C_1,C_2			370		145[a]
C_3	153	4.23	646		133
i-C_4	109	4.69	511	62	
n-C_4	154	4.86	748	88	
C_5^+	4,515	6.82	30,788[b]	2,580	_____
Totals			33,652	2,730	2,960
Feed	5,250	6.29	33,652	3,000	—

[a]Assume lb C_1/lb C_2 = 0.5 (i.e., C_1, C_2 = 23.3 mol wt)
[b]Lb/hr C_5^+ obtained by difference from total feed less other products.

Naphtha Desulfurizer

Assume crude oil = 1.0% S

Then, from curve for miscellaneous crudes

Naphtha contains 0.05% S (240°F MBP)

Calculate amount of sulfur produced. Assume K_w for desulfurized feed is 11.8. This combined with 240°F MBP gives a naphtha gravity of 56°API (see reformer calculations).

56°API = 6.29 lb/gal
wt S in naphtha = S_N
S_N = 4,000 X 42 X 6.29 X 0.0005 = 528 lb/day
max H_2S formed = $\frac{34}{32}$ X 528 = 561 lb/day
theoretical H_2 required = 561-528
= 33 lb/day
= 16.5 mol/day
= 6.26 Mscf/day

Makeup H_2 required [see ref. 8, p. 202] is about 100 to 150 scf/bbl or

$$4,000 \times 0.15 = 600 \text{ Mscf/day}$$

Hydrogen from catalytic reformer is 2,682 Mscf/day, which is more than adequate.

Summary of Investment Costs and Utilities

Item	BPSD	$(X10^3) 1982 G.C.	CW gpm	lb/hr stm	kW	Fuel MBtu/hr
Desalter	30,000	750	83		150	
Crude unit	30,000	6,800	125	7,500	625	88
Vac unit	18,000	4,000	750	6,750	150	30
Naph desulf	4,000	2,200	833	1,000	133	17
Reformer	3,000	4,600	1,250	3,750	375	38
Init cat (Desulf)		Included				
Init cat (Reform)		600				
Subtotal		18,950	3,041	19,000	1,633	173
CW system, 3,500 gpm[a]		265			88	
Stm system, 22,800 lb/hr[b]		1,140				27
		20,355	3,041	19,000	1,721	200
Storage[c]	12 days[d]	9,900				
Subtotal		30,255				
Offsites[c]	(20%)	6,050				
Subtotal		36,305				
Location factor	1.3					
Spec cost factor	1.04[e]					
Contingency	1.15[e]					
Escalation	$(1.15)^2$					
Total		74,650[f]				

Note: Values shown are to be multiplied by 1,000 when notes as $(X10^3).
[a]Add 15% excess capacity to calculated cooling water circulation.
[b]Add 20% excess capacity to calculated steam supply.
[c]Individual values for utilities in the storage and offsite categories are accounted for by notes a and b.
[d]360,000 barrels ar average cost of $27.50/bbl.
[e]These factors are compounded.
[f]This is the projected cost at the location in St. Louis in 1984. No paid-up royalties are included.

Calculation of Direct Annual Operating Costs

After completing the investment and yield calculations the annual operating costs of the refinery can be determined. Operating costs can be considered to include three major categories:
1. Costs which vary as a function of plant throughput and on-stream time—these include water makeup to the boilers and cooling tower, electric power, fuel, running royalties and catalyst consumption.
2. Costs which are a function of the plant investment—these include insurance, local taxes, maintenance (both material and labor) and miscellaneous supplies.
3. Costs which are determined by the size and complexity of the refinery—these include operating, clerical, technical and supervisory personnel. The following section illustrates development of the above costs.

On-Stream Time

Refineries generally have an on-stream (full capacity) factor of about 92 to 96%. For this example, a factor of 93.15% (340 days per year) is used.

Water Makeup

1. To cooling tower ($30°F \Delta t$):

    1% evaporation for $10°F \Delta t$
    1/2% windage loss
    1% blowdown to control solids concentration
    Cooling tower makeup = 3 X 1% + 1/2% + 1% = 4 1/2%
                    makeup = 0.045 X 3,500 gpm = 157.5 gpm
2. To boiler:

    Average boiler blowdown to control solids concentration can be
      assumed to be 5%
    Boiler makeup = 0.05 X 22,800
                  = 1,140 lb/hr = 2.3 gpm
    Total makeup water = 160 gpm
    Average cost to provide makeup water is approximately 20¢/1,000
      gallons
    Therefore, annual water makeup cost is
      160 X 1,440 X 340 X 0.20 X 10^{-3} = $15,670

Power

Industrial power costs range from 2.5¢/kWh (in locations where there is hydroelectric power) to 6.0¢/kWh. For this example, use 3¢/kWh.

Power cost = 1,721 X 24 X 340 X 0.03
            = 421,290 per year

Fuel

In this example, no separate charge will be made for fuel, since it is assumed that the refinery will use some of the heavy gas oil products for fuel. The amount of gas oil consumed must be calculated, so that this quality can be deducted from the products available for sale.

From the summary tabulation of utilities, we require 200 MMBtu/hr for full-load operation. This fuel is supplied by combustion of reformer "off-gas" supplemented with heavy gas oil. Some of the reformer off-gas is consumed in the hydrodesulfurizer and this quantity (hydrogen portion only) must be deducted from available fuel. A fuel balance is made as shown below to determine the amount of heavy gas oil consumed as fuel.

Step 1:

From reformer calculations the available fuel gas is:

Component	Total (lb/hr)	HDS* usage (lb/hr)	Available for fuel (lb/hr)
H_2	589	132	457
C_1	123		123
C_2	246		246
C_3	646	___	646
	1,604	132	1,472

*From desulfurizer calculations, hydrogen makeup was 600 Mscfd.

In petroleum work, "standard conditions" are 60°F, and 14.696 psia. At these conditions, 1 pound mole = 379.5 scf. Thus, hydrogen consumed in HDS unit is

$$\frac{600,000}{24 \times 379.5} \times 2 = 132 \text{ lb/hr}.$$

Step 2:

Calculated lower heating value (LHV) of available fuel gas is:

Component	Total (lb/hr)	LHV* (Btu/lb)	LHV (MMBtu/hr)
H_2	457	51,600	23.6
C_1	123	21,500	2.6
C_2	246	20,420	5.0
C_3	646	19,930	12.9
	1,472		44.1

*From Maxwell [11] or Perry [12], or other convenient source.

Step 3:

Heavy gas oil required for fuel. Assume 5.5 MMBtu/bbl LHV.

$$\frac{200 - 44.1}{5.5} \times 24 = 680 \text{ BPSD}$$

Step 4:

Heavy gas oil remaining for sale.

$$4,000 - 680 = 3,320 \text{ BPSD}$$

Royalties

The desulfurizer and reformer are proprietary processes and, therefore, royalties must be paid. On a "running" basis, these are:

Desulfurizer: 1.5¢/bbl feed

   Annual cost = 0.015 X 4,000 X 340 = $20,400

Reformer: 7.0¢/bbl feed

   Annual cost = 0.07 X 3,000 X 340 = $71,400

Total royalties: $91,800/yr

Catalyst Consumption:

Catalyst consumption costs are as follows:

Desulfurizer: 0.002 lb/bbl; $1/lb
  Annual cost = 4,000 X 340 X 0,002 X 1 = $ 2,700

Reformer: 0.004 lb/bbl; $5 per lb

  Annual cost = 3,000 X 340 X 0.004 X 5 = <u>$20,400</u>

Total catalyst cost: $23,100/yr

Insurance

This cost usually is 0.5% of the plant investment per year.

$74,650,000 X 0.005 = $373,250/yr

Local Taxes

Usually 1% of the plant investment per year.

$74,650,000 X 0.01 = $746,500/yr

Maintenance

This cost varies between 3% and 8% of plant investment per year. For this example, use an average value of 5.5%. This includes material and labor.

$74,650,000 X 0.055 = $4,105,750/yr

Miscellaneous Supplies

This item includes miscellaneous chemicals used for corrosion control, drinking water, office supplies, etc. An average value is 0.15% of the plant investment per year.

$74,650,000 X 0.0015 = $119,980/yr

Plant Staff and Operators

The number of staff personnel and operators depends on plant complexity and location. For this example, the following staff could be considered typical of a modern refinery:

	No. per shift	Payroll total
Plant Manager		1
Process engineer		1
Mechanical engineer		1
Clerk		2
Stenographer		1
Chemist		1
Operating superintendent		1
Process operators	2	9
Utility operators	1	5
		22

Assume average annual salary plus payroll burden is $35,000 per person. Total annual cost for staff and operators is thus

$$22 \times \$35,000 = \$770,000$$

Note that maintenance personnel are not listed above since this cost was included with the maintenance item. Also note that it takes about 4 1/2 men on the payroll for each shift job to cover vacations, holidays, illness, and fishing time.

Summary of Direct Annual Operating Costs

Item	$/yr($\times 10^3$)
Makeup water	16
Power	421
Fuel[a]	—
Royalties	92
Catalyst	23
Insurance	373
Local taxes	747
Maintenance	4,106
Miscellaneous supplies	120
Plant staff and operators	770
Subtotal	6,668
Contingency (10%)	667
Total[b]	7,335

[a]Fuel quantity is deducted from available heavy gas oil for sale.
[b]Additional items such as corporate overhead, research and development, sales expense, etc. are omitted from this sample.

Calculation of Income Before Income Tax

Sales:

Product	BPD	MBPY	$/bbl	$/yr(X10^3)
Gasoline				
LSR	1,000			
Reformate	2,730			
	3,730	1,268	39.00	49,452
Light gas oil	4,000	1,360	38.00	51,680
Heavy gas oil	3,320	1,129	36.00	40,644
Vac gas oil	6,000	2,040	35.00	71,400
Vac residual	12,000	4,080	28.00	114,240
Total:				327,416
Crude Cost:	30,000	10,200	29.50	300,900
Direct operating costs:				7,335
Income before income tax:				19,181

Calculation of Return on Investment

Investment = $74,650,000

Item	$/yr(X10^3)
Income before tax	19,181
Less depreciation allowance*	4,977
Taxable income	14,204
Income tax at 50%	7,102
Income after tax	7,102
Plus depreciation allowance	4,977
Cash flow	12,079
Return on investment (% per year)	16.18
Payout period (years)	6.2

True rate of return (discounted cash flow)
Basis: 20-year life, no salvage value (see
Appendix E)

$$TRR = DCF = i = \frac{S}{I} - \frac{i}{(1 + i)^T - 1}$$

$$i = \frac{12,079}{74,650} - \frac{i}{(1 + i)^{20} - 1}$$

By trial-and-error solution:

$$i = 0.109 = 10.9 = 11\%$$

Note: Interest on capital during construction
period and average feedstock and product inventories are not considered in above product.
These items would result in an increase in investment and a decrease in the rate of return.
*15 year, straight-line.

PROBLEMS

1. Calculate the chemical and utility requirements for a crude unit consisting
   of both atmospheric and vacuum fractionation towers charging 100,000
   BPCD of the assigned crude oil and producing typical side, overhead,
   and bottom streams. Using assigned utility and chemical costs, calculate
   the cost per barrel of charge and total annual costs for chemicals and
   utilities.

2. Calculate the DCFROR on a 100,000 BPCD crude unit charging the
   assigned crude oil and producing typical side, overhead, and bottom
   streams. Use product values from OGJ and for those streams not listed
   base on equivalent heating value of closest comparable listed product.
   Construction costs should be brought to current value by use of Nelson
   Indices. Include total operating and materials costs. Use 30 year useful
   life with salvage value equal to cost of removal. Construction began 2
   years before unit start-up. Land costs are $400,000 and the labor re-
   quirement is 3 persons per shift.

NOTES

1. H. J. Lang, Chem. Eng. *55*(6), 112 (1948).
2. W. Williams, Jr., Chem. Eng. *54*(12), 124-125 (1947).
3. A. B. Woodler and J. W. Woodcock, Eur. Chem. News, pp. 8-9
    (10 September 1965).
4. K. M. Guthrie, Chem. Eng. *77*(13), 140-156 (1970).
5. W. L. Nelson, Oil Gas J. *63*(20), 1939-1940 (1965).
6. J. D. Chase, Chem. Eng. *77*(7), 113-118 (1970).
7. K. M. Guthrie, Chem. Eng. *76*(6), 114-142 (1969).
8. W. L. Nelson, Oil Gas J. *71*(17), 88-89 (1973).
9. Ibid, *70*(49), 56-57 (1972).
10. Ibid, *72*(30), 160-162 (1974).
11. J. B. Maxwell, *Data Book on Hydrocarbons* (D. Van Nostrand, New
    York, 1950).
12. R. H. Perry, C. H. Chilton, and S. D. Kirkpatrick, *Chemical
    Engineers' Handbook*, 4th ed. (McGraw-Hill Book Company, New
    York, 1963).

Chapter 16
ECONOMIC EVALUATION

Economic evaluations are generally carried out to determine if a proposed investment meets the profitability criteria of the company or to evaluate alternatives. There are a number of methods of evaluation and a good summary of the advantages and disadvantages of each is given in Perry's "Chemical Engineers' Handbook" [1]. Most companies do not rely upon one method alone but utilize several to obtain a more objective viewpoint.

As this is primarily concerned with cost estimation procedures, there will be no attempt to go into the theory of economics but equations will be presented which are used for the economic evaluation calculations. There is a certain amount of basic information needed to undertake the calculations for an economic evaluation of a project [2, 3, 4, 5].

16.1 DEFINITIONS

Depreciation

Depreciation arises from two causes, deterioration and obsolescence. These two causes do not necessarily operate at the same rate and the one having the faster rate determines the economic life of the project. Depreciation is an expense and there are several permissible ways of allocating it. For engineering purposes depreciation is usually calculated by the "straight-line method" for the economic life of the project. Frequently economic lives of 10 years or less are assumed for projects of less than $250,000.

Working Capital

The working capital (WC) consists of feed and product inventories, cash for wages and materials, accounts receivable, and spare parts. A reasonable figure is the sum of the above items for a 30-day period.

Annual Cash Flow

The annual cash flow (ACF) is the sum of the earnings after taxes and the depreciation for a one-year period.

Sensitivity Analysis

Uncertainties in the costs of equipment, labor, operation, and raw materials as well as in future prices received for products can have a major effect on the evaluation of investments. It is important in appraising the risks involved to know how the outcome would be affected by errors in estimation, and a senstivity analysis is made to show the changes in the rate of return due to errors of estimation of investment costs and raw material and product prices. These will be affected by the type of cost analysis performed (rough estimate or detailed analysis), stability of the raw material and product markets, and the economic life of the project. Each company will have its own bases for sensitivity analyses but, when investment costs are derived from the installed-cost figures in this book, the following values are reasonable:

	Decrease by	Increase by
Investment cost	15%	20%
Raw materials costs	3	5
Product prices	5	5
Operating volumes	2	2

Product and Raw Material Cost Estimation

It is very important that price estimation and projections for raw materials and products be as realistic as possible. Current posted prices may not be representative of future conditions, or even of the present value to be received from an addition to the quantities available on the market. A more realistic method is to use the average of the published low over the past several months.

16.2 RETURN ON ORIGINAL INVESTMENT

This method is also known as the engineer's method, du Pont method, or the capitalized earning rate. It does not take into account the time value of money, but because of this, offers a more realistic comparison of returns during the latter years of the investment. The return on original investment is defined as:

$$ROI = \frac{\text{Average yearly profit}}{\text{Original fixed investment + working capital}} \times 100$$

The return on investment should be reported to two significant figures.

16.3 PAYOUT TIME

The payout time is also referred to as the cash recovery period and years to
pay out. It is calculated by the following formula and is expressed to the
nearest one-tenth year:

$$\text{Payout time} = \frac{\text{Original depreciable fixed investment}}{\text{Annual cash flow}}$$

If the annual cash flow varies, the payout time can be calculated by
adding the cash income after income taxes for consecutive years until the
sum is equal to the total investment. The results can be reported to a frac-
tional year by indicating at what point during the year the cash flow will
completely offset the depreciable investment.

16.4 DISCOUNTED CASH FLOW RATE OF RETURN

This method is called the investors' return on investment, internal rate of
return, profitability index, and interest rate of return as well as discounted
cash flow. A trial-and-error solution is necessary to calculate the average
rate of interest earned on the company's outstanding investment in the project. It can also be considered as the maximum interest rate at which funds
could be borrowed for investment in the project with the project breaking
even at the end of its expected life.

The discounted cash flow is basically the ratio of the average annual
profit during construction and earning life to the sum of the average fixed
investment, working capital, and the interest charged on the fixed and
working capital that reflects the time value of money. This ratio is expressed
as a percentage rather than a fraction. Discounted cash flow is discussed
in detail, with an example of its use, in Appendix E.

In order to compare investments having different lives or with variations in return during their operating lives, it is necessary to convert
rates of return to a common time basis for comparison. While any time may
be taken for this comparison, the plant startup time is usually taken as the
most satisfactory. Expenditures prior to startup and income and expenditures after startup are converted to their worth at startup. The discussion
to follow is based upon the predicted startup time being the basis of
calculation.

Expenditures Prior to Startup

The expenditures prior to startup can be placed in two categories: those that occur uniformly over the priod of time before startup and lump-sum payments that occur in-an-instant at some point before the startup time.

Construction costs are generally assumed to be disbursed uniformly between the start of construction and the startup time, although equivalent results can be obtained if they are considered to be a lump-sum disbursement taking place halfway between the start of construction and startup. The present worth of construction costs that are assumed to occur uniformly over a period of years T prior to startup can be calculated using either continuous interest compounding or discrete (annual) interest compounding.

Continuous interest compounding:

$$P_0 = \left(\frac{CC}{T}\right)\left(\frac{e^{iT} - 1}{i}\right)$$

Discrete (annual) interest compounding:

$$P_0 = \left(\frac{CC}{T}\right)\frac{\ln[1/(1 + i)]}{[1/(1 + i)]^T - [1/(1 + i)]^{T-1}}$$

where

P = worth at startup time
CC = total construction cost
T = length of construction period in years before startup
i = annual interest rate.

The cost of the land is a lump-sum disbursement and, in the equation given, it is assumed that the land payment coincides with the start of construction. If the disbursement is made at some other time, then the proper value should be substituted for T.

Continuous interest compounding:

$$P_0 = (LC)\, e^{iT}$$

Discrete (annual) interest compounding:

$$P_0 = (LC)(1 + i)^T$$

where

> LC = land cost
> T = years before startup time that payment was made
> i = annual interest rate.

Expenditures at Startup

Any costs which occur at startup time do not have to be factored, but have a present worth equal to their cost. The major investment at this time is the working capital, but there also may be some costs involved with the startup of the plant that would be invested at this time.

Income After Startup

The business income is normally spread throughout the year, and a realistic interpretation is that 1/365th of the annual earnings is being received at the end of each day. The present-worth factors for this type of incremental income are essentially equal to the continuous-income present-worth factors [6]. Even though the present worth of the income should be computed on a continuous-income basis, it is a matter of individual policy as to whether continuous or discrete compounding of interest is used [6, 7]. The income for each year can be converted to the reference point by the appropriate equation.

Continuous-income, continuous-interest:

$$P_0 = (ACF) \left(\frac{e^i - 1}{i} \right) e^{-in}$$

Continuous-income, with interest compounded annually:

$$P_0 = (ACF) \frac{[1/(1 + 1)]^n - [1/(1 + i)]^{n - 1}}{\ln[1/(1 + i)]}$$

where

> ACF = annual cash flow for year N
> n = years after startup
> i = annual interest rate.

For the special case where the income occurs uniformly over the life of the project after startup, the calculations can be simplified.

Uniform continuous-income, continuous-interest:

$$P_0 = (ACF) \left(\frac{e^i - 1}{i}\right) \left(\frac{e^{in} - 1}{ie^{in}}\right)$$

Uniform continuous-income, with interest compounded annually:

$$P_0 = (ACF) \frac{[1/(1 + i)]^n - 1}{\ln[1/(1 + i)]}$$

There are certain costs which are assumed not to depreciate and which are recoverable at the end of the normal service life of the project. Among these are the cost of land, working capital, and salvage value of equipment. These recoverable values must be corrected to their present worth at the startup time.

Continuous interest:

$$P_0 = (SV + LC + WC)e^{-in}$$

Interest compounded annually:

$$P_0 = (SV + LC + WC)[1/(1 + i)]^n$$

where

SV = salvage value, \$
LC = land cost, \$
WC = working capital, \$
 i = annual interest rate, decimal/yr
 n = economic life, yr.

For many studies, the salvage value is assumed equal to dismantling costs and is not included in the recoverable value of the project.

It is necessary to use a trial-and-error solution to calculate the discounted cash flow rate of return because the interest rate must be determined that makes the present value at the startup time of all earnings equal to that of all investments. An example of a typical balance for continuous-income and continuous-interest with uniform annual net income is:

$$(LC)(e^{iT}) + (CC) \left(\frac{e^{iT} - 1}{i}\right)\left(\frac{1}{T}\right) + (WC)$$

$$- (ACF) \left(\frac{e^i - 1}{i}\right)\left(\frac{e^{in} - 1}{ie^{in}}\right) - (SV + LC + WC) \left(\frac{1}{e^{in}}\right) = 0$$

All of the values are known except i, the effective interest rate. An interest rate is assumed and if the results give a positive value, the trial rate of return is too high and the calculations should be repeated using a lower value for i. If the calculated value is negative, the trial rate is too low and a higher rate of return should be tested. Continue the trial calculations until the rate of return is found which gives a value close to zero.

The return on investment should be reported only to two significant figures.

16.5 CASE-STUDY PROBLEM: ECONOMIC EVALUATION

The estimated 1982 construction costs of the refinery process units and their utility requirements are listed in Table 16.1.

The cooling water and steam systems and water makeup requirements were calculated according to the guidelines set up in Chapter 15.

The estimated refinery construction start is expected to be August, 1983, and the process startup date is anticipated to be in August, 1985. Inflation rates of 10% per year are used to bring the costs to their values in 1985.

The working capital is assumed to be equal to 10% of the construction costs. A review of the refinery manning requirements indicates that approximately 139 people will be required to operate the refinery, exclusive of maintenance personnel. The maintenance personnel are included in the 4.5% annual maintenance costs. An average annual salary of $35,000, including fringe benefits, is used.

A 20-year life will be assumed for the refinery with a dismantling cost equal to salvage value. Straight-line depreciation will be used.

The Federal tax rate is 48% and the state rate is 7%.

Investment costs and utility requirements are summarized in Table 16.1, operating costs in Table 16.2.

A summary of overall costs and realizations is given in Table 16.3, annual costs and revenues in Table 16.4, and total investment costs in Table 16.5, together with payout time and rates of return on investment. The effects of changes in investment and raw material cost, product prices, and operating volume are shown in Table 16.6.

TABLE 16.1
Investment Costs and Utilities

Item	BPCD	%[a] on-stream	BPSD	$(×10³) 1982	Power, MWh/day	Water, gpm[b] Cooling	Water, gpm[b] Process	Steam, Mlb/hr	Fuel, MBtu/hr
Desalter	100,000	96.9	103,200	1,500	12	277	87	25.0	292
Atm crude still	100,000	96.9	103,200	17,000	50	417		21.0	92
Vac pipe still	56,060	96.9	57,850	8,500	11	2,336		26.0	158
Coker	23,760	96.1	24,720	51,000	50	1,650		4.5	75
Hydrotreater	18,040	96.8	18,640	5,500	36	3,758		31.2	312
Cat reformer[c]	24,930	96.8	25,750	23,100	75	10,387			135
Cat cracker	32,300	95.7	33,750	39,000	194	11,200			323
Alkylation unit	7,460	97.2	7,670	14,000	28	19,200		3.4	194
Hydrocracker[c]	16,930	97.1	17,440	49,365	245	6,620		66.3	253
H₂ unit, MMscfd	25.6	97.1	26.4	10,600	20	6,100	94		260
Gas plt, MMscfd	36.2	96.9	38.1	13,500	124	30,920			34
Amine treater, gpm	566	96.9	584	4,400	8	2,490			
Claus sulfur, LT/day	87	96.9	90	3,100	2		49	(−23.0)	
Stretford unit, LT/day	10	96.9	10	3,100	21	90			
Subtotal				243,665	876	95,445	230	154.4	2,128
CW system (+15%)				8,250	66		5,500	30.9	222
Steam system (+20%)				9,250			18		
Subtotal				261,165	942	110,000	5,748	185.3	2,350
Storage				153,320					
Subtotal				414,485					
Offsites (15%)				62,172					
Subtotal				476,657					
Location (1.1)									
Special costs (1.04) } (0.3156)(476,657) =				207,441					
Contingency (1.15) }									
Total (1982)				684,098					

a W. L. Nelson, Oil Gas J. 69 (11), 86 (1971).
b Process water includes makeup water.
c Catalyst cost for initial charge included in unit construction cost; $5,150,000 for reformer and $2,616,000 for hydrocracker.

TABLE 16.2

Summary of Operating Costs

Item	$/yr($\times 10^3$)
Royalties	1,635
Chemicals and catalysts	5,398
Water makeup	453
Power	10,315
Fuel[a,b]	217,536
Insurance	3,420
Local taxes	6,841
Maintenance	30,784
Miscellaneous supplies	1,026
Plant staff and operators	4,865
Total[c]:	282,273

[a]Fuel quantity is deducted from refinery fuel gas, propane, heavy fuel oil, and some distillate fuel.
[b]Fuel cost (alternative value basis):

Fuel gas	1,138 MMBtu/hr@ $2.75/MMBtu	75,108
Propane	2,973 BPCD @ $23.10/bbl	68,676
No. 2 FO	840 BPCD @ $37.80/bbl	31,752
Hvy FO	1,680 BPCD @ $25.00/bbl	42,000
		$217,536

($217,536)(365) = $79,400,640/yr
[c]Additional items, such as corporate overhead, research and development vary among refineries and companies. These items are omitted here.

16.6 CASE-STUDY PROBLEM: ECONOMIC SOLUTION

Storage Costs

Based on 50 bbl of storage per BPCD crude oil processed.
Assume 21 days storage provided for n-butane.
n-butane: 5,260 BPD X 21 days = 110,460 bbl

Total Storage = 5,000,000 bbl
        Spheroid = 110,460

        General = 4,889,540 bbl

Cost:	$($X10^3$)
General @ $30/bbl	
(4,889,540) ($30) =	146,690
Spheroid @ $60/bbl	
(110,460) ($60) =	6,630
Total storage costs: =	153,320

TABLE 16.3

Refinery Annual Summary (1982 Prices)

	BPCD	($/bbl)	$/yr($\times 10^3$)
Inputs:		Costs	
North Slope crude	100,000	24.35	888,775
i-Butane	2,340	25.40	21,694
n-Butane	2,460	23.10	20,741
Methane, Mscf/day	2,660	2.75[a]	6,685
			937,895
Products:		Realizations	
Gasoline	65,640	40.00	958,344
Jet fuel	12,870	40.00	187,902
Distillate fuel (No. 2)	14,720	37.80	203,092
Coke, ton/day	1,259	20.00[b]	9,191
Sulfur, LT/day	96	45.50[c]	3,854
			1,362,383

[a]$/Mscf
[b]$/ton
[c]$/long ton

TABLE 16.4

Costs and Revenues

		$/yr($\times 10^3$)
Gross income:		1,362,383
Production Costs:		
Raw materials	937,895	
Operating costs	282,273	
Depreciation	45,522	
		1,269,690
Income Before Tax:		96,693
Less Federal and State income tax		49,932
Net income		46,761
Cash flow		92,283

TABLE 16.5

Total Investment

Construction costs	$910,534,000
Land cost	18,211,000
Working capital	91,053,000
Total:	$1,019,798,000
Return on original investment	4.6%
Payout time	9.9 years
Discounted cash flow rate of return	5.9%

Investment Cost

1982 investment cost = $684,098,000 (Table 16.1)
Inflation rate estimated at 10% per year
Completion date scheduled for August, 1985
Estimated completed cost = $(684,098,000)(1.10)^3 = \$910,534,000$

Royalty Costs

Unit	BPCD	$/bbl	$/CD
Hydrorefiner	18,040	0.015	270
Cat Reformer	24,930	0.045	1,121
FCC	32,300	0.04	1,292
Alkylation	7,460	0.15	1,119
Hydrocracker	16,930	0.04	677

$(4,481)(365) = \$1,635,500/yr$

TABLE 16.6

Sensitivity Analysis
(in percent)

DCFRR on original basis = 5.9%

	Investment cost		Raw materials cost		Product prices		Operating volume	
	+20	−15	+5	−3	+5	−5	+2	−2
DCFRR	3.9	6.9	2.9	6.9	8.9	1.9	5.9	4.9
Change in DCFRR	−2.0	+1.0	−3.0	+1.0	+3.0	−4.0	0	−1.0

Note: In the above analysis, the annual costs for insurance, local taxes, and maintenance were not changed with varying investment costs. Royalties, and catalyst replacement costs were not changed with operating volumes.

Chemical and Catalysts Costs

	$/CD
Desalter: 200 lb/day X $1.30/lb	260
Hydrorefiner: 18,040 BPD X 0.02	361
Catalytic reformer: 24,930 BPD X 0.10	2,493
FCC: 32,200 BPD X 0.15	4,845
Alkylation: 2,240 lb HF/day X $1.24/lb	2,778
1,490 lb NaOH/day X $0.19/lb	283
Hydrocracker: 16,930 BPD X 0.20	3,386
Hydrogen unit	429
Amine treater: 54 lb/day X $0.95/lb	51
Stretford unit	32
	14,918

$$(14,918)(365) = \$5,445,070/yr$$

Water Makeup Cost

1. Cooling tower makeup = 5%
   Makeup = (0.05)(110,000 gpm) = 5,500 gpm

2. To boiler
   Boiler makeup = 5%
   Total steam produced = 185,000 lb/hr
   Makeup = (185,000)(0.05) = 9,250 lb/hr = 18 gpm

3. Process water = 230 gpm

Total makeup water = 5,748 gpm

Annual water makeup cost:
      (5,748 gpm)(1,440 min/day)(365 days/hr)($0.15/kgal) = $453,172

Power Costs

Power usage = 942,000 kWh/day @ $0.03 per kWh
Annual cost = (942,000)(365)($0.03) = $10,314,900

Fuel

Fuel requirements = 2,350 MMBtu/hr

Fuel gas = 53,930 lb/hr @ 21,100 BTu/lb
  Available hat = 1,138 MMBtu/hr
Heavy fuel oil = 1,680 BPCD = 26,730 lb/hr
  LHV = 16,700 Btu/lb (Maxwell)
  Available heat = 440.4 MMBtu/hr
Total from fuel gas and heavy fuel oil = 1.138 + 440 = 1,578 MMBtu/hr

Balance needed = 2,350 − 1,578 = 772 MMBtu/hr
If propane* is added to fuel gas:
  (29,430 lb/hr)(19,930 Btu/lb) = 578 MMBtu/hr
Balance needed from No. 2 distillate fuel: (772 − 578) = 194 MMBtu/hr 32°API
  No. 2 distillate fuel LHV = 18,310 Btu/lb = 132,100 Btu/gal†
  LHV = 5,548,200 Btu/bbl

Number of barrels of No. 2 fuel oil required = 35.0 bbl/hr
                                             = 840 BPCD
              No. 2 fuel oil available for sale = 15,560 − 840
                                             = 14,720 BPCD

Insurance Costs

Average of 0.5% of plant investment per year:
  ($910,534,000)(.005) = $4,553,000

Local Taxes

Average of 1% of plant investment per year:
  ($910,534,000(0.01)) = $9,105,000

Maintenance Costs

Average of 4.5% of plant investment per year, including material and
labor:
  ($910,534,000)(0.045) = $40,974,000

Miscellaneous Supplies Costs

Average of 0.15% of plant investment per year:
  ($910,534,000)(0.0015) = $1,366,000

PROBLEMS

1. Determine the discounted cash flow rate of return, return on investment,
    and payout time for a refinery with the following specifications:

    Construction cost = $210,000,000
    Working capital = $21,000,000
    Land cost = $4,200,000
    Annual raw materials cost = $283,000,000
    Annual gross income = $433,000,000
    Annual operating costs = $31,300,000

---

*19,930 Btu/lb C_3
†See Appendix B, Table B.3.

Assume a three-year construction time, a 20-year useful life with straight-line depreciation, and Federal and state income taxes of 48% and 8% respectively. Dismantling costs will equal salvage value.

2. For the conditions listed in problem 1, what would be the effect of the DCFRR if the construction took (a) four years? (b) two years?

3. What would be the DCFRR, payout time and return on investment for the refinery listed in problem 1, if the refinery was operated at 95% capacity? Operating costs will be decreased by 1%.

4. Using the specifications of problem 1, make a sensitivity analysis as described in Section 14.1. What factors will be of greatest concern with respect to accurate forecasting?

5. If the declining balance method of depreciation is used instead of the straight-line method for the refinery in problem 1, what would the rates of return be during the first ten years of operation?

NOTES

1. R. H. Perry, C. H. Chilton, and S. D. Kirkpatrick, *Chemical Engineers' Handbook*, 4th ed. (McGraw-Hill Book Company, New York, 1963), pp. 26-34.
2. W. H. Dickinson, AIEE Trans., paper 60-55, pp. 110-124, April, 1960.
3. J. Happel, Chem. Eng. Progr. *51*(12), 533-539 (1955).
4. F. J. Stermole, *Economic Evaluation and Investment Decision Methods*, 3rd ed. (Investment Evaluation Corp., Golden, Colorado, 1982).
5. J. B. Weaver and R. J. Reilly, Chem. Eng. Progr. *52*(10), 405-412 (1956).
6. R. H. Congelliere, Chem. Eng. *77*(25), 109-112 (1970).
7. F. C. Helen and J. H. Black, *Cost and Optimization Engineering*, 2nd ed. (McGraw-Hill Book Company, New York, 1983).

Appendix A
DEFINITIONS OF REFINING TERMS

ACID HEAT—A test which is indicative of unsaturated components in petroleum distillates. The test measures the amount of reaction of unsaturated hydrocarbons with sulfuric acid (H_2SO_4).

ALKYLATE—The product of an alkylation process.

ALKYLATE BOTTOMS—A thick, dark-brown oil containing high molecular-weight polymerization products of alkylation reactions.

ALKYLATION—A polymerization process uniting olefins and isoparaffins; particularly, the reacting of butylene and isobutane using sulfuric or hydrofluoric acid as a catalyst to produce a high-octane, low-sensitivity blend-agent for gasoline.

ALUMINUM-CHLORIDE TREATING—A quality improvement process for steam-cracked naphthas using aluminum chloride ($AlCl_3$) as a catalyst. The process improves the color and odor of the naphtha by the polymerization of undesirable olefins into resins. The process is also used when production of resins is desirable.

ANILINE POINT—The minimum temperature for complete miscibility of equal volumes of aniline and the test sample. The test is considered an indication of the paraffinicity of the sample. The aniline point is also used as a classification of the ignition quality of diesel fuels.

API GRAVITY—An arbitrary gravity scale defined as:

$$°API = (141.5/\text{Specific gravity } 60\text{-}60 \text{ F}) - 131.5$$

This scale allows representation of the specific gravity of oils, which on the 60/60°F scale varies only over a range of 0.776 by a scale which ranges from less than 0 (heavy residual oil) to 340 (methane).

ASTM DISTILLATION—A standardized laboratory batch distillation for naphthas and middle distillates carried out at atmospheric pressure without fractionation.

B-B—Butane-butylene fraction.

BFOE—Barrels fuel oil equivalent based on net heating value (LHV) of 6,050,000 Btu per BFOE.

BARREL—42 gallons.

BARRELS PER CALENDAR DAY (BPCD)—Average flow rates based on operating 365 days per year.

BARRELS PER STREAM DAY (BPSD)—Flow rates based on actual on-stream time of a unit or group of units. This notation equals barrels per calendar day divided by the service factor.

BATTERY LIMITS (BL)—The periphery of the area surrounding any process unit which includes the equipment for that particular process.

BITUMEN—That portion of petroleum, asphalt, and tar products which will dissolve completely in carbon disulfide (CS_2). This property permits a complete separation from foreign products not soluble in carbon disulfide.

BLENDING—One of the final operations in refining, in which two or more different components are mixed together to obtain the desired range of properties in the final product.

BLENDING OCTANE NUMBER—When blended into gasoline in relatively small quantities, high octane materials behave as though they had octane numbers higher than shown by laboratory tests on the pure material. The effective octane number of the material in the blend is known as the blending octane number.

BLOCKED OPERATION—Operation of a unit, e.g., a pipestill, under periodic change of feed (one charge stock is processed at a time rather than mixing charge stocks) or internal conditions in order to obtain a required range of raw products. Blocked operation is demanded by critical specifications of various finished products. This frequently results in a more efficient operation because each charge stock can be processed at its optimum operating conditions.

BRIGHT STOCK—Heavy lube oils (frequently the vacuum still bottoms) from which asphaltic compounds, aromatics, and waxy paraffins have been removed. Bright stock is one of the feeds to a lube oil blending plant.

BROMINE INDEX—Measure of amount of bromine-reactive material in a sample; ASTM D-2710.

BROMINE NUMBER—A test which indicates the degree of unsaturation in the sample (olefins and diolefins); ASTM D-1159.

BOTTOMS—In general, the higher-boiling residue which is removed from the bottom of a fractionating tower.

CABP—Cubic average boiling point.

CAFFEINE NUMBER—A value related to the amount of carcinogenic compounds (high molecular-weight aromatics referred to as tars) in an oil.

CATALYST—A substance that assists a chemical reaction to take place but which is not itself chemically changed as a result.

CATALYST/OIL RATIO (C/O)—The weight of circulating catalyst fed to the reactor of a fluid-bed catalytic cracking unit divided by the weight of hydrocarbons charged during the same interval.

CATALYTIC CYCLE STOCK—That portion of a catalytic cracker reactor effluent which is not converted to naphtha and lighter products. This material, generally $340°F^+$ ($170°C^+$), may either be completely or partially recycled. In the latter case the remainder will be blended to products or processed further.

CETANE NUMBER—The percentage of pure cetane in a blend of cetane and alpha-methyl-naphthalene which matches the ignition quality of a diesel fuel sample. This quality, specified for middle distillate fuels, is synonymous with the octane number of gasolines.

CFR—Combined feed ratio. Ratio of total feed (including recycle) to fresh feed.

CHARACTERIZATION FACTOR—An index of feed quality, also useful for correlating data on physical properties. The Watson (UOP) characterization factor is defined as the cube root of the mean average boiling point in degrees Rankine divided by the specific gravity.

CLAY TREATING—An elevated temperature and pressure process usually applied to thermally cracked naphthas to improve stability and color. The stability is increased by the adsorption and polymerization of reactive diolefins in the cracked naphtha. Clay treating is used for treating jet fuel to remove surface-active agents which adversely affect the Water Separator Index specifications.

CLOUD POINT—The temperature at which solidifiable compounds present in the sample begin to crystallize or separate from the solution under a method of prescribed chilling. Cloud point is a typical specification of middle distillate fuels; ASTM D-2500.

CONRADSON CARBON—A test used to determine the amount of carbon residue left after the evaporation and pyrolysis of an oil under specified conditions. Expressed as weight %; ASTM D-189.

CORRELATION INDEX—The U.S. Bureau of Mines factor for evaluating individual fractions from crude oil. The CI scale is based upon straight-chain hydrocarbons having a CI value of 0 and benzene having a value of 100.

CRACKING—The breaking down of higher molecular-weight hydrocarbons to lighter components by the application of heat. Cracking in the presence of a suitable catalyst produces an improvement in yield and quality over simple thermal cracking.

CRUDE ASSAY DISTILLATION—See 15/5 distillation.

CUT—That portion of crude oil boiling within certain temperature limits. Usually the limits are on a crude assay true boiling point basis.

CUT POINT—A temperature limit of a cut, usually on a true boiling point basis although ASTM distillation cut points are not uncommon.

DIESEL INDEX (DI)—A measure of the ignition quality of a diesel fuel. Diesel index is defined as:

$$DI = (°API \times Aniline\ Point)/100$$

The higher the diesel index, the more satisfactory the ignition quality of the fuel. By means of correlations unique to each crude and manufacturing process, this quality can be used to predict cetane number (if no standardized test for the latter is available).

DOCTOR TEST—A method for determining the presence of mercaptan sulfur in petroleum products. This test is used for products in which a "sweet" odor is desirable for commercial reasons, especially naphthas; ASTM D-484.

DRY GAS—All C_1 to C_3 material, whether associated with a crude or produced as a byproduct of refinery processing. Convention often includes hydrogen in dry-gas yields.

DEWAXING—The removal of wax from lubricating oils, either by chilling and filtering, solvent extraction, or selective hydrocracking.

END POINT—Upper temperature limit of a distillation.

ENDOTHERMIC REACTION—A reaction in which heat must be added to maintain reactants and products at a constant temperature.

EXOTHERMIC REACTION—A reaction in which heat is evolved. Alkylation, polymerization, and hydrogenation reactions are exothermic.

FBP—The final boiling point of a cut, usually on an ASTM distillation basis.

FOE—Fuel oil equivalent [6.05×10^6 Bbu (LHV)].

FVT—The final vapor temperature of a cut. Boiling ranges expressed in this manner are usually on a crude assay true boiling point basis.

FIFTEEN-FIVE (15/5) DISTILLATION—A laboratory batch distillation performed in a fifteen theoretical plate fractionating column with a five-to-one reflux ratio. A good fractionation results in accurate boiling temperatures. For this reason, this distillation is referred to as the true boiling point (TBP) distillation. This distillation corresponds very closely to the type of fractionation obtained in a refinery.

FIXED CARBON—The organic portion of the residual coke obtained on the evaporation to dryness of hydrocarbon products in the absence of air.

FLASH POINT—The temperature to which a product must be heated under prescribed conditions to release sufficient vapor to form a mixture with air that can be readily ignited. Flash point is used generally as an indication of the fire and explosion potential of a product; ASTM D-56, D-92, D-93, E-134, D-1310.

FLUX—The addition of a small percentage of a material to a product in order to meet some specification of the final blend.

FOOTS OIL—Oil and low melting materials removed from slack wax to produce a finished wax.

FREE CARBON—The organic material in tars which is in soluble in carbon disulfide.

FUEL OIL EQUIVALENT (FOE)—The heating value of a standard barrel of fuel oil, equal to 6.05×10^6 Btu (LHV). On a yield chart, dry gas and refinery fuel gas are usually expressed in FOE barrels.

GAS OIL—The material boiling within the general range of 330 to 750°F FVT (166 to 400°C). This range usually includes kerosine, diesel fuel, heating oils, and light fuel oils. Actual initial and final cut points are determined by the specifications of the desired products.

GHV—Gross heating value of fuels. The heat produced by complete oxidation of materials at 60°F (25°C) to carbon dioxide and liquid water at 60°F (25°C).

HEART CUT RECYCLE—That unconverted portion of the catalytically cracked material which is recycled to the catalytic cracker. This recycle is usually in the boiling range of the feed and, by definition, contains no bottoms. Recycle allows less severe operation and suppresses the further cracking of desirable products.

HEMPEL DISTILLATION—U.S. Bureau of Mines (now Department of Energy) Routine Method Distillation. Results are frequently used interchangeably with TBP Distillation.

IBP—Initial boiling point of a cut, usually on an ASTM basis.

IVT—Initial vaporization temperature of a cut, usually based on a crude assay distillation.

ISOMERATE—The product of an isomerization process.

ISOMERIZATION—The rearrangement of straight-chain hydrocarbon molecules to form branched-chain products. Pentanes and hexanes, which are difficult to reform, are isomerized using aluminum chloride or precious metal catalysts to form gasoline-blending components of fairly high octane value. Normal butane may be isomerized to provide a portion of the isobutane feed needed for alkylation processes.

KEROSINE—A middle-distillate product composed of material of 300 to 550°F boiling range. The exact cut is determined by various specifications of the finished kerosine.

LAMP SULFUR—The total amount of sulfur present per unit of liquid product. The analysis is made by burning a sample so that the sulfur content will be converted to sulfur dioxide, which can be quantitatively measured. Lamp sulfur is a critical specification of all motor, tractor, and burner fuels; ASTM D-1266.

LEAD SUSCEPTIBILITY—The variation of the octane number of a gasoline as a function of the tetraethyl lead content.

LHSV—Liquid hour space velocity; volume of feed per hour per volume of catalyst.

LHV—Lower heating value of fuels (net heat of combustion). The heat produced by complete oxidation of materials at 60°F (25°C) to carbon dioxide and water vapor at 60°F (25°C).

LIGHT ENDS—Hydrocarbon fractions in the butane and lighter boiling range.

LNG—Liquefied natural gas or methane.

LIQUEFIED PETROLEUM GAS (LPG)—Liquefied light ends gases used for home heating and cooking. This gas is usually 95% propane, the remainder being split between ethane and butane. It can be any one of several specified mixtures of propane and butane which are sold as liquids under pressure at ambient temperatures.

MABP—Molal average boiling point:

$$MABP = \sum_{i=1}^{n} X_i T_{bi}$$

MeABP—Mean average boiling point:

MeABP = (MABP + CABP)/2

MID-BOILING POINT—That temperature, usually based on a crude assay distillation, at which one-half of the material of a cut has been vaporized.

MID-PERCENT POINT—The vapor temperature at which one-half of the material of a cut has been vaporized. Mid-percent point is often used to characterize a cut in place of temperature limits.

MIDDLE DISTILLATES—Atmospheric pipestill cuts boiling in the range of 300 to 700°F vaporization temperature. The exact cut is determined by the specifications of the product.

MOTOR OCTANE NUMBER (MON, ASTM ON, F-2)—The percentage by volume of isooctane in a mixture of isooctane and n-heptane that knocks with the same intensity as the fuel being tested. A standardized test engine operating under standardized conditions (900 rpm) is used. This test approximates cruising conditions of an automobile; ASTM D-2723.

NAPHTHA—A pipestill cut in the range of C_5 to 420°F. Naphthas are subdivided, according to the actual pipestill cuts, into light, intermediate, heavy, and very heavy virgin naphthas. A typical pipestill operation would be:

C_5-160°F - light virgin naphtha

160-280°F - intermediate virgin naphtha

280-330°F - Heavy virgin naphtha

330-420°F - very heavy virgin naphtha

Naphthas, the major constituents of gasoline, generally need processing to make suitable quality gasoline.

NEUTRALIZATION NUMBER—The quantity of acid or base which is required to neutralize all basic or acidic components present in a specified quantity of sample. This is a measure of the amount of oxidation of a product in storage or in service; ASTM D-664, D-974.

NEUTRAL OILS—Pale or red oils of low viscosity produced from dewaxed oils by distillation.

OLEFIN SPACE VELOCITY—Volume of olefin charged per hour to an alkylation reactor divided by the volume of acid in the reactor.

PERFORMANCE RATING—A method of expressing the quality of a high-octane gasoline relative to isooctane. This rating is used for fuels which are of better quality than isooctane.

PETROLATUM—Microcrystalline wax.

PIPESTILL—A heater or furnace containing tubes through which oil is pumped while being heated or vaporized. Pipestills are fired with waste gas, natural gas, or heavy oils and, by providing for rapid heating under conditions of high pressure and temperature, are useful for thermal cracking as well as distillation operations.

POLYMERIZATION—The combination of two or more unsaturated molecules to form a molecule of higher molecular weight. Propylenes and butylenes are the primary feed material for refinery polymerization processes which use solid or liquid phosphoric acid catalysts.

POSTED OCTANE NUMBER—Arithmetic average of research octane number and motor octane number.

POUR POINT—The lowest temperature at which a petroleum oil will flow or pour when it is chilled without disturbance at a controlled rate. Pour point is a critical specification of middle distillate products used in cold climates; ASTM D-99.

POUR BLENDING INDEX (PBI)—An empirical quantity related to pour point, which allows volumetric blending of pour points of various blend components. This method of blending is most accurate for blending of similar fractions of the same crude.

RAFFINATE—The residue recovered from an extraction process. An example is the furfural extraction of aromatics from high molecular weight distillates. The raffinate is relatively free of aromatics which have poor viscosity-temperature characteristics (low VI).

RAMSBOTTOM CARBON—Recommended to replace Conradson Carbon; ASTM D-524. Carbon residue expressed in weight %.

RECONSTITUTED CRUDE—A crude to which has been added a specific crude fraction for the purpose of meeting some product volume unattainable with the original crude.

REDUCED CRUDE—A crude whose API gravity has been reduced by distillation of the lighter lower-boiling constituents.

REFORMATE—A reformed naphtha which is upgraded in octane by means of catalytic or thermal reforming.

REFORMING—The conversion of naphtha fractions to products of higher octane value. Thermal reforming is essentially a light cracking process applied to heavy naphthas to produce increased yields of hydrocarbons in the gasoline boiling range. Catalytic reforming is applied to various straight-run and cracked naphtha fractions and consists primarily of dehydrogenation of naphthenes to aromatics. A number of catalysts, including platinum and platinum-rhenium, are used. A high partial pressure of hydrogen is maintained to prevent the formation of excessive coke.

REID VAPOR PRESSURE (RVP)—The vapor pressure at 100°F of a product determined in a volume of air four times the liquid volume. Reid vapor pressure is an indication of the vapor-lock tendency of a motor gasoline as well as explosion and evaporation hazards; ASTM D-323. It is usually expressed in psig.

RESEARCH OCTANE NUMBER (RON, CFRR, F-1)—The percentage by volume of isooctane in a blend of isooctane and n-heptane that knocks with the same intensity as the fuel being tested. A standardized test engine operating under standardized conditions (600 rpm) is used. Results are comparable to those obtained in an automobile engine operated at low speed; ASTM D-2722.

ROAD OCTANE NUMBER—The percentage by volume of isooctane which would be required in a blend of isooctane and n-heptane to give incipient knock in an automobile engine operating under the same conditions of engine load, speed, and degree of spark advance as that of the fuel being tested.

SAPONIFICATION NUMBER—An indication of the fraction of fat or fatty oils in a given product. The number of milligrams of potassium hydroxide required to saponify one gram of the sample.

SCF—Volume of gas as "standard cubic feet". Standard conditions in petroleul and natural gas usage refer to a pressure base of 14.696 psia and a temperature base of 60°F (15°C).

SELECTIVITY—The difference between the research octane number and the motor octane number of a given gasoline. Alkylate is an excellent low-sensitivity and reformate a high-sensitivity gasoline component.

SERVICE FACTOR—A quantity which relates the actual on-stream time of a process unit to the total time available for use. Service factors include both expected and unexpected unit shutdowns.

SEVERITY—The degree of intensity of the operating conditions of a process unit. Severity may be indicated by clear research octane number of the product (reformer), percent disappearance of the feed (catalytic cracking), or operating conditions alone.

SLACK WAX—Wax produced in the dewaxing of lube oil base stocks. This wax still contains some oil and must be deoiled to produce a finish wax product.

SMOKE POINT—A test measuring the burning quality of jet fuels, kerosine, and illuminating oils. It is defined as the height of the flame in millimeters beyond which smoking takes place; ASTM D-1322.

SOUR OR SWEET CRUDE—A rather general method for classifying crudes according to sulfur content. Various definitions are available:

Sour Crude—A crude which contains sulfur in amounts greater than 0.5 to 1.0 wt%, or which contains 0.05 cu ft or more of hydrogen sulfide (H_2S) per 100 gal, except West Texas crude, which is always considered sour regardless of its hydrogen sulfide content. Arabian crudes are high-sulfur crudes which are not always considered sour because they do not contain highly active sulfur compounds.

Sweet Crude—As evident from the above definitions, a sweet crude contains little or no dissolved hydrogen sulfide and relatively small amounts of mercaptans and other sulfur compounds.

SPACE VELOCITY—The volume (or weight) of gas and/or liquid passing through a given catalyst or reactor space per unit time, divided by the volume (or weight) of catalyst through which the fluid passes. High space velocities correspond to short reaction times. See LHSV and WHSV.

STRAIGHT-RUN GASOLINE—An uncracked gasoline fraction distilled from crude. Straight-run gasolines contain primarily paraffinic hydrocarbons and have lower octane values than cracked gasolines from the same crude feedstock.

SYNTHETIC CRUDE—Wide boiling-range product of catalytic cracking or some other chemical structure change operation.

SWEETENING—The removal or conversion to innocuous substances of sulfur compounds in a petroleum product by any of a number of processes (doctor treating, caustic and water washing, etc.).

TA—Total alkylate.

TAIL GAS—Light gases (C_1 to C_3 and H_2) produced as byproducts of refinery processing.

TBP DISTILLATION—See fifteen-five (15/5) distillation.

TETRAETHYL LEAD—An antiknock additive for gasoline.

THEORETICAL PLATE—A theoretical contacting unit useful in distillation, absorption, and extraction calculations. Vapors and liquid leaving any such unit are required to be in equilibrium under the cnditions of temperature and pressure which apply. An actual fractionator tray or plate is generally less effective than a theoretical plate. The ratio of the number of theoretical plates required to perform a given distillation separation to the number of actual plates used gives the overall tray efficiency of the fractionator.

TOPPING—Removal by distillation of the light products from crude oil, leaving in the still bottoms all of the heavier constituents.

TREAT GAS—Light gases, usually high in hydrogen content, which are required for refinery hydrotreating processes such as hydrodesulfurization. The treat gas for hydrodesulfurization is usually the tail gas from catalytic reforming.

TUBE STILL—See pipestill.

U.S. BUREAU OF MINES ROUTINE METHOD DISTILLATION—See Hempel distillation.

VAPOR LOCK INDEX—A measure of the tendency of a gasoline to generate excessive vapors in the fuel line thus causing displacement of liquid fuel and subsequent interruption of normal engine operation. The vapor lock index generally is related to RVP and % distilled at 158°F (70°C).

VIRGIN STOCKS—Petroleum oils which have not been cracked or otherwise subjected to any treatment which would produce appreciable chemical change in their components.

VISCOSITY—The property of liquids under flow conditions which causes them to resist instantaneous change of shape or instantaneous rearrangement of their parts due to internal friction. Viscosity is generally measured as the number of seconds, at a definite temperature, required for a standard quantity of oil to flow through a standard apparatus. Common viscosity scales in use are Saybolt Universal, Saybolt Furol, and Kinematic (Stokes).

VOLATILITY FACTOR—An empirical quantity which indicates good gasoline performance with respect to volatility. It involves actual automobile operating conditions and climatic factors. The volatility factor is generally defined as a function of RVP, % of 158°F (70°C), and % off 212°F (100°C). This factor is an attempt to predict the vapor lock tendency of a gasoline.

WABP—Weight average boiling point.

$$WABP = \sum_{i=1}^{n} X_{wi} T_{bi}$$

where

X_{wi} = weight fraction of component i.

WHSV—Weight hour space velocity; weight of feed per hour per weight of catalyst.

WICK CHAR—A test used as an indication of the burning quality of a kerosine or illuminating oil. It is defined as the weight of deposits remaining on the wick after a specified quantity of sample is burned.

Appendix B
PHYSICAL PROPERTIES

TABLE B.1
Density Conversion Table

Specific gravity 60/60°F	°API	Density in vacuo lb/bbl	lb/gal	lb/hr* from bbl/day	Specific gravity 60/60°F	°API	Density in vacuo lb/bbl	lb/gal	lb/hr* from bbl/day
1.165	−10.0	407.8	9.71	16.99	1.092	−2.0	382.6	9.11	15.94
1.163	−9.8	407.1	9.69	16.95	1.090	−1.8	382.0	9.09	15.92
1.161	−9.6	406.5	9.68	16.94	1.089	−1.6	381.4	9.08	15.89
1.159	−9.4	405.8	9.66	16.91	1.087	−1.4	380.8	9.07	15.87
1.157	−9.2	405.1	9.65	16.88	1.085	−1.2	380.3	9.05	15.85
1.155	−9.0	404.5	9.63	16.85	1.084	−1.0	379.7	9.04	15.82
1.153	−8.8	403.8	9.61	16.82	1.082	−0.8	379.1	9.03	15.80
1.151	−8.6	403.2	9.60	16.80	1.080	−0.6	378.5	9.01	15.77
1.149	−8.4	402.5	9.58	16.77	1.079	−0.4	377.9	9.00	15.75
1.147	−8.2	401.9	9.57	16.74	1.077	−0.2	377.4	8.98	15.72
1.145	−8.0	401.2	9.55	16.72	1.076	0.0	376.8	8.97	15.70
1.143	−7.8	400.6	9.54	16.69	1.074	.2	376.2	8.96	15.67
1.142	−7.6	399.9	9.52	16.66	1.073	.4	375.6	8.94	15.65
1.140	−7.4	399.3	9.51	16.64	1.071	.6	375.1	8.93	15.63
1.138	−7.2	398.6	9.49	16.61	1.070	.8	374.5	8.92	15.60
1.136	−7.0	398.0	9.48	16.58	1.068	1.0	373.9	8.90	15.53
1.134	−6.8	397.3	9.46	16.55	1.066	.2	373.4	8.89	15.56
1.132	−6.6	396.7	9.45	16.53	1.065	.4	372.8	8.88	15.53
1.131	−6.4	396.1	9.43	16.50	1.063	.6	372.3	8.86	15.51
1.129	−6.2	395.4	9.42	16.47	1.062	.8	371.7	8.85	15.49
1.127	−6.0	394.8	9.40	16.45	1.060	2.0	371.1	8.84	15.46
1.125	−5.8	394.2	9.39	16.42	1.053	.2	370.6	8.82	15.44
1.123	−5.6	393.6	9.37	16.40	1.057	.4	370.0	8.81	15.42
1.122	−5.4	392.9	9.36	16.37	1.055	.6	369.5	8.80	15.40
1.120	−5.2	392.3	9.34	16.35	1.054	.8	368.9	8.78	15.37
1.118	−5.0	391.7	9.33	16.32	1.052	3.0	368.4	8.77	15.35
1.116	−4.8	391.1	9.31	16.30	1.051	.2	367.8	8.76	15.32
1.115	−4.6	390.5	9.30	16.27	1.049	.4	367.3	8.75	15.30
1.113	−4.4	389.8	9.23	16.24	1.047	.6	366.8	8.73	15.28
1.111	−4.2	389.2	9.27	16.22	1.046	.8	366.2	8.72	15.26
1.109	−4.0	388.6	9.25	16.19	1.044	4.0	365.7	8.71	15.24
1.108	−3.8	388.0	9.24	16.17	1.043	.2	365.1	8.69	15.21
1.106	−3.6	387.4	9.22	16.14	1.041	.4	364.6	8.68	15.19
1.104	−3.4	386.8	9.21	16.12	1.040	.6	364.0	8.67	15.17
1.102	−3.2	386.2	9.19	16.09	1.038	.8	363.5	8.66	15.15
1.101	−3.0	385.6	9.18	16.07	1.037	5.0	363.0	8.64	15.12
1.099	−2.8	385.0	9.16	16.04	1.035	.2	362.4	8.63	15.10
1.097	−2.6	384.4	9.15	16.02	1.034	.4	361.9	8.62	15.08
1.096	−2.4	383.8	9.14	15.99	1.032	.6	361.4	8.60	15.06
1.094	−2.2	383.2	9.12	15.97	1.031	.8	360.9	8.59	15.04

TABLE B.1 (Continued)

Specific gravity 60/60° F	°API	Density in vacuo lb/bbl	lb/gal	lb/hr* from bbl/day	Specific gravity 60/60° F	°API	Density in vacuo lb/bbl	lb/gal	lb/hr* from bbl/day
1.029	6.0	360.3	8.58	15.01	0.973	14.0	340.5	8.11	14.19
1.028	.2	359.8	8.57	14.99	0.971	.2	340.1	8.10	14.17
1.026	.4	359.3	8.55	14.97	0.970	.4	339.6	8.09	14.15
1.025	.6	358.8	8.54	14.95	0.969	.6	339.1	8.08	14.13
1.023	.8	358.3	8.53	14.93	0.967	.8	338.7	8.06	14.11
1.022	7.0	357.7	8.52	14.90	0.966	15.0	338.2	8.05	14.09
1.020	.2	357.2	8.51	14.88	0.965	.2	337.8	8.04	14.07
1.019	.4	356.7	8.49	14.86	0.963	.4	337.3	8.03	14.05
1.017	.6	356.2	8.48	14.84	0.962	.6	336.8	8.02	14.03
1.016	.8	355.7	8.47	14.82	0.961	.8	336.4	8.01	14.02
1.014	8.0	355.2	8.46	14.80	0.959	16.0	335.9	8.00	14.00
1.013	.2	354.7	8.44	14.78	0.958	.2	335.5	7.99	13.98
1.011	.4	354.2	8.43	14.76	0.957	.4	335.0	7.98	13.96
1.010	.6	353.7	8.42	14.74	0.955	.6	334.6	7.96	13.94
1.009	.8	353.2	8.41	14.72	0.954	.8	334.1	7.95	13.92
1.007	9.0	352.7	8.40	14.70	0.953	17.0	333.7	7.94	13.90
1.006	.2	352.2	8.38	14.67	0.952	.2	333.2	7.93	13.88
1.004	.4	351.7	8.37	14.65	0.950	.4	332.8	7.92	13.87
1.003	.6	351.2	8.36	14.63	0.949	.6	332.3	7.91	13.85
1.001	.8	350.7	8.35	14.61	0.948	.8	331.9	7.90	13.83
1.000	10.0	350.2	8.34	14.59	0.947	18.0	331.4	7.89	13.81
0.999	10.2	349.7	8.33	14.57	0.945	.2	331.0	7.88	13.79
0.997	10.4	349.2	8.31	14.55	0.944	.4	330.5	7.87	13.77
0.996	10.6	348.7	8.30	14.53	0.943	.6	330.1	7.86	13.75
0.994	10.8	348.2	8.29	14.51	0.942	.8	329.7	7.85	13.74
0.993	11.0	347.7	8.28	14.49	0.940	19.0	329.2	7.84	13.72
0.992	.2	347.2	8.27	14.47	0.939	.2	328.0	7.83	13.70
0.990	.4	346.7	8.26	14.45	0.938	.4	328.4	7.82	13.68
0.989	.6	346.2	8.24	14.43	0.937	.6	327.9	7.81	13.66
0.987	.8	345.8	8.23	14.41	0.935	.8	327.5	7.80	13.65
0.986	12.0	345.3	8.22	14.39	0.934	20.0	327.1	7.79	13.63
0.985	.2	344.8	8.21	14.37	0.933	.2	326.6	7.78	13.61
0.983	.4	344.3	8.20	14.35	0.932	.4	326.2	7.77	13.59
0.982	.6	343.8	8.19	14.33	0.930	.6	325.8	7.76	13.57
0.981	.8	343.4	8.18	14.31	0.929	.8	325.3	7.75	13.55
0.979	13.0	342.9	8.16	14.29	0.928	21.0	324.9	7.74	13.54
0.978	.2	342.4	8.15	14.27	0.927	.2	324.5	7.73	13.52
0.977	.4	341.9	8.14	14.25	0.925	.4	324.0	7.72	13.50
0.975	.6	341.5	8.13	14.23	0.924	.6	323.6	7.71	13.48
0.974	.8	341.0	8.12	14.21	0.923	.8	323.2	7.70	13.47

TABLE B.1 (Continued)

Specific gravity 60/60°F	°API	Density in vacuo lb/bbl	lb/gal	lb/hr* from bbl/day	Specific gravity 60/60°F	°API	Density in vacuo lb/bbl	lb/gal	lb/hr* from bbl/day
0.922	22.0	322.8	7.69	13.45	0.876	30.0	306.8	7.30	12.78
0.921	.2	322.4	7.68	13.43	0.875	.2	306.4	7.30	12.77
0.919	.4	321.9	7.67	13.41	0.874	.4	306.0	7.29	12.75
0.918	.6	321.5	7.66	13.40	0.873	.6	305.7	7.28	12.74
0.917	.8	321.1	7.65	13.38	0.872	.8	305.3	7.27	12.72
0.916	23.0	320.7	7.64	13.36	0.871	31.0	304.9	7.26	12.70
0.915	.2	320.3	7.63	13.35	0.870	.2	304.5	7.25	12.69
0.914	.4	319.9	7.62	13.33	0.869	.4	304.2	7.24	12.67
0.912	.6	319.5	7.61	13.31	0.868	.6	303.8	7.23	12.66
0.911	.8	319.0	7.60	13.29	0.867	.8	303.4	7.22	12.64
0.910	24.0	318.6	7.59	13.27	0.865	32.0	303.0	7.21	12.62
0.909	.2	318.2	7.58	13.26	0.864	.2	302.7	7.20	12.61
0.908	.4	317.8	7.57	13.24	0.863	.4	302.3	7.19	12.60
0.907	.6	317.4	7.56	13.22	0.862	.6	301.9	7.19	12.58
0.905	.8	317.0	7.55	13.21	0.861	.8	301.6	7.18	12.57
0.904	25.0	316.6	7.54	13.19	0.860	33.0	301.2	7.17	12.55
0.903	.2	316.2	7.53	13.17	0.859	.2	300.8	7.16	12.53
0.902	.4	315.8	7.52	13.16	0.858	.4	300.5	7.15	12.52
0.901	.6	315.4	7.51	13.14	0.857	.6	300.1	7.14	12.50
0.900	.8	315.0	7.50	13.12	0.856	.8	299.7	7.14	12.49
0.898	26.0	314.6	7.49	13.11	0.855	34.0	299.4	7.13	12.47
0.897	.2	314.2	7.48	13.09	0.854	.2	299.0	7.12	12.46
0.896	.4	313.8	7.47	13.07	0.853	.4	298.7	7.11	12.45
0.895	.6	313.4	7.46	13.06	0.852	.6	298.3	7.10	12.43
0.894	.8	313.0	7.45	13.04	0.851	.8	297.9	7.09	12.41
0.893	27.0	312.6	7.44	13.02	0.850	35.0	297.6	7.09	12.40
0.892	.2	312.2	7.43	13.01	0.849	.2	297.2	7.08	12.38
0.891	.4	311.8	7.42	12.99	0.848	.4	296.9	7.07	12.37
0.889	.6	311.4	7.41	12.97	0.847	.6	296.5	7.06	12.35
0.888	.8	311.0	7.40	12.96	0.846	.8	296.2	7.05	12.34
0.887	28.0	310.6	7.40	12.95	0.845	36.0	295.8	7.04	12.32
0.886	.2	310.3	7.39	12.93	0.844	.2	295.4	7.04	12.31
0.885	.4	309.9	7.38	12.91	0.843	.4	295.1	7.03	12.30
0.884	.6	309.5	7.37	12.90	0.842	.6	294.8	7.02	12.28
0.883	.8	309.1	7.36	12.88	0.841	.8	294.4	7.01	12.27
0.882	29.0	308.7	7.35	12.86	0.840	37.0	294.0	7.00	12.25
0.881	.2	308.3	7.34	12.85	0.839	.2	293.7	6.99	12.24
0.879	.4	307.9	7.33	12.83	0.838	.4	293.4	6.99	12.21
0.878	.6	307.6	7.32	12.82	0.837	.6	293.0	6.98	12.21
0.877	.8	307.2	7.31	12.80	0.836	.8	292.7	6.97	12.20

TABLE B.1 (Continued)

Specific gravity 60/60°F	°API	Density in vacuo lb/bbl	lb/gal	lb/hr* from bbl/day	Specific gravity 60/60°F	°API	Density in vacuo lb/bbl	lb/gal	lb/hr* from bbl/day
0.835	38.0	292.3	6.96	12.18	0.797	46.0	279.1	6.64	11.63
0.834	.2	292.0	6.95	12.17	0.796	.2	278.3	6.64	11.62
0.833	.4	291.6	6.94	12.15	0.795	.4	278.5	6.63	11.60
0.832	.6	291.3	6.94	12.14	0.795	.6	278.2	6.63	11.59
0.831	.8	291.0	6.93	12.12	0.794	.8	277.9	6.62	11.58
0.830	39.0	290.6	6.92	12.11	0.793	47.0	277.6	6.61	11.57
0.829	.2	290.3	6.91	12.10	0.792	.2	277.3	6.60	11.55
0.828	.4	290.0	6.90	12.08	0.791	.4	277.0	6.59	11.54
0.827	.6	289.6	6.89	12.07	0.790	.6	276.7	6.59	11.53
0.826	.8	289.2	6.89	12.05	0.789	.8	276.3	6.58	11.51
0.825	40.0	288.9	6.88	12.04	0.788	48.0	276.0	6.57	11.50
0.824	.2	288.6	6.87	12.02	0.787	.2	275.7	6.56	11.49
0.823	.4	288.2	6.86	12.01	0.787	.4	275.4	6.56	11.47
0.822	.6	287.9	6.85	12.00	0.786	.6	275.1	6.55	11.46
0.821	.8	287.6	6.84	11.93	0.785	.8	274.1	6.54	11.45
0.820	41.0	287.2	6.84	11.97	0.784	49.0	274.5	6.54	11.44
0.819	.2	286.9	6.83	11.95	0.783	.2	274.2	6.53	11.42
0.818	.4	286.6	6.82	11.94	0.782	.4	273.9	6.52	11.41
0.817	.6	286.2	6.81	11.92	0.781	.6	273.6	6.51	11.40
0.817	.8	285.9	6.81	11.91	0.781	.8	273.3	6.51	11.39
0.816	42.0	285.6	6.80	11.90	0.780	50.0	273.0	6.50	11.37
0.815	.2	285.3	6.79	11.89	0.779	.2	272.7	6.49	11.36
0.814	.4	284.9	6.79	11.87	0.778	.4	272.4	6.49	11.35
0.813	.6	284.6	6.78	11.86	0.777	.6	272.1	6.48	11.34
0.812	.8	284.3	6.77	11.85	0.776	.8	271.8	6.47	11.32
0.811	43.0	283.9	6.76	11.83	0.775	51.0	271.5	6.46	11.31
0.810	.2	283.6	6.75	11.82	0.775	.2	271.2	6.46	11.30
0.809	.4	283.3	6.74	11.80	0.774	.4	270.9	6.45	11.29
0.808	.6	283.0	6.74	11.79	0.773	.6	270.6	6.44	11.27
0.807	.8	282.6	6.73	11.77	0.772	.8	270.3	6.44	11.26
0.806	44.0	282.3	6.72	11.76	0.771	52.0	270.0	6.43	11.25
0.805	.2	282.0	6.71	11.75	0.770	.2	269.7	6.42	11.24
0.804	.4	281.7	6.70	11.74	0.769	.4	269.4	6.41	11.22
0.804	.6	281.4	6.70	11.72	0.769	.6	269.1	6.41	11.21
0.803	.8	281.0	6.69	11.71	0.768	.8	268.8	6.40	11.20
0.802	45.0	280.7	6.69	11.70	0.767	53.0	268.5	6.39	11.19
0.801	.2	280.4	6.68	11.68	0.766	.2	268.3	6.39	11.18
0.800	.4	280.1	6.67	11.67	0.765	.4	268.0	6.38	11.17
0.799	.6	279.8	6.66	11.66	0.764	.6	267.7	6.37	11.15
0.798	.8	279.5	6.65	11.65	0.764	.8	267.4	6.37	11.14

TABLE B.1 (Continued)

Specific gravity 60/60°F	°API	Density in vacuo lb/bbl	lb/gal	lb/hr* from bbl/day	Specific gravity 60/60°F	°API	Density in vacuo lb/bbl	lb/gal	lb/hr* from bbl/day
0.763	54.0	267.1	6.36	11.13	0.731	62.0	256.1	6.10	10.67
0.762	.2	266.8	6.35	11.12	0.730	.2	255.8	6.09	10.66
0.761	.4	266.5	6.34	11.10	0.730	.4	255.5	6.08	10.65
0.760	.6	266.2	6.34	11.09	0.729	.6	255.3	6.08	10.64
0.760	.8	265.9	6.33	11.08	0.728	.8	255.0	6.07	10.62
0.759	55.0	265.7	6.33	11.07	0.728	63.0	254.7	6.07	10.61
0.758	.2	265.4	6.32	11.06	0.727	.2	254.5	6.06	10.60
0.757	.4	265.1	6.31	11.05	0.726	.4	254.2	6.05	10.59
0.756	.6	264.8	6.30	11.03	0.725	.6	254.0	6.05	10.58
0.756	.8	264.5	6.30	11.02	0.724	.8	253.7	6.04	10.57
0.755	56.0	264.3	6.29	11.01	0.724	64.0	253.4	6.03	10.56
0.754	.2	264.0	6.29	11.00	0.723	.2	253.2	6.03	10.55
0.753	.4	263.7	6.28	10.99	0.722	.4	252.9	6.02	10.54
0.752	.6	263.4	6.27	10.97	0.722	.6	252.7	6.02	10.53
0.752	.8	263.1	6.27	10.96	0.721	.8	252.4	6.01	10.52
0.751	57.0	262.9	6.26	10.95	0.720	65.0	252.2	6.00	10.51
0.750	.2	262.6	6.26	10.94	0.719	.2	251.9	6.00	10.50
0.749	.4	262.3	6.24	10.93	0.719	.4	251.6	5.99	10.48
0.748	.6	262.0	6.24	10.92	0.718	.6	251.4	5.98	10.47
0.748	.8	261.7	6.23	10.90	0.717	.8	251.1	5.98	10.46
0.747	58.0	261.5	6.23	10.89	0.716	66.0	250.9	5.97	10.45
0.746	.2	261.2	6.22	10.88	0.716	.2	250.6	5.97	10.44
0.745	.4	260.9	6.21	10.87	0.715	.4	250.4	5.96	10.43
0.744	.6	260.6	6.20	10.86	0.714	.6	250.1	5.95	10.42
0.744	.8	260.4	6.20	10.85	0.714	.8	249.9	5.95	10.41
0.743	59.0	260.1	6.19	10.84	0.713	67.0	249.6	5.94	10.40
0.742	.2	259.8	6.19	10.82	0.712	.2	249.4	5.94	10.39
0.741	.4	259.6	6.18	10.81	0.711	.4	249.1	5.93	10.38
0.740	.6	259.3	6.17	10.80	0.711	.6	248.9	5.93	10.37
0.740	.8	259.0	6.17	10.79	0.710	.8	248.6	5.92	10.36
0.739	60.0	258.7	6.16	10.78	0.709	68.0	248.4	5.91	10.35
0.738	.2	258.5	6.15	10.77	0.709	.2	248.1	5.91	10.34
0.737	.4	258.2	6.15	10.76	0.708	.4	247.9	5.90	10.33
0.737	.6	257.9	6.14	10.75	0.707	.6	247.6	5.90	10.32
0.736	.8	257.7	6.14	10.74	0.706	.8	247.4	5.89	10.31
0.735	61.0	257.4	6.13	10.72	0.706	69.0	247.1	5.88	10.30
0.734	.2	257.1	6.12	10.71	0.705	.2	246.9	5.88	10.29
0.734	.4	256.9	6.12	10.70	0.704	.4	246.6	5.87	10.28
0.733	.6	256.6	6.11	10.69	0.704	.6	246.4	5.87	10.27
0.732	.8	256.3	6.10	10.68	0.703	.8	246.1	5.86	10.26

TABLE B.1 (Continued)

Specific gravity 60/60°F	°API	Density in vacuo lb/bbl	lb/gal	lb/hr* from bbl/day	Specific gravity 60/60°F	°API	Density in vacuo lb/bbl	lb/gal	lb/hr* from bbl/day
0.702	70.0	245.9	5.85	10.25	0.646	.5	226.2	5.39	9.42
0.701	.5	245.3	5.84	10.22	0.645	88.0	225.7	5.38	9.40
0.699	71.0	244.7	5.83	10.20	0.643	.5	225.2	5.36	9.38
0.697	.5	244.1	5.81	10.17	0.642	89.0	224.7	5.35	9.36
0.695	72.0	243.5	5.80	10.15	0.640	.5	224.2	5.34	9.34
0.694	.5	242.9	5.78	10.12	0.639	90.0	223.7	5.33	9.32
0.692	73.0	242.3	5.77	10.10	0.637	.5	223.2	5.31	9.30
0.609	.5	241.7	5.75	10.07	0.636	91.0	222.7	5.30	9.28
0.689	74.0	241.1	5.74	10.05	0.635	.5	222.2	5.29	9.26
0.687	.5	240.5	5.73	10.02	0.633	92.0	221.7	5.28	9.24
0.685	75.0	239.9	5.71	10.00	0.632	.5	221.2	5.27	9.22
0.684	.5	239.4	5.70	9.97	0.630	93.0	220.7	5.26	9.20
0.682	76.0	238.8	5.69	9.95	0.629	.5	220.2	5.24	9.18
0.680	.5	238.2	5.67	9.92	0.628	94.0	219.7	5.23	9.16
0.679	77.0	237.6	5.66	9.90	0.626	.5	219.2	5.22	9.14
0.677	.5	237.1	5.64	9.88	0.625	95.0	218/8	5.21	9.12
0.675	78.0	236.5	5.63	9.85	0.623	.5	218.3	5.20	9.10
0.674	.5	235.9	5.62	9.83	0.622	96.0	217.8	5.19	9.03
0.672	79.0	235.4	5.60	9.81	0.621	.5	217.3	5.17	9.06
0.671	.5	234.8	5.59	9.78	0.619	97.0	216.8	5.16	9.04
0.669	80.0	234.3	5.58	9.76	0.618	.5	216.4	5.15	9.02
0.668	.5	233.7	5.56	9.74	0.617	98.0	215.9	5.14	9.00
0.666	81.0	233.2	5.55	9.72	0.615	.5	215.4	5.13	8.98
0.664	.5	232.6	5.54	9.69	0.614	99.0	215.0	5.12	8.96
0.663	82.0	232.1	5.53	9.67	0.613	.5	214.5	5.11	8.94
0.661	.5	231.5	5.51	9.65	0.611	100.0	214.0	5.10	8.92
0.660	83.0	231.0	5.50	9.62	0.610	.5	213.6	5.09	8.90
0.658	.5	230.4	5.49	9.60	0.609	101.0	213.1	5.07	8.83
0.657	84.0	229.9	5.48	9.58	0.607	.5	212.7	5.06	8.86
0.655	.5	229.4	5.46	9.56	0.606	102.0	212.2	5.05	8.84
0.654	85.0	228.9	5.45	9.54	0.605	.5	211.7	5.04	8.82
0.652	.5	228.3	5.44	9.51	0.603	103.0	211.3	5.03	8.80
0.651	86.0	227.8	5.43	9.49	0.602	.5	210.8	5.02	8.78
0.649	.5	227.3	5.41	9.47	0.601	104.0	210.4	5.01	8.77
0.648	87.0	226.8	5.40	9.45	0.600	.5	209.9	5.00	8.75
					0.598	105.0	209.5	4.99	8.73

Source: From ASTM D-1250.

*Multiply barrels/day by the factor in this column corresponding to the API gravity to obtain pounds/hour.

TABLE B.2

Physical Constants of Paraffin Hydrocarbons and Other Components of Natural Gas
[NGPA Publication 2145-75[(1)]]

Component	Notes	Methane	Ethane	Propane	Iso-Butane	N-Butane	Iso-Pentane	N-Pentane
Molecular Weight	*	16.043	30.070	44.097	58.124	58.124	72.151	72.151
Boiling Point @ 14.696 psia, °F		−258.69	−127.48	− 43.67	10.90	31.10	82.12	96.92
Freezing Point @ 14.696 psia, °F		−296.46[d]	−297.89[d]	−305.84[d]	−255.29	−217.05	−255.83	−201.51
Vapor Pressure @ 100°F, psia		(5000)	(800)	190	72.2	51.6	20.44	15.570
Density of Liquid @ 60°F & 14.696 psia								
Specific Gravity @ 60°F/60°F	a,b	0.3[i]	0.3564[h]	0.5077[h]	0.5631[h]	0.5844[h]	0.6247	0.6310
°API	* a,b	340[i]	265.5[h]	147.2[h]	119.8[h]	110.6[h]	95.0	92.7
Lb/gal @ 60°F , wt in vacuum	*	2.5[i]	2.971[h]	4.233[h]	4.695[h]	4.872[h]	5.208	5.261
Lb/gal @ 60°F, wt in air	* c	2.5[i]	2.962[h]	4.223[h]	4.686[h]	4.865[h]	5.199	5.251
Density of Gas @ 60°F & 14.696 psia								
Specific Gravity, Air = 1.00, ideal gas	*	0.5539	1.0382	1.5225	2.0068	2.0068	2.4911	2.4911
Lb/M cu ft, ideal gas	*	42.28	79.24	116.20	153.16	153.16	190.13	190.13
Volume Ratio @ 60°F and 14.696 psia								
Gal/lb mol	*	6.4[i]	10.12[h]	10.42[h]	12.38[h]	11.93[h]	13.85	13.71
Cu ft gas/gal liquid, ideal gas	*	59[i]	37.5[h]	36.43[h]	30.65[h]	31.81[h]	27.39	27.67
Gas vol/liquid vol, ideal gas	*	443[i]	280.5[h]	272.51[h]	229.30[h]	237.98[h]	204.93	207.00
Critical Conditions								
Temperature, °F		−116.63	90.09	206.01	274.98	305.65	369.10	385.7
Pressure, psia		667.8	707.8	616.3	529.1	550.7	490.4	488.6
Gross Heat of Combustion @ 60°F								
Btu/lb liquid	*	—	22,214[d]	21,513[d]	21,091[d]	21,139[d]	20,889	20,928
Btu/lb gas	*	23,885	22,323	21,665	21,237	21,298	21,040	21,089
Btu/cu ft, ideal gas	*p	1009.7	1768.8	2517.5	3252.7	3262.1	4000.3	4009.6
Btu/gal liquid	*	—	65,998[d]	91,065[d]	99,022[d]	102,989[d]	108,790	110,102
Cu ft air to burn 1 cu ft gas — ideal gas	*	9.54	16.70	23.86	31.02	31.02	38.18	38.18
Flammability Limits @ 100°F & 14.696 psia								
Lower, vol % in air		5.0	2.9	2.1	1.8	1.8	1.4	1.4
Upper, vol % in air		15.0	13.0	9.5	8.4	8.4	(8.3)	8.3
Heat of Vaporization @ 14.696 psia								
Btu/lb @ boiling point		219.22	210.41	183.05	157.53	165.65	147.13	153.59
Specific Heat @ 60°F & 14.696 psia								
C_P gas — Btu/lb, °F, ideal gas		0.5266	0.4097	0.3881	0.3872	0.3867	0.3827	0.3883
C_V gas — Btu/lb, °F, ideal gas	*	0.4027	0.3436	0.3430	0.3530	0.3525	0.3552	0.3608
$N = C_P/C_V$	*	1.308	1.192	1.131	1.097	1.097	1.078	1.076
C_P liquid — Btu/lb, °F	*	—	0.9256	0.5920	0.5695	0.5636	0.5353	0.5441
Octane Number								
Motor clear		—	+ .05[f]	97.1	97.6	89.6[j]	90.3	62.6[j]
Research clear		—	+1.6[j,f]	+1.8[j,f]	+0.10[j,f]	93.8[j]	92.3	61.7[j]
Refractive Index n_D @ 68°F		—	—	—	—	1.3326[h]	1.35373	1.35748

NOTES

a. Air saturated hydrocarbons.
b. Absolute values from weights in vacuum.
c. The apparent values for weight in air are shown for users' convenience. All other mass data in this table are on an absolute mass (weight in vacuum) basis.
d. At saturation pressure (triple point).
f. The + sign and number following signify the octane number corresponding to that of 2,2,4 trimethylpentane with the indicated number of ml of TEL added.
h. Saturation pressure and 60°F.
i. Apparent value for methane at 60°F.
j. Average value from octane numbers of more than one sample.
m. Density of liquid, gm/ml at normal boiling point.
n. Heat of sublimation.
p. Gross heat values are reported on a dry basis at 60°F and 14.696 psia based on ideal gas calculations. To convert to water saturation basis, multiply by 0.9826.
s. Extrapolated to room temperature from higher temperature.
* Calculated values. 1969 atomic weights used. See "Constants for Use in Calculations."
( ) Estimated values.

N-Hexane	N-Heptane	N-Octane	N-Nonane	N-Decane	Carbon Dioxide	Hydrogen Sulfide	Nitrogen	Oxygen	Air	Water
86.178	100.205	114.232	128.259	142.286	44.010	34.076	28.013	31.999	28.964	18.015
155.72	209.17	258.22	303.47	345.48	−109.3[z]	− 76.6[24]	−320.4[2]	−297.4[2]	−317.6[2]	212.0
−139.58	−131.05	− 70.18	− 64.28	− 21.36	−	−117.2[7]	−346.0[24]	−361.8[24]	−	32.0
4.956	1.620	0.537	0.179	0.0597	−	394.0[6]	−	−	−	0.9492[12]
0.6640	0.6882	0.7068	0.7217	0.7342	0.827[h(6)]	0.79[h(6)]	0.808[m(3)]	1.14[m(3)]	0.856[m(8)]	1.000
81.6	74.1	68.7	64.6	61.2	39.6[h]	47.6[h]	43.6[m]	−7.4[m]	33.8[m]	10.0
5.536	5.738	5.893	6.017	6.121	6.89[h]	6.59[h]	6.74[m]	9.50[m]	7.14[m]	8.337
5.526	5.728	5.883	6.008	6.112	6.89[h]	6.58[h]	6.73[m]	9.50[m]	7.13[m]	8.328
2.9753	3.4596	3.9439	4.4282	4.9125	1.5195	1.1765	0.9672	1.1048	1.0000	0.6220
227.09	264.05	301.01	337.98	374.94	115.97	89.79	73.82	84.32	76.32	47.47
15.57	17.46	19.39	21.32	23.24	6.38[h]	5.17[h]	4.16[m]	3.37[m]	4.06[m]	2.16
24.38	21.73	19.58	17.80	16.33	59.5[h]	73.3[h]	91.3[m]	112.7[m]	93.5[m]	175.6
182.37	162.56	146.45	133.18	122.13	444.8[h]	548.7[h]	682.7[m]	843.2[m]	699.5[m]	1313.8
453.7	512.8	564.22	610.68	652.1	87.9[23]	212.7[17]	−232.4[(-*)]	−181.1[17]	−221.3[2]	705.6[17]
436.9	396.8	360.6	332	304	1071[17]	1306[17]	493.0[24]	736.9[24]	547[2]	3208[17]
20,784	20,681	20,604	20,544	20,494	−	−	−	−	−	−
20,944	20,840	20,762	20,701	20,649	−	−	−	−	−	−
4756.2	5502.8	6249.7	6996.5	7742.1	−	637[16]	−	−	−	−
115,060	118,668	121,419	123,613	125,444	−	−	−	−	−	−
45.34	52.50	59.65	66.81	73.97	−	7.16	−	−	−	−
1.2	1.0	0.96	0.87[a]	0.78[a]	−	4.30[5]	−	−	−	−
7.7	7.0	−	2.9	2.6	−	45.50	−	−	−	−
143.95	136.01	129.53	123.76	118.68	238.2[n(14)]	235.6[7]	87.8[14]	91.6[14]	92[3]	970.3[12]
0.3864	0.3875	(0.3876)	0.3840	0.3835	0.1991[13]	0.238[4]	0.2482[13]	0.2188[13]	0.2400[9]	0.4446[13]
0.3633	0.3677	0.3702	0.3685	0.3695	0.1539	0.1797	0.1773	0.1567	0.1714	0.3343
1.063	1.054	1.047	1.042	1.038	1.293	1.325	1.400	1.396	1.400	1.330
0.5332	0.5283	0.5239	0.5228	0.5208	−	−	−	−	−	1.0009[7]
26.0	0.0	−	−	−	−	−	−	−	−	−
24.8	0.0	−	−	−	−	−	−	−	−	−
1.37486	1.38764	1.39743	1.40542	1.41189	−	−	−	−	−	1.3330[8]

REFERENCES

1. Values for hydrocarbons were selected or calculated from API Project 44 and are identical to or consistent with ASTM DS 4A "Physical Constants of Hydrocarbons C1 - C10," 1971, American Society for Testing Materials, 1916 Race Street, Philadelphia.
2. International critical tables.
3. Hodgman, Handbook of Chemistry & Physics, 31 edition (1949).
4. West, J. R., Chemical Engineering Progress, 44, 287 (1948).
5. Jones, Chemical Review, 22, 1 (1938).
6. Sage & Lacey, API Research Project 37, Monograph (1955).
7. Perry, Chemical Engineers Handbook, 4th edition (1963).
8. Matteson and Hanna, Oil and Gas Journal, 41, No. 2, 33 (1942).
9. Keenan & Keyes, Thermodynamic Properties of Air (1947).
12. Keenan & Keyes, Thermodynamic Properties of Steam (Twenty-ninth Printing 1956).
13. American Petroleum Institute, Project 44.
14. Dreisback, Physical Properties of Chemical Compounds, American Chemical Society, 1961.
16. Maxwell, J. B., Data Book on Hydrocarbons, Van Nostrand Co. (1950).
17. Kobe, K. A. & R. E. Lynn, Jr., Chemical Review, 52, 117-236 (1953).
23. Din., "Thermodynamic Functions of Gases," Butterworths (1956).
24. Thermodynamic Research Center Data Project, Texas A&M University, (formerly MCA Research Project).

CONSTANTS FOR USE IN CALCULATIONS

Atomic Weights based on 1969 values, Pure Applied Chemistry 20(4) (1969)
      C - 12.011   H - 1.0080   N - 14.0067   O - 15.9994   S - 32.06

Ideal Gas
      1 mol = 379.49 cu ft @ 14.696 psia and 60°F
      1 mol = 22.414 liters @ 14.696 psia and 32°F

Conversion Factors
      1 cu ft = 28.317 liters   1 gal = 3785.41 milliliters   1 cu ft = 7.4805 gal
      760 mm Hg = 14.696 lb/sq in = 1 atm
      1 lb = 453.59 gms   0°F = 459.67° Rankine

Density of Water @ 60°F = 8.3372 lb/gal = 0.999015 g/cc (weight in vacuum)
Specific Gravity @ 60°F/60°F x 0.999015 = Density @ 60°F, g/cc

$$°F = 9/5 \ °C + 32 \qquad °API = \frac{141.5}{Sp\ Gr\ @\ 60°F/60°F} - 131.5$$

CALCULATED VALUES

Density of Liquid @ 60°F and 14.696 psia

Lb/gal @ 60°F (weight-in-vacuum) = sp gr @ 60°F/60°F (wt-in-vac) x 8.3372 lb/gal (wt-in-vac)
Lb/gal @ 60°F (weight-in-air) — See ASTM DS 4A, Page 61.
Gal/lb mol @ 60°F = mol wt ÷ [lb/gal @ 60°F (wt-in-vac)]

Density of Gas @ 60°F and 14.696 psia (Ideal Gas)

Sp Gr @ 60°F = mol wt ÷ 28.964
Lb/M cu ft = (mol wt x 1000) ÷ 379.49
Cu ft vap/gal liq = [lb/gal @ 60°F (wt-in-vac) x 379.49] ÷ mol wt

Ratio gas vol/liq vol

Gas vol/liq vol = [lb/gal @ 60°F (wt-in-vac) x 379.49 x 7.4805] ÷ mol wt

Heat of Combustion @ 60°F

The heat of combustion in Btu/gal was calculated as follows: The gross heat of combustion in Btu/lb gas less the heat of vaporization at 60°F is the calculated heat of combustion for the liquid state in Btu/lb. The heat of vaporization at 60°F was calculated from the normal boiling point, critical temperature, and the heat of vaporization at the normal boiling point using the method of Fishtine (see Reid and Sherwood, The Properties of Gases and Liquids, 2nd edition, page 148). The heat of combustion in Btu/lb liquid multiplied by the weight in vacuum, lb/gal, yields the heat of combustion in Btu/gal at the saturation pressure and 60°F.

Cu ft of air to burn 1 cu ft gas (Ideal Gas) — $C_a H_b$

$$\frac{cu\ ft\ air}{cu\ ft\ gas} = \frac{(a + b/4)}{0.2095} \qquad \text{See ASTM DS 4A, Page 63.}$$

$(H_2S + 1.5O_2 \rightarrow H_2O + SO_2)$

Specific Heat @ 60°F and 14.696 psia (Ideal Gas)

C_v (gas) = C_p (gas) − (1.98719 ÷ mol wt) (For hydrocarbons)

C_v values for nonhydrocarbon components are calculated from C_p and N values.

TABLE B.3

Heats of Combustion of Residual Fuel Oils

Gravity		Density, lb per gal	Total heat of combustion at constant volume, Qv			Net heat of combustion at constant pressure, Qp		
°API at 60°F	Specific at 60°/60°F		Cal per g	Btu per lb	Btu per gal	Cal per g	Btu per lb	Btu per gal
0	1.0760	8.962	9,970	17,950	160,900	9,470	17,050	152,800
1.0	1.0679	8.895	10,010	18,010	160,200	9,500	17,100	152,100
2.0	1.0599	8.828	10,040	18,070	159,500	9,530	17,150	151,400
3.0	1.0520	8.762	10,080	18,140	158,900	9,560	17,210	150,800
4.0	1.0443	8.698	10,110	18,200	158,300	9,590	17,260	150,100
5.0	1.0366	8.634	10,140	18,250	157,600	9,620	17,320	149,500
6.0	1.0291	8.571	10,180	18,320	157,000	9,650	17,370	148,900
7.0	1.0217	8.509	10,210	18,380	156,400	9,670	17,410	148,200
8.0	1.0143	8.448	10,240	18,430	155,700	9,690	17,450	147,400
9.0	1.0071	8.388	10,270	18,490	155,100	9,720	17,500	146,800
10.0	1.0000	8.328	10,300	18,540	154,400	9,740	17,540	146,100
11.0	0.9930	8.270	10,330	18,590	153,700	9,770	17,580	145,500
12.0	0.9861	8.212	10,360	18,640	153,000	9,790	17,620	144,800
13.0	0.9792	8.155	10,390	18,690	152,400	9,810	17,670	144,100
14.0	0.9725	8.099	10,410	18,740	151,800	9,840	17,710	143,500
15.0	0.9659	8.044	10,440	18,790	151,100	9,860	17,750	142,800
16.0	0.9593	7.989	10,470	18,840	150,500	9,880	17,790	142,200
17.0	0.9529	7.935	10,490	18,890	149,900	9,900	17,820	141,500
18.0	0.9465	7.882	10,520	18,930	149,200	9,920	17,860	140,800
19.0	0.9402	7.830	10.540	18,980	148,600	9,940	17,900	140,200
20.0	0.9340	7.778	10,570	19,020	147,900	9,960	17,930	139,500
21.0	0.9279	7.727	10,590	19,060	147,300	9,980	17,960	138,900
22.0	0.9218	7.676	10,620	19,110	146,600	10,000	18,000	138,200
23.0	0.9159	7.627	10,640	19,150	146,000	10,020	18,030	137,600
24.0	0.9100	7.578	10.660	19,190	145,400	10,040	18,070	137,000
25.0	0.9024	7.529	10.680	19,230	144,800	10,050	18,100	136,300
26.0	0.8984	7.481	10,710	19,270	144,100	10,070	18,130	135,700
27.0	0.8927	7.434	10,730	19,310	143,500	10,090	18,160	135,100
28.0	0.8871	7.387	10,750	19,350	142,900	10,110	18,190	134,500
29.0	0.8816	7.341	10,770	19,380	142,300	10,120	18,220	133,800
30.0	0.8762	7.305	10,790	19,420	141,800	10,140	18,250	133,300
31.0	0.8708	7.260	10,810	19,450	141,200	10,150	18,280	132,700
32.0	0.8654	7.215	10,830	19,490	140,600	10,170	18,310	132,100
33.0	0.8602	7.171	10,850	19,520	140,000	10,180	18,330	131,500
34.0	0.8550	7.128	10,860	19,560	139,400	10,200	18,360	130,900

TABLE B.3 (Continued)

Gravity		Density, lb per gal	Total heat of combustion at constant volume, Qv			Net heat of combustion at constant pressure, Qp		
°API at 60°F	Specific at 60°/60°F		Cal per g	Btu per lb	Btu per gal	Cal per g	Btu per lb	Btu per gal
35.0	0.8498	7.085	10,880	19,590	138,800	10,210	18,390	130,300
36.0	0.8448	7.043	10,900	19,620	138,200	10,230	18,410	129,700
37.0	0.8398	7.001	10,920	19,650	137,600	10,240	18,430	129,100
38.0	0.8348	6.960	10,940	19,680	137,000	10,260	18,460	128,500
39.0	0.8299	6.920	10,950	19,720	136,400	10,270	18,840	127,900
40.0	0.8251	6.879	10,970	19,750	135,800	10,280	18,510	127,300
41.0	0.8203	6.839	10,990	19,780	135,200	10,300	18,530	126,700
42.0	0.8155	6.799	11,000	19,810	134,700	10,310	18,560	126,200
43.0	0.8109	6.760	11,020	19,830	134,100	10,320	18,580	125,600
44.0	0.8063	6.722	11,030	19,860	133,500	10,330	18,600	125,000
45.0	0.8017	6.684	11,050	19,890	132,900	10,340	18,620	124,400
46.0	0.7972	6.646	11,070	19,920	132,400	10,360	18,640	123,900
47.0	0.7927	6.609	11,080	19,940	131,900	10,370	18,660	123,300
48.0	0.7883	6.572	11,100	19,970	131,200	10,380	18,680	122,800
49.0	0.7839	6.536	11,110	20,000	130,700	10,390	18,700	122,200

Source: K.M. Guthrie, ed., "Petroleum Products Handbook." Copyright 1960, McGraw-Hill Book Company. Used with permission of McGraw-Hill Book Company.

Appendix C
CATALYST COSTS

Many of the modern refining processes utilize relatively expensive catalysts to achieve economical reaction rates and yields. Although these catalysts are not consumed in the process reactions, they are subject to deactivation and degradation as a result of carbon deposition and feedstock contamination. Most of the catalysts can be reactivated several times to extend their useful life. However, after being used for varying lengths of time (typically two to six years), it is necessary to completely replace the used catalyst. Thus, data on both initial catalyst cost and replacement costs are required to determine investment and operating costs for these processes. Data for this purpose are given below.

The amount of catalyst required and the cost for any specific application varies over a wide range. This variation reflects differences in catalyst composition and process operating conditions. It must also be recognized that the initial costs of the catalysts used for reforming, isomerization, and hydrocracking are dependent on the current costs of the noble metal constituents (platinum, rhenium, palladium). When new catalyst is purchased, the used material is returned to the catalyst manufacturer for recovery of these constituents, and a corresponding credit is applied to the cost of the new catalyst.

The costs shown below are only order-of-magnitude approximations as of the year 1982. The process licensors must be consulted for accurate costs.

Process	Initial Catalyst $/BPD Feed	Replacement Catalyst cents/bbl of Feed
Catalytic Reforming	200	10
Isomerization	150	5
Hydrodesulfurizing	10	2
Hydrocracking	150	20
FCC (Gas Oil Feed)	10	15
FCC (Resid. Feed)	20	50

Appendix D
U.S. BUREAU OF MINES ROUTINE ANALYSIS OF SELECTED CRUDE OILS

CRUDE PETROLEUM ANALYSIS

Bureau of Mines ...Bartlesville............ Laboratory
Sample ....55151...............

IDENTIFICATION

Ten Section field
Stevens, Upper Miocene
7,800-8,400 feet

California
Kern County

GENERAL CHARACTERISTICS

Gravity, specific, ...0.854......   Gravity, ° API, ..34.2...........   Pour point, ° F., .below 5.............
Sulfur, percent, ...0.45..........   Color, brownish black.................
Viscosity, Saybolt Universal at ....100°F., 43 sec.........................   Nitrogen, percent, ..0.289.............

DISTILLATION, BUREAU OF MINES ROUTINE METHOD

STAGE 1—Distillation at atmospheric pressure, ..756........ mm. Hg

First drop, ...82.... ° F.

Fraction No.	Cut temp. ° F.	Percent	Sum. percent	Sp. gr. 60/60° F.	° API. 60° F.	C. I.	Refractive index, n_D at 20° C.	Specific dispersion	S. U. visc., 100° F.	Cloud test, ° F.
1......	122	2.6	2.6	0.644	88.2					
2......	167	2.3	4.9	.683	75.7	14	1.38469	122.3		
3......	212	5.0	9.9	.725	63.7	24	1.40300	124.0		
4......	257	7.9	17.8	.751	56.9	27	1.41569	128.6		
5......	302	6.2	24.0	.772	51.8	29	1.42785	133.6		
6......	347	4.9	28.9	.791	47.4	32	1.43863	135.5		
7......	392	4.6	33.5	.808	43.6	33	1.44778	140.5		
8......	437	5.2	38.7	.825	40.0	36	1.45638	144.7		
9......	482	4.9	43.6	.837	37.6	36	1.46441	152.5		
10......	527	6.2	49.8	.852	34.6	39	1.47262	151.8		

STAGE 2—Distillation continued at 40 mm. Hg

11......	392	4.3	54.1	0.867	31.7	42	1.47941	156.0	41	10
12......	437	5.2	59.3	.872	30.8	40	1.48461	163.8	49	30
13......	482	5.3	64.6	.890	27.5	46	1.49418	170.9	66	55
14......	527	3.2	67.8	.897	26.3	46			105	70
15......	572	5.4	73.2	.915	23.1	51			200	80
Residuum.	......	25.0	98.2	.984	12.3					

Carbon residue, Conradson: Residuum, 10.5 percent; crude, 3.0.... percent.

APPROXIMATE SUMMARY

	Percent	Sp. gr.	° API	Viscosity
Light gasoline................................	9.9	0.694	72.4	
Total gasoline and naphtha...................	33.5	0.752	56.7	
Kerosine distillate...........................	5.2	.825	40.0	
Gas oil......................................	18.5	.855	34.0	
Nonviscous lubricating distillate.............	8.5	.873-.896	30.6-26.4	50-100
Medium lubricating distillate.................	4.8	.896-.915	26.4-23.1	100-200
Viscous lubricating distillate................	2.7	.915-.926	23.1-21.3	Above 200
Residuum....................................	25.0	.984	12.3	
Distillation loss..............................				

CRUDE PETROLEUM ANALYSIS

Bureau of Mines ....Laramie........................ Laboratory
Sample ..PC-65-28..........................

IDENTIFICATION

Elk Hills field (Naval Reserve No. 1)
Sub-Scales No. 1 sandstone - Pilocene
3,333-3,370 feet

California
Kern County
SE1/4SW1/4, sec. 8,
T 31 S, R 24 E

GENERAL CHARACTERISTICS

Gravity, specific, ..0.896.......   Gravity, ° API, ...26.4..............   Pour point, ° F., ..below 5......
Sulfur, percent, ...51.........   Color, .greenish black.....
Viscosity, Saybolt Universal at 100°F, 61 sec; at 77°F, 83 sec   Nitrogen, percent, ..0.398...............

DISTILLATION, BUREAU OF MINES ROUTINE METHOD

STAGE 1—Distillation at atmospheric pressure, ...760...... mm. Hg
First drop, ...154.... ° F.

Fraction No.	Cut temp. ° F.	Percent	Sum. percent	Sp. gr. 60/60° F.	° API 60° F.	C. I.	Refractive index, n_D at 20° C.	Specific dispersion	S. U. visc., 100° F.	Cloud test, ° F.
1......	122									
2......	167	0.6	0.6	0.697	71.5		1.39020	118.3		
3......	212	3.3	3.9	.739	60.0	30	1.40622	121.1		
4......	257	5.8	9.7	.762	54.2	32	1.41838	122.5		
5......	302	5.2	14.9	.782	49.5	34	1.42933	125.8		
6......	347	5.1	20.0	.805	44.3	38	1.44218	137.8		
7......	392	4.4	24.4	.825	40.0	41	1.45419	141.2		
8......	437	5.7	30.1	.840	37.0	43	1.46360	145.8		
9......	482	7.4	37.5	.855	34.0	45	1.47338	155.8		
10.....	527	5.5	43.0	.871	31.0	48	1.48395	170.5		

STAGE 2—Distillation continued at 40 mm. Hg

Fraction No.	Cut temp. ° F.	Percent	Sum. percent	Sp. gr. 60/60° F.	° API 60° F.	C. I.	Refractive index, n_D at 20° C.	Specific dispersion	S. U. visc., 100° F.	Cloud test, ° F.
11......	392	1.5	44.5	.890	27.5	53	1.49180	166.4	41	below 5
12......	437	5.8	50.3	.892	27.1	50	1.49533	174.4	47	below 5
13......	482	5.3	55.6	.906	24.7	53	1.49684	171.6	63	below 5
14......	527	4.8	60.4	.931	20.5	62			110	below 5
15......	572	5.3	65.7	.938	19.4	62			280	below 5
Residuum..		33.8	99.5	.984	12.3					

Carbon residue, Conradson: Residuum, ..11.6.. percent; crude, ..4.3.. percent.

APPROXIMATE SUMMARY

	Percent	Sp. gr.	° API	Viscosity
Light gasoline......................................	3.9	0.733	61.5	
Total gasoline and naphtha	24.4	.782	49.5	
Kerosine distillate				
Gas oil	23.9	.863	32.5	
Nonviscous lubricating distillate	8.6	.894-.926	26.8-21.3	50-100
Medium lubricating distillate	3.8	.926-.935	21.3-19.8	100-200
Viscous lubricating distillate	5.0	.935-.942	19.8-18.7	Above 200
Residuum...	33.8	.984	12.3	
Distillation loss	.5			

CRUDE PETROLEUM ANALYSIS

Bureau of Mines .....Bartlesville.......... Laboratory
Sample ....55126......................

IDENTIFICATION

Torrance Field
Del Amo, Miocene
3,100-5,000 feet

California
Los Angeles County

GENERAL CHARACTERISTICS

Gravity, specific, ..0.911........   Gravity, ° API, ..23.8.................   Pour point, ° F., below 5
Sulfur, percent, ...1.84.............   Color, brownish black
Viscosity, Saybolt Universal at 100°F.,160sec; 130°F.,96 sec   Nitrogen, percent, 0.555

DISTILLATION, BUREAU OF MINES ROUTINE METHOD

STAGE 1—Distillation at atmospheric pressure, ..746...... mm. Hg
First drop, ...81... ° F.

Fraction No.	Cut temp. °F.	Percent	Sum. percent	Sp. gr., 60/60° F.	° API. 60° F.	C. I.	Refractive index, n_D at 20° C.	Specific dispersion	S. U. visc., 100° F.	Cloud test, ° F.
1.....	122	0.3	0.3	0.675	78.1					
2.....	167	1.1	1.4	.683	75.7	14	1.39514	126.2		
3.....	212	2.2	3.6	.725	63.7	24	1.41771	130.2		
4.....	257	3.7	7.3	.755	55.9	29	1.42942	134.5		
5.....	302	3.8	11.1	.777	50.6	32	1.43986	138.3		
6.....	347	3.3	14.4	.796	46.3	34	1.44908	141.7		
7.....	392	3.5	17.9	.813	42.6	36	1.45776	142.1		
8.....	437	4.3	22.2	.830	39.0	38	1.46618	148.3		
9.....	482	4.6	26.8	.843	36.4	39	1.47548	151.0		
10.....	527	7.4	34.2	.861	32.8	43				

STAGE 2—Distillation continued at 40 mm. Hg

Fraction No.	Cut temp. °F.	Percent	Sum. percent	Sp. gr., 60/60° F.	° API. 60° F.	C. I.	Refractive index, n_D at 20° C.	Specific dispersion	S. U. visc., 100° F.	Cloud test, ° F.
11.....	392	1.6	35.8	0.872	30.8	44	1.48248	–	43	10
12.....	437	5.4	41.2	.884	28.6	46	1.48731	150.6	49	30
13.....	482	6.0	47.2	.901	25.6	51	1.49769	–	70	45
14.....	527	5.5	52.7	.910	24.0	52	1.50544	–	125	65
15.....	572	4.6	57.3	.927	21.1	57			250	80
Residuum.	.....	41.9	99.2	1.004	9.4					

Carbon residue, Conradson: Residuum, 13.2 percent; crude, 6.1 percent.

APPROXIMATE SUMMARY

	Percent	Sp. gr.	° API	Viscosity
Light gasoline	3.6	0.708	68.4	
Total gasoline and naphtha	17.9	0.769	52.5	
Kerosine distillate	–	–	–	
Gas oil	21.0	.855	34.0	
Nonviscous lubricating distillate	8.5	.855-.906	28.4-24.7	50-100
Medium lubricating distillate	5.7	.906-.921	24.7-22.1	100-200
Viscous lubricating distillate	4.2	.921-.934	22.1-20.0	Above 200
Residuum	41.9	1.004	9.4	
Distillation loss	.8			

U. S. GOVERNMENT PRINTING OFFICE   16—57835-3

CRUDE PETROLEUM ANALYSIS

Bureau of Mines ......Bartlesville............ Laboratory
Sample ....55126........................

IDENTIFICATION

Torrance Field
Del Amo, Miocene
3,100-5,000 feet

California
Los Angeles County

GENERAL CHARACTERISTICS

Gravity, specific, ..0.911..........   Gravity, ° API, ..23.8..................   Pour point, ° F., below 5
Sulfur, percent, ...1.84............   Color, brownish black
Viscosity, Saybolt Universal at 100°F.,160sec; 130°F.,96 sec   Nitrogen, percent, 0.555

DISTILLATION, BUREAU OF MINES ROUTINE METHOD

STAGE 1—Distillation at atmospheric pressure, ..746........ mm. Hg
First drop, ...81.... ° F.

Fraction No.	Cut temp. ° F.	Percent	Sum, percent	Sp. gr. 60/60° F.	° API 60° F.	C. I.	Refractive index, n_D at 20° C.	Specific dispersion	S. U. visc., 100° F.	Cloud test, ° F.
1.....	122	0.3	0.3	0.675	78.1					
2.....	167	1.1	1.4	.683	75.7	14	1.39514	126.2		
3.....	212	2.2	3.6	.725	63.7	24	1.41771	130.2		
4.....	257	3.7	7.3	.755	55.9	29	1.42942	134.5		
5.....	302	3.8	11.1	.777	50.6	32	1.43986	138.3		
6.....	347	3.3	14.4	.796	46.3	34	1.44908	141.7		
7.....	392	3.5	17.9	.813	42.6	36	1.45776	142.1		
8.....	437	4.3	22.2	.830	39.0	38	1.46618	148.3		
9.....	482	4.6	26.8	.843	36.4	39	1.47548	151.0		
10.....	527	7.4	34.2	.861	32.8	43				

STAGE 2—Distillation continued at 40 mm. Hg

Fraction No.	Cut temp. ° F.	Percent	Sum, percent	Sp. gr. 60/60° F.	° API 60° F.	C. I.	Refractive index, n_D at 20° C.	Specific dispersion	S. U. visc., 100° F.	Cloud test, ° F.
11.....	392	1.6	35.8	0.872	30.8	44	1.48248	–	43	10
12.....	437	5.4	41.2	.884	28.6	46	1.48731	150.6	49	30
13.....	482	6.0	47.2	.901	25.6	51	1.49769	–	70	45
14.....	527	5.5	52.7	.910	24.0	52	1.50544	–	125	65
15.....	572	4.6	57.3	.927	21.1	57			250	80
Residuum.	.....	41.9	99.2	1.004	9.4					

Carbon residue, Conradson: Residuum, 13.2 percent; crude, 6.1 percent.

APPROXIMATE SUMMARY

	Percent	Sp. gr.	° API	Viscosity
Light gasoline	3.6	0.708	68.4	
Total gasoline and naphtha	17.9	0.769	52.5	
Kerosine distillate	–	–	–	
Gas oil	21.0	.855	34.0	
Nonviscous lubricating distillate	8.5	.855-.906	28.4-24.7	50-100
Medium lubricating distillate	5.7	.906-.921	24.7-22.1	100-200
Viscous lubricating distillate	4.2	.921-.934	22.1-20.0	Above 200
Residuum	41.9	1.004	9.4	
Distillation loss	.8			

CRUDE PETROLEUM ANALYSIS

Bureau of Mines ...Bartlesville............ Laboratory

Sample ...57069.....................

IDENTIFICATION

Bridgeport field
Bridgeport, Pennsylvanian
906-938 feet

Illinois
Lawrence County

GENERAL CHARACTERISTICS

Gravity, specific, ...0.847......   Gravity, ° API, ..35.6............   Pour point, ° F., below 5

Sulfur, percent, ...0.21........   Color, brownish green

Viscosity, Saybolt Universal at ...100°F, 46 sec.   Nitrogen, percent, 0.138

DISTILLATION, BUREAU OF MINES ROUTINE METHOD

STAGE 1—Distillation at atmospheric pressure, ... mm. Hg

First drop, ... ° F.

Fraction No.	Cut temp. ° F.	Percent	Sum, percent	Sp. gr., 60/60° F.	° API. 60° F.	C. I.	Refractive index, n_D at 20° C.	Specific dispersion	S. U. visc., 100° F.	Cloud test, ° F.
1......	122	2.3	2.3	0.638	90.3					
2......	167	2.2	4.5	.671	79.4	8.0	1.37182	121.0		
3......	212	4.6	9.1	.712	67.2	18	1.39622	116.9		
4......	257	5.7	14.8	.738	60.2	21	1.40852	123.8		
5......	302	5.3	20.1	.756	55.7	22	1.41885	128.7		
6......	347	5.3	25.4	.776	50.9	24	1.42922	129.4		
7......	392	4.4	29.8	.792	47.2	26	1.43930	135.2		
8......	437	4.7	34.5	.809	43.4	28	1.44784	133.7		
9......	482	4.8	39.3	.823	40.4	30	1.45582	134.3		
10.....	527	5.9	45.2	.837	37.6	32	1.46433	142.6		

STAGE 2—Distillation continued at 40 mm. Hg

Fraction No.	Cut temp. ° F.	Percent	Sum, percent	Sp. gr., 60/60° F.	° API. 60° F.	C. I.	Refractive index, n_D at 20° C.	Specific dispersion	S. U. visc., 100° F.	Cloud test, ° F.
11.....	392	3.7	48.9	0.852	34.6	35	1.47330	150.0	40	10
12.....	437	5.7	54.6	.858	33.4	34	1.47744	--	47	25
13.....	482	4.5	59.1	.873	30.6	37	1.48469	--	61	50
14.....	527	4.7	63.8	.885	28.4	40	1.49563	--	89	65
15.....	572	5.5	69.3	.901	25.6	45			170	75
Residuum.		28.3	97.6	.960	15.9					

Carbon residue, Conradson: Residuum, 8.4 percent; crude, 2.7 percent.

APPROXIMATE SUMMARY

	Percent	Sp. gr.	° API	Viscosity
Light gasoline	9.1	0.683	75.7	
Total gasoline and naphtha	29.8	0.739	60.0	
Kerosine distillate	9.5	.816	41.9	
Gas oil	13.7	.847	35.6	
Nonviscous lubricating distillate....................	9.2	.861-.887	32.8-27.7	50-100
Medium lubricating distillate	6.4	.887-.907	27.7-24.5	100-200
Viscous lubricating distillate	.7	.907-.909	24.5-24.2	Above 200
Residuum	28.3	.960	15.9	
Distillation loss	2.4			

U. S. GOVERNMENT PRINTING OFFICE   16—57835-3

CRUDE PETROLEUM ANALYSIS

Bureau of Mines .Bartlesville............... Laboratory

Sample .61084............................

IDENTIFICATION

Marcotte field
Arbuckle, Cambro-Ordovician
3,757-3,761 feet

Kansas
Rooks County

GENERAL CHARACTERISTICS

Gravity, specific, .0.897..........   Gravity, ° API, ..26.3...............   Pour point, ° F., ...15..................

Sulfur, percent, ..0.77..............   Color, .brownish black.................

Viscosity, Saybolt Universal at .100.°F., .242 sec.; .130.°F., .122 sec...   Nitrogen, percent, .0.19.................

DISTILLATION, BUREAU OF MINES ROUTINE METHOD

STAGE 1—Distillation at atmospheric pressure, ...746..... mm. Hg

First drop, .185.......... ° F.

Fraction No.	Cut temp. ° F.	Percent	Sum, percent	Sp. gr., 60/60° F.	° API 60° F.	C. I.	Refractive index, n_D at 20° C.	Specific dispersion	S. U. visc., 100° F.	Cloud test, ° F.
1	122									
2	167									
3	212									
4	257	2.3	2.3	0.730	62.3	–	1.40764	126.2		
5	302	2.5	4.8	.749	57.4	18	1.41629	124.0		
6	347	3.4	8.2	.770	52.3	22	1.42613	126.1		
7	392	3.6	11.8	.790	47.6	25	1.43574	127.7		
8	437	4.6	16.4	.804	44.5	26	1.44520	129.3		
9	482	4.5	20.9	.821	40.9	29	1.45305	132.8		
10	527	6.1	27.0	.836	37.8	31	1.46130	139.3		

STAGE 2—Distillation continued at 40 mm. Hg

Fraction No.	Cut temp. ° F.	Percent	Sum, percent	Sp. gr., 60/60° F.	° API 60° F.	C. I.	Refractive index, n_D at 20° C.	Specific dispersion	S. U. visc., 100° F.	Cloud test, ° F.
11	392	2.7	29.7	0.848	35.4	33	1.46983	143.7	39	below 5
12	437	6.4	36.1	.860	33.0	35	1.47567	148.0	46	15
13	482	8.2	44.3	.876	30.0	39	1.48456	155.4	61	30
14	527	4.6	48.9	.893	27.0	44	1.49161	161.2	98	60
15	572	6.7	55.6	.897	26.3	43	1.49562	168.5	155	75
Residuum.	. . . . . .	43.6	99.2	.974	13.8					

Carbon residue, Conradson: Residuum, 8.5 percent; crude, 4.1 percent.

APPROXIMATE SUMMARY

	Percent	Sp. gr.	° API	Viscosity
Light gasoline.	–	–	–	
Total gasoline and naphtha	11.8	0.764	53.7	
Kerosine distillate	9.1	.812	42.8	
Gas oil	14.9	.847	35.6	
Nonviscous lubricating distillate	11.0	.866-.893	31.9-27.0	50-100
Medium lubricating distillate	8.8	.893-.899	27.0-25.9	100-200
Viscous lubricating distillate	–	–	–	Above 200
Residuum	43.6	.974	13.8	
Distillation loss	.8			

CRUDE PETROLEUM ANALYSIS

Bureau of Mines ..Bartlesville.......... Laboratory
Sample ..60052...................

IDENTIFICATION

Black Bay, West field
9200', Miocene
9,178-9,185 feet

Louisiana
Piaquemines Parish

GENERAL CHARACTERISTICS

Gravity, specific, ...0.853.....   Gravity, ° API, ...34.4............   Pour point, ° F., ..below 5.....
Sulfur, percent, ...0.19.........   Color, ..brownish green.....
Viscosity, Saybolt Universal at ...100°F., 46 sec.................   Nitrogen, percent, ...0.04..........

DISTILLATION, BUREAU OF MINES ROUTINE METHOD

STAGE 1—Distillation at atmospheric pressure, .758......... mm. Hg
First drop, ..113....... ° F.

Fraction No.	Cut temp. ° F.	Percent	Sum, percent	Sp. gr., 60/60° F.	° API, 60° F.	C. I.	Refractive index, n_D at 20° C.	Specific dispersion	S. U. visc., 100° F.	Cloud test, ° F.
1......	122									
2......	167									
3......	212	2.6	2.6	0.706	68.9	–	1.39971	129.4		
4......	257	3.1	5.7	.739	60.0	21	1.41235	132.0		
5......	302	3.7	9.4	.762	54.2	25	1.42308	135.4		
6......	347	4.2	13.6	.780	49.9	26	1.43298	137.1		
7......	392	5.8	19.4	.796	46.3	28	1.44076	138.4		
8......	437	4.9	24.3	.807	43.8	27	1.44701	139.1		
9......	482	7.6	31.9	.820	41.1	28	1.45389	140.8		
10......	527	9.1	41.0	.834	38.2	30	1.46161	143.0		

STAGE 2—Distillation continued at 40 mm. Hg

Fraction No.	Cut temp. ° F.	Percent	Sum, percent	Sp. gr., 60/60° F.	° API, 60° F.	C. I.	Refractive index, n_D at 20° C.	Specific dispersion	S. U. visc., 100° F.	Cloud test, ° F.
11......	392	6.0	47.0	0.846	35.8	32	1.46906	148.8	40	below 5
12......	437	8.3	55.3	.854	34.2	32	1.47238	147.4	46	20
13......	482	6.8	62.1	.866	31.9	34	1.47868	144.0	58	50
14......	527	5.8	67.9	.881	29.1	38	1.48434	–	81	60
15......	572	6.1	74.0	.892	27.1	40			135	70
Residuum.	......	24.5	98.5	.940	19.0					

Carbon residue, Conradson: Residuum, .4.6. percent; crude, .1.2... percent.

APPROXIMATE SUMMARY

	Percent	Sp. gr.	° API	Viscosity
Light gasoline............................	2.6	0.706	68.9	
Total gasoline and naphtha	19.4	0.765	53.5	
Kerosine distillate	12.5	.815	42.1	
Gas oil....................................	21.9	.843	36.4	
Nonviscous lubricating distillate	12.9	.858-.884	33.4-28.6	50-100
Medium lubricating distillate	7.3	.884-.898	28.6-26.1	100-200
Viscous lubricating distillate	-	-	-	Above 200
Residuum	24.5	.940	19.0	
Distillation loss	1.5			

CRUDE PETROLEUM ANALYSIS

Bureau of Mines ...Bartlesville........... Laboratory
Sample ...54060...................

IDENTIFICATION

Bayou des Allemands field   Louisiana
Miocene   Lafourche Parish

GENERAL CHARACTERISTICS

Gravity, specific, 0.845   Gravity, ° API, 36.0   Pour point, ° F., 35
Sulfur, percent, 0.20   Color, brownish green
Viscosity, Saybolt Universal at 100°F, 49 sec.   Nitrogen, percent, 0.040

DISTILLATION, BUREAU OF MINES ROUTINE METHOD

STAGE 1—Distillation at atmospheric pressure, 743 mm. Hg
First drop, 86 ° F.

Fraction No.	Cut temp. ° F.	Percent	Sum, percent	Sp. gr., 60/60° F.	° API, 60° F.	C. I.	Refractive index, n_D at 20° C.	Specific dispersion	S. U. visc., 100° F.	Cloud test, ° F.
1	122	0.5	0.5	0.670	79.7					
2	167	1.2	1.7	.675	78.1	11				
3	212	1.6	3.3	.722	64.5	23	1.39163	137.0		
4	257	2.7	6.0	.748	57.7	26	1.41725	141.7		
5	302	3.1	9.1	.765	53.5	26	1.42648	142.3		
6	347	3.9	13.0	.778	50.4	25	1.43374	140.7		
7	392	4.7	17.7	.789	47.8	24	1.43962	138.0		
8	437	5.7	23.4	.801	45.2	24	1.44529	137.6		
9	482	8.0	31.4	.814	42.3	25	1.45193	137.4		
10	527	10.7	42.1	.825	40.0	26	1.45884	142.9		

STAGE 2—Distillation continued at 40 mm. Hg

11	392	5.0	47.1	0.845	36.0	31	1.46614	142.6	40	15
12	437	10.0	57.1	.854	32	32	1.46870	139.9	45	30
13	482	7.8	64.9	.863	32.5	33	1.47403	140.4	56	50
14	527	7.0	71.9	.874	30.4	35			81	65
15	572	6.5	78.4	.889	27.7	39			145	85
Residuum		20.8	99.2	.931	20.5					

Carbon residue, Conradson: Residuum, 3.7 percent; crude, 0.8 percent.

APPROXIMATE SUMMARY

	Percent	Sp. gr.	° API	Viscosity
Light gasoline	3.3	0.697	71.5	
Total gasoline and naphtha	17.7	0.759	54.9	
Kerosine distillate	24.4	.816	41.9	
Gas oil	14.2	.850	35.0	
Nonviscous lubricating distillate	14.1	.858-.878	33.4-29.7	50-100
Medium lubricating distillate	8.0	.878-.895	29.7-26.6	100-200
Viscous lubricating distillate	-	-	-	Above 200
Residuum	20.8	.931	20.5	
Distillation loss	.8			

CRUDE PETROLEUM ANALYSIS

Bureau of Mines ..Bartlesville............... Laboratory
Sample ..54064..........................

IDENTIFICATION

Sho-Vel-Tum field
Camp area   Oklahoma
Springer, Pennsylvanian   Carter County
6,295-6,385 feet

GENERAL CHARACTERISTICS

Gravity, specific, .0.887............   Gravity, ° API, ...28.0.............   Pour point, ° F., ...10..........
Sulfur, percent, ...1.41...............   Color, .brownish black................
Viscosity, Saybolt Universal at 100°F. 115 sec; 130°F. 81 sec........   Nitrogen, percent, ..0.318............

DISTILLATION, BUREAU OF MINES ROUTINE METHOD

Stage 1—Distillation at atmospheric pressure, ...745.... mm. Hg
First drop, ...84........ ° F.

Fraction No.	Cut temp. ° F.	Percent	Sum, percent	Sp. gr., 60/60° F.	° API. 60° F.	C. I.	Refractive index, n_D at 20° C.	Specific dispersion	S. U. visc., 100° F.	Cloud test, ° F.
1......	122	1.3	1.3	0.648	86.9					
2......	167	1.5	2.8	.674	78.4	9.4				
3......	212	3.3	6.1	.712	67.2	18	1.39123	127.3		
4......	257	4.3	10.4	.739	60.0	21	1.40995	127.9		
5......	302	4.0	14.4	.758	55.2	23	1.42105	130.1		
6......	347	4.1	18.5	.779	50.1	26	1.43181	134.3		
7......	392	3.7	22.2	.798	45.8	29	1.44175	136.9		
8......	437	4.1	26.3	.814	42.3	31	1.45087	141.2		
9......	482	4.8	31.1	.831	38.8	33	1.46025	145.7		
10.....	527	6.0	37.1	.848	35.4	37	1.46939	155.6		

Stage 2—Distillation continued at 40 mm. Hg

Fraction No.	Cut temp. ° F.	Percent	Sum, percent	Sp. gr., 60/60° F.	° API. 60° F.	C. I.	Refractive index, n_D at 20° C.	Specific dispersion	S. U. visc., 100° F.	Cloud test, ° F.
11......	392	1.1	38.2	0.862	32.7	39	1.47778	153.8	43	10
12......	437	4.7	42.9	.873	30.6	41	1.48216	156.6	46	25
13......	482	4.6	47.5	.882	28.9	42	1.48952	161.8	58	40
14......	527	5.3	52.8	.898	26.1	46			88	55
15......	572	5.3	58.1	.911	23.8	49			175	70
Residuum.	......	40.9	99.0	.982	12.6					

Carbon residue, Conradson: Residuum, ..11.4 percent; crude, .5.2.. percent.

APPROXIMATE SUMMARY

	Percent	Sp. gr.	° API	Viscosity
Light gasoline	6.1	0.689	73.9	
Total gasoline and naphtha	22.2	0.746	58.2	
Kerosine distillate	4.1	.814	42.3	
Gas oil	16.0	.844	36.2	
Nonviscous lubricating distillate	8.6	.854-.871	34.2-31.0	50-100
Medium lubricating distillate	6.1	.871-.891	31.0-27.3	100-200
Viscous lubricating distillate	1.1	.891-.894	27.3-26.8	Above 200
Residuum..	40.9	.982	12.6	
Distillation loss	1.0			

CRUDE PETROLEUM ANALYSIS

Bureau of Mines ..... Bartlesville .... Laboratory

Sample ..... 59172 ....

IDENTIFICATION

Sho-Vel-Tum field
Tatums area
Pennsylvanian

Oklahoma
Garvin County

GENERAL CHARACTERISTICS

Gravity, specific, ...0.928.......... Gravity, ° API, ...21.0.............. Pour point, ° F., ...below 5..............

Sulfur, percent, ...1.68............ Color, ...brownish black.............

Viscosity, Saybolt Universal at 100°F., 550 sec; 130°F., 440 sec. Nitrogen, percent, ...0.482..............

DISTILLATION, BUREAU OF MINES ROUTINE METHOD

STAGE 1—Distillation at atmospheric pressure, ...743..... mm. Hg

First drop, ...147...... ° F.

Fraction No.	Cut temp. °F.	Percent	Sum. percent	Sp. gr. 60/60° F.	° API 60° F.	C. I.	Refractive index, n_D at 20° C.	Specific dispersion	S. U. visc., 100° F.	Cloud test, °F.
1......	122									
2......	167									
3......	212	2.6	2.6	0.695	72.1	-	1.38886	124.3		
4......	257	2.8	5.4	.737	60.5	20	1.40876	124.0		
5......	302	2.9	8.3	.758	55.2	23	1.41971	128.8		
6......	347	3.5	11.8	.778	50.3	25	1.43164	131.0		
7......	392	2.7	14.5	.799	45.6	29	1.44187	135.5		
8......	437	3.1	17.6	.817	41.7	32	1.45080	136.7		
9......	482	4.4	22.0	.832	38.6	-	1.46062	143.8		
10......	527									

STAGE 2—Distillation continued at 40 mm. Hg

11......	392	6.1	28.1	0.859	33.2	-	1.47659	156.3	38	below 5
12......	437	4.5	32.6	.878	29.7	43	1.48547	160.2	47	do.
13......	482	3.9	36.5	.890	27.5	46	1.49374	157.4	62	do.
14......	527	4.8	41.3	.909	24.2	-	1.50253	161.8	105	do.
15......	572									
Residuum.	......	55.9	97.2	1.012	8.3					

Carbon residue, Conradson: Residuum, ..8.2 percent; crude 5.0..... percent.

APPROXIMATE SUMMARY

	Percent	Sp. gr.	° API	Viscosity
Light gasoline.....................................	2.6	0.695	72.1	
Total gasoline and naphtha	14.5	0.755	55.9	
Kerosine distillate	3.1	.817	41.7	
Gas oil	13.6	.854	34.2	
Nonviscous lubricating distillate	7.3	.880-.907	29.3-24.5	50-100
Medium lubricating distillate	2.8	.907-.919	24.5-22.5	100-200
Viscous lubricating distillate	-	-	-	Above 200
Residuum..	55.9	1.012	8.3	
Distillation loss	2.8			

CRUDE PETROLEUM ANALYSIS

Bureau of Mines ...Bartlesville........... Laboratory
Sample ..64036...............................

Sho-Vel-Tum field
Sholem Alechem area
Pennsylvanian
3,468-3,488 feet

IDENTIFICATION

Oklahoma
Stephens County

GENERAL CHARACTERISTICS

Gravity, specific, ..0.893.........   Gravity, ° API, ..27.0...............   Pour point, ° F., ..below 5..........
Sulfur, percent, ...1.34...............   Color, ..greenish black.............
Viscosity, Saybolt Universal at .100°F.,.131.sec;.130°F,.84.sec....   Nitrogen, percent, .0.243.............

DISTILLATION, BUREAU OF MINES ROUTINE METHOD

STAGE 1—Distillation at atmospheric pressure, ..732........ mm. Hg
First drop, ...77......... ° F.

Fraction No.	Cut temp. ° F.	Percent	Sum, percent	Sp. gr., 60/60° F.	° API. 60° F.	C. I.	Refractive index, n_D at 20° C.	Specific dispersion	S. U. visc., 100° F.	Cloud test, ° F.
1......	122									
2......	167									
3......	212	3.2	3.2	0.704	69.5	-	1.39260	128.2		
4......	257	3.5	6.7	.737	60.5	20	1.40952	131.7		
5......	302	3.7	10.4	.759	54.9	23	1.42062	133.4		
6......	347	3.8	14.2	.778	50.4	25	1.43156	135.0		
7......	392	3.5	17.7	.798	45.8	29	1.44148	137.7		
8......	437	4.0	21.7	.814	42.3	31	1.45105	140.0		
9......	482	4.7	26.4	.832	38.6	34	1.46032	148.1		
10......	527	5.1	31.5	.847	35.6	36	1.46909	154.9		

STAGE 2—Distillation continued at 40 mm. Hg

11......	392	3.7	35.2	0.870	31.1	43	1.47931	158.1	42	5
12......	437	4.7	39.9	.877	29.8	43	1.48496	164.2	50	25
13......	482	5.6	45.5	.895	26.6	48	1.49389	-	73	50
14......	527	3.8	49.3	.905	24.9	49			115	60
15......	572									
Residuum.		48.9	98.2	.972	14.1					

Carbon residue, Conradson: Residuum, 9.2 percent; crude, 4.9 percent.

APPROXIMATE SUMMARY

	Percent	Sp. gr.	° API	Viscosity
Light gasoline	3.2	0.704	69.5	
Total gasoline and naphtha	17.7	0.757	55.5	
Kerosine distillate	4.0	.814	42.3	
Gas oil	15.8	.852	34.6	
Nonviscous lubricating distillate	8.2	.877-.901	29.8-25.5	50-100
Medium lubricating distillate	3.6	.901-.909	25.5-24.2	100-200
Viscous lubricating distillate	-	-	-	Above 200
Residuum	48.9	.972	14.1	
Distillation loss	1.8			

CRUDE PETROLEUM ANALYSIS

Bureau of Mines ...Bartlesville............... Laboratory

Sample ...62066................................

IDENTIFICATION

Hastings, East field
Frio, Oligocene
6,020-6,050 feet

Texas
Brazoria County

GENERAL CHARACTERISTICS

Gravity, specific, 0.871   Gravity, ° API, 31.0   Pour point, °F., below 5

Sulfur, percent, 0.15   Color, greenish black

Viscosity, Saybolt Universal at 77°F., 62 sec.; 100°F., 55 sec.   Nitrogen, percent, 0.02

DISTILLATION, BUREAU OF MINES ROUTINE METHOD

STAGE 1—Distillation at atmospheric pressure, 743 mm. Hg

First drop, 145 ° F.

Fraction No.	Cut temp. ° F.	Percent	Sum, percent	Sp. gr., 60/60° F.	° API, 60° F.	C. I.	Refractive index, n_D at 20° C.	Specific dispersion	S. U. visc., 100° F.	Cloud test, ° F.
1	122									
2	167	1.1	1.1	0.748	57.7	-	1.40061	126.1		
3	212	1.8	2.9	.753	56.4	37	1.40946	129.6		
4	257	1.7	4.6	.757	55.4	30	1.41686	131.3		
5	302	2.7	7.3	.770	52.3	28	1.42613	139.7		
6	347	3.4	10.7	.789	47.8	31	1.43860	142.8		
7	392	5.1	15.8	.813	42.6	36	1.45011	147.6		
8	437	5.9	21.7	.829	39.2	38	1.45805	149.8		
9	482	9.8	31.5	.846	35.8	41	1.46806	153.5		
10	527	10.7	42.2	.860	33.0	42	1.47690	158.6		

STAGE 2—Distillation continued at 40 mm. Hg

Fraction No.	Cut temp. ° F.	Percent	Sum, percent	Sp. gr., 60/60° F.	° API, 60° F.	C. I.	Refractive index, n_D at 20° C.	Specific dispersion	S. U. visc., 100° F.	Cloud test, ° F.
11	392	4.4	46.6	0.871	31.0	44	1.48289	158.7	42	Below 5
12	437	8.7	55.3	.880	29.3	44	1.48436	156.3	49	do.
13	482	6.7	62.0	.891	27.3	46	1.48938	155.1	68	do.
14	527	5.9	67.9	.904	25.0	49	1.49414	153.0	110	do.
15	572	6.6	74.5	.910	24.0	49			225	10
Residuum		23.0	97.5	.942	18.7					

Carbon residue, Conradson: Residuum, 4.3 percent; crude, 1.1 percent.

APPROXIMATE SUMMARY

	Percent	Sp. gr.	° API	Viscosity
Light gasoline	2.9	0.751	56.9	
Total gasoline and naphtha	15.8	0.783	49.2	
Kerosine distillate	-	-	-	
Gas oil	35.6	.855	34.0	
Nonviscous lubricating distillate	12.1	.880-.901	29.3-25.6	50-100
Medium lubricating distillate	6.4	.901-.908	25.6-24.3	100-200
Viscous lubricating distillate	4.6	.908-.913	24.3-23.5	Above 200
Residuum	23.0	.942	18.7	
Distillation loss	2.5			

CRUDE PETROLEUM ANALYSIS

Bureau of Mines .....Bartlesville..... Laboratory
Sample .....51051.....

IDENTIFICATION

Cedar Lake field   Texas
San Andres, Permian   Gaines County
4,580-4,765 feet

GENERAL CHARACTERISTICS

Gravity, specific, ...0.863.....   Gravity, ° API, ...32.5.....   Pour point, ° F., ...below 5.....
Sulfur, percent, ...2.12.....   Color, ...greenish black.....
Viscosity, Saybolt Universal at .....100°F., 45 sec.....   Nitrogen, percent, ...0.09.....

DISTILLATION, BUREAU OF MINES ROUTINE METHOD

STAGE 1—Distillation at atmospheric pressure, ...746..... mm. Hg
First drop, ...88..... ° F.

Fraction No.	Cut temp. ° F.	Percent	Sum, percent	Sp. gr., 60/60° F.	° API. 60° F.	C. I.	Refractive index, n_D at 20° C.	Specific dispersion	S. U. visc., 100° F.	Cloud test, ° F.
1	122	1.8	1.8	0.656	84.2		1.39023	139.5		
2	167	6.4	8.2	.697	71.5	20	-	-		
3	212	3.1	11.3	.740	69.5	31	1.42215	146.8		
4	257	4.7	16.0	.761	54.4	32	1.43240	150.7		
5	302	6.4	22.4	.777	50.6	32	1.43892	147.8		
6	347	4.6	27.0	.790	47.6	31	1.44401	145.3		
7	392	4.8	31.8	.801	45.2	30	1.45084	145.9		
8	437	4.3	36.1	.815	42.1	31	1.46039	151.1		
9	482	5.4	41.5	.831	38.8	33	1.47221	162.4		
10	527	6.8	48.3	.849	35.2	37				

STAGE 2—Distillation continued at 40 mm. Hg

Fraction No.	Cut temp. ° F.	Percent	Sum, percent	Sp. gr., 60/60° F.	° API. 60° F.	C. I.	Refractive index, n_D at 20° C.	Specific dispersion	S. U. visc., 100° F.	Cloud test, ° F.
11	392	1.4	49.7	0.862	32.7	39			40	15
12	437	5.7	55.4	.873	30.6	41			44	30
13	482	5.6	61.0	.889	27.7	45			56	50
14	527	5.3	66.3	.899	25.9	47			82	65
15	572	5.5	71.8	.916	23.0	55			150	85
Residuum		27.0	98.8	.987	11.9					

Carbon residue, Conradson: Residuum, ...10.7..... percent; crude, ...3.3..... percent.

APPROXIMATE SUMMARY

	Percent	Sp. gr.	° API	Viscosity
Light gasoline	11.3	0.702	70.1	
Total gasoline and naphtha	31.8	0.754	56.2	
Kerosine distillate	4.3	.815	42.1	
Gas oil	19.5	.852	34.6	
Nonviscous lubricating distillate	9.5	.881-.903	29.1-25.2	50-100
Medium lubricating distillate	6.7	.903-.925	25.2-21.5	100-200
Viscous lubricating distillate	-	-	-	Above 200
Residuum	27.0	.987	11.9	
Distillation loss	1.2			

CRUDE PETROLEUM ANALYSIS

Bureau of Mines ...Bartlesville........... Laboratory
Sample ...56112...........................

IDENTIFICATION

Corsicana field
Wolf City, Upper Cretaceous
1,088-1,116 feet

Texas
Navarro County

GENERAL CHARACTERISTICS

Gravity, specific, .0.834............   Gravity, ° API, .38.2...............   Pour point, ° F., ...below 5...............
Sulfur, percent, ...0.24...............   Color, ..brownish green..................
Viscosity, Saybolt Universal at ..100°F., 43 sec................................   Nitrogen, percent, ..0.000...............

DISTILLATION, BUREAU OF MINES ROUTINE METHOD

STAGE 1—Distillation at atmospheric pressure, ..751........ mm. Hg
First drop, ...106...... ° F.

Fraction No.	Cut temp. ° F.	Percent	Sum, percent	Sp. gr., 60/60° F.	° API, 60° F.	C. I.	Refractive index, n_D at 20° C.	Specific dispersion	S. U. visc., 100° F.	Cloud test, ° F.
1	122									
2	167	1.7	1.7	0.666	81.0					
3	212	3.2	4.9	.701	70.4	12	1.38738	127.2		
4	257	6.2	11.1	.724	63.9	14	1.40466	128.3		
5	302	5.9	17.0	.742	59.2	15	1.41441	122.9		
6	347	7.6	24.6	.762	54.2	18	1.42379	132.2		
7	392	6.8	31.4	.778	50.4	19	1.43298	128.6		
8	437	5.7	37.1	.795	46.5	22	1.44101	125.8		
9	482	7.1	44.2	.809	43.4	23	1.44869	130.9		
10	527	7.4	51.6	.826	39.8	26	1.45727	134.7		

STAGE 2—Distillation continued at 40 mm. Hg

Fraction No.	Cut temp. ° F.	Percent	Sum, percent	Sp. gr., 60/60° F.	° API, 60° F.	C. I.	Refractive index, n_D at 20° C.	Specific dispersion	S. U. visc., 100° F.	Cloud test, ° F.
11	392	4.5	56.1	0.844	36.2	31	1.46666	136.0	41	10
12	437	6.2	62.3	.854	34.2	32	1.47114	139.4	47	25
13	482	5.1	67.4	.866	31.9	34	1.47746	141.3	59	45
14	527	4.8	72.2	.884	28.6	40	1.48359	147.2	84	60
15	572	5.5	77.7	.894	26.8	41	1.49112	142.5	155	75
Residuum.		21.6	99.3	.951	17.2					

Carbon residue, Conradson: Residuum, ..8.3. percent; crude, ..2.0.. percent.

APPROXIMATE SUMMARY

	Percent	Sp. gr.	° API	Viscosity
Light gasoline	4.9	0.689	73.9	
Total gasoline and naphtha	31.4	0.743	58.9	
Kerosine distillate	12.8	.803	44.7	
Gas oil	16.5	.839	37.2	
Nonviscous lubricating distillate	10.3	.857-.886	33.6-28.2	50-100
Medium lubricating distillate	6.7	.886-.889	28.2-25.9	100-200
Viscous lubricating distillate	-	-	-	Above 200
Residuum	21.6	.951	17.3	
Distillation loss	.7			

Appendix E
ECONOMIC EVALUATION EXAMPLE PROBLEM

UNDERSTANDING "TRUE RATE OF RETURN"

The continuous inflation of our economy and escalation of prices makes imperative the consideration of the "time-value" of money when evaluating new investments. The easily understood simple payout and simple percent return which were the yardsticks of engineers and management 10 to 15 years ago have been replaced with new evaluation techniques which account for the "time-value" of money.

These procedures, although widely used, have many sophisticated names and definitions. All too frequently, the managerial and technical people who use these tools do not clearly understand the actual physical significance of the terms.

A brief review of these definitions, explained with a numerical example, is the best method of providing a clear understanding of the "true rate of return" (TRR). The terms listed below are all synonymous with true rate of return.

Average annual rate of return (AARR)
Internal rate of return (IRR)
Interest rate of return (IRR)
Discounted cash flow rate of return (DCFRR)
Investors method rate of return (IMRR)
Profitability index (PI)

A common definition of the terms listed above is: "The true rate of return is the [highest, constant] percentage of the outstanding investment which can be realized as a profit [or saving] each year [for the life of the facility] and at the same time provide funds [from the operating income] that will exactly amortize [recover] the investment over the useful life of the facility." This definition is easier to understand if the bracketed words are omitted.

This rate can be calculated for the case of a single lump investment and constant annual cash flow by the following equation:

$$i = \frac{S}{I} - \frac{i}{(1 + i)^T - 1}$$

where

i = annual rate of return (as a fraction)

S = annual cash flow (net after-tax income plus depreciation)

I = investment

T = life of facility (years).

An example will provide tangible significance of the above definition.

Example:

Initial (and only) investment (I) $1,000,000

Useful life of facility   (T) 3 years

Annual cash flow   (S) $500,000

(After-tax income + depreciation allowed for tax calculation)

Calculate i by solution of the equation:

Given $\frac{S}{I}$ = 0.5, and T = 3

$$i = 0.5 - \frac{i}{(1 + i)^3 - 1} = 0.234$$

Year	Amt. outstand. investment	Cash flow	Profit realized (23.4%XA)	Cash flow used to amort. inv. (B - C)	Remain. inv. (A - D)
	A	B	C	D	E
1	$1,000,000	$ 500,000	$234,000	$ 266,000	$734,000
2	734,000	500,000	172,000	328,000	406,000
3	406,000	500,000	94,000	406,000	
		$1,500,000	$500,000	$1,000,000	

From the above example, it can be seen that the total cash flow minus the total profit, taken at a constant percentage (23.4) of outstanding investment, is equal to a total cash allowance which will exactly recover the investment over the project life.

$$\Sigma B - \Sigma C = \Sigma D = \text{investment}$$

The above statement simply says that over the life of the project, the total of the annual net income minus the investment is equal to the total profit. The value of this obvious statement is that it makes the bankers' and account-ants' definition of TRR, AARR, IRR, DCFRR, IMRR, and PI understandable. This definition is: "The TRR is the particular interest rate that equates the total present value cash investments to the total present value cash flow incomes."

This definition leads to the more common way of determining the TRR. Our previous example can be solved by determining the discount rate at which total cash flows for three years will equal the investment.

Year	Cash flow	Discounted factor (23.4%)	Discounted cash flow
1	500,000	$1/1.234$	406,000
2	500,000	$1/(1.234)^2$	328,000
3	500,000	$1/(1.234)^3$	266,000
			1,000,000

A third approach defines the TRR as: "The interest rate required for an annuity fund started with the same sum of capital to make the same pay-ments at the same time as the proposed investment." Illustrating this defini-tion with our example, we have the following values:

Year	Principal + interest	Less payments	Balance inv.
1	1,000,000 X 1.234 = 1,234,000	500,000	734,000
2	734,000 X 1.234 = 906,000	500,000	406,000
3	406,000 X 1.234 = 500,000	500,000	—

The above procedure is suitable for evaluation of projects which have varying cash flows and distributed investments.

The examples are based on periodic interest rates and the assumption that the annual cash flow all occurs at the end of the year. For most evaluations, these conditions are sufficiently accurate. A review of alternative procedures to allow for continuous interest and continuous cash flow has been given by Souders [3].

For those cases where cash flow is constant and the investment is all made at the start of the project, a simple graphical solution for the TRR is possible [2].

A recent article by Reul [1] reviews the relative merits the several types of investment evaluation techniques. As very clearly illustrated, there is really no acceptable alternative to the true rate of return for comparing the economic performance of various investments.

NOTES

1. R. Reul, Chem. Eng. p. 212 (22 Apr. 1968).
2. R. Salmon, Chem. Eng. p. 79 (1 Apr. 1963).
3. M. Souders, Chem. Eng. Progr. 62(3), 79 (1966).

Appendix F
PHOTOGRAPHS

Photo 1. Overall refinery view.

Photo 2. Crude oil distillation unit. Left: atmospheric distillation tower; right: vacuum distillation tower.

Photo 3. Delayed coking unit.

1 – Reactor
2 – Heater
3 – Stabilizer
4 – Control House

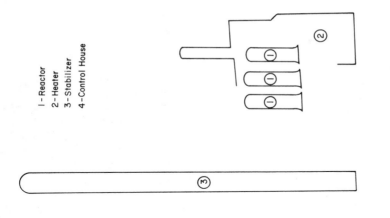

Photo 4.   2,000 BPD platformer.   (Photo courtesy of the UOP Process Division).

Photo 5. Powerformer catalytic reforming unit.

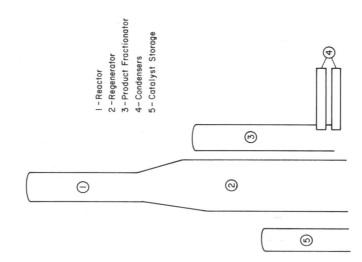

1 – Reactor
2 – Regenerator
3 – Product Fractionator
4 – Condensers
5 – Catalyst Storage

Photo 6.   20,000 BPD fluid catalytic cracker.   (Photo courtesy of the UOP Process Division).

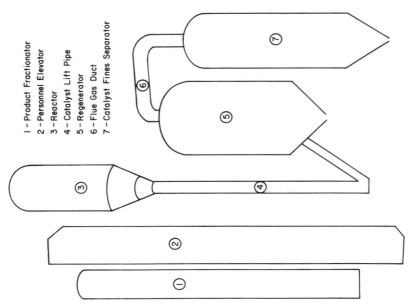

1 – Product Fractionator
2 – Personnel Elevator
3 – Reactor
4 – Catalyst Lift Pipe
5 – Regenerator
6 – Flue Gas Duct
7 – Catalyst Fines Separator

Photo 7. Fluid catalytic cracker. (Photo courtesy of Fluor Engineers and Constructors, Inc.)

1 – Reactor - Kiln Structure
2 – Feed Heaters

Photo 8.   12,000 BPD TCC unit.   (Photo by Ray Manley, courtesy of the Stearns Roger Corp.).

Photo 9.   24,000 BPD H-Oil unit.   (Photo courtesy of Fluor Engineers and Constructors, Inc.).

1-Acid Coolers
2-Reactor
3-Combination Fractionator
4-Reboiler Heater
5-Control House

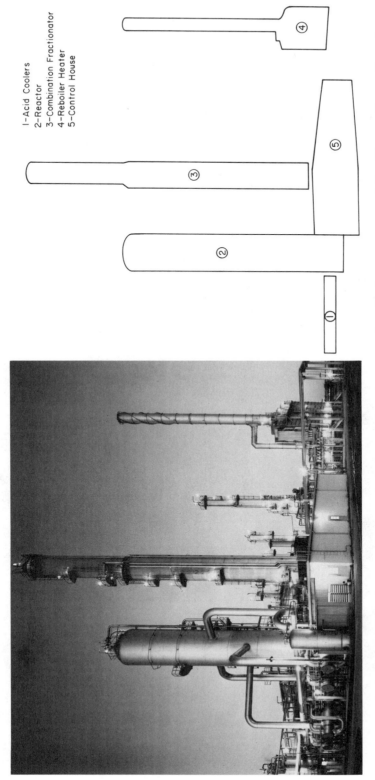

Photo 10. 10,000 BPD hydrofluoric acid alkylation unit. (Photo by Ray Manley, courtesy of the Stearns Roger Corp.)

Photo 11.   Sulfuric acid alkylation unit.

Photo 12. Light ends distillation units.

Appendix G
YIELD AND COST DATA FIGURES

The figures relating to yield and cost data which are included in the body of the text are reproduced in this appendix full size for engineering use.

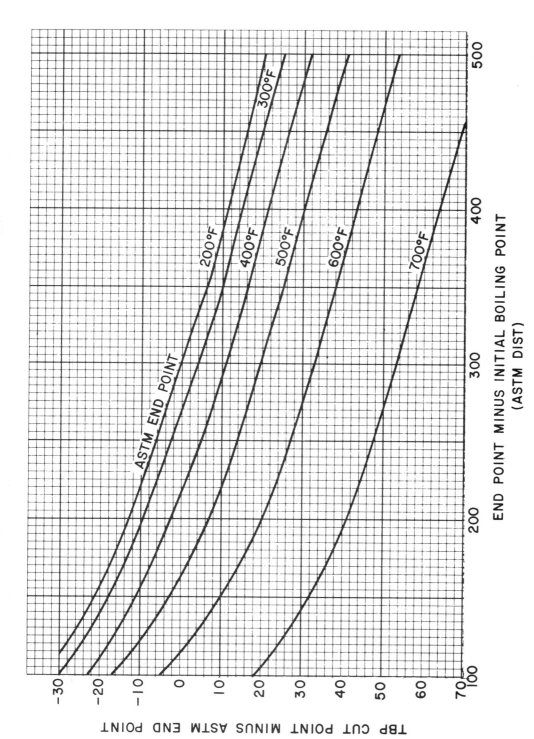

FIG. 3.1. TBP cut point vs ASTM end point.

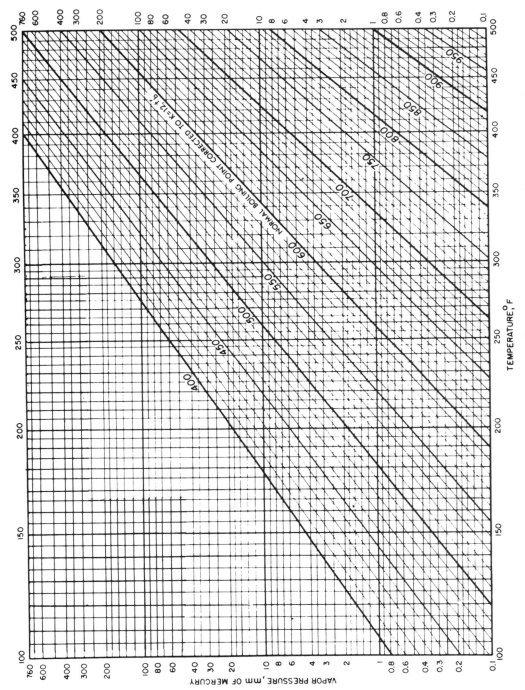

FIG. 3.6.   Vapor pressure of pure hydrocarbons and narrow-boiling petroleum fractions (low-temperature range). From "API Technical Data Book" (American Petroleum Institute, Division of Refining, 1964).

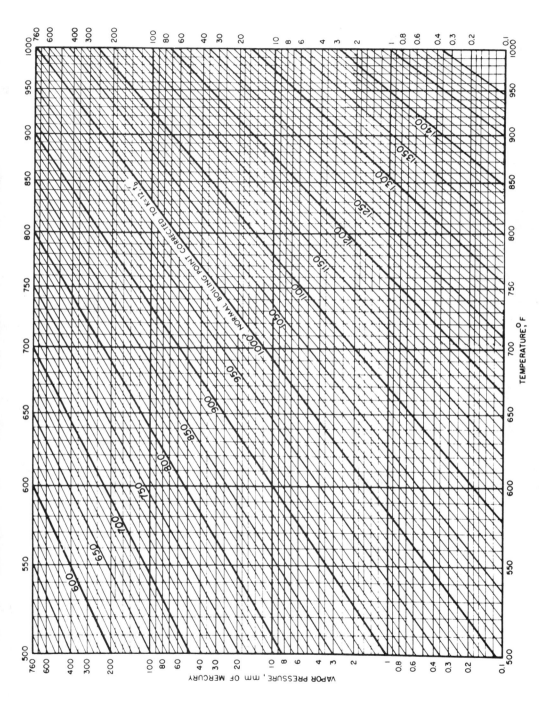

FIG. 3.7. Vapor pressure of pure hydrocargons and narrow-boiling petroleum fractions (high-temperature range) From "API Technical Data Book" (American Petroleum Institute, Division of Refining, 1964).

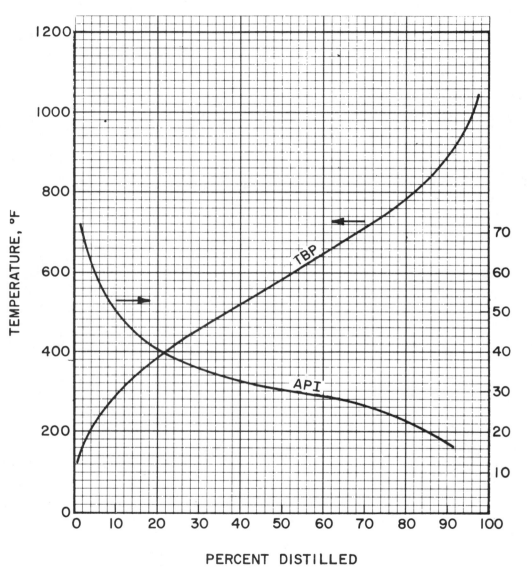

FIG. 3.9. TBP and gravity curves   Crude: Hastings Field, Texas; gravity:
3.17°API; sulfur: 0.15 wt %.

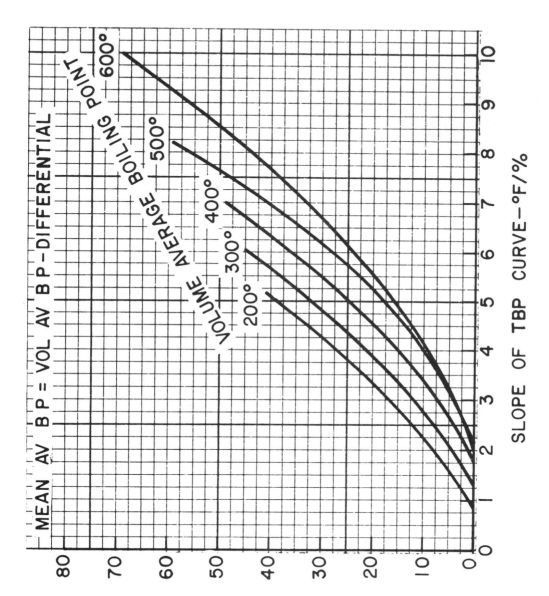

FIG. 4.1a. Mean average boiling point of petroleum fractions.

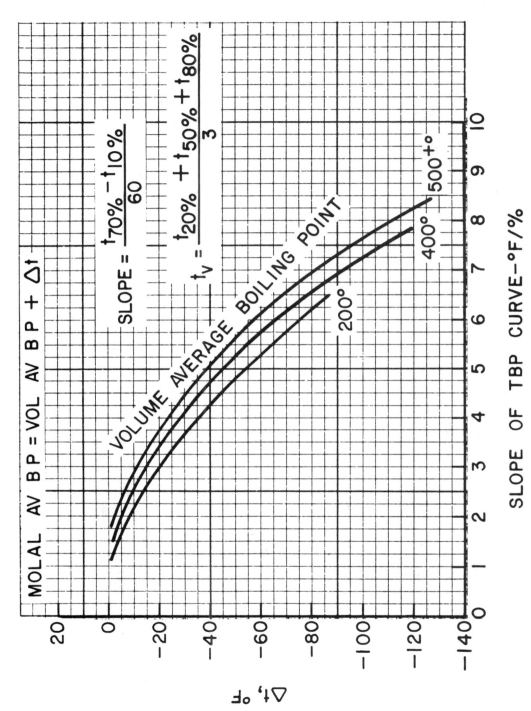

FIG. 4.1b. Molal average boiling point of petroleum fractions.

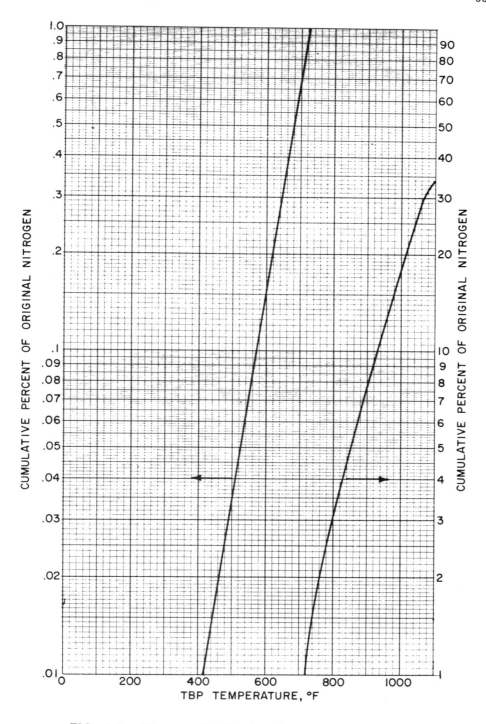

FIG. 4.2. Nitrogen distributions in crude oil fractions.

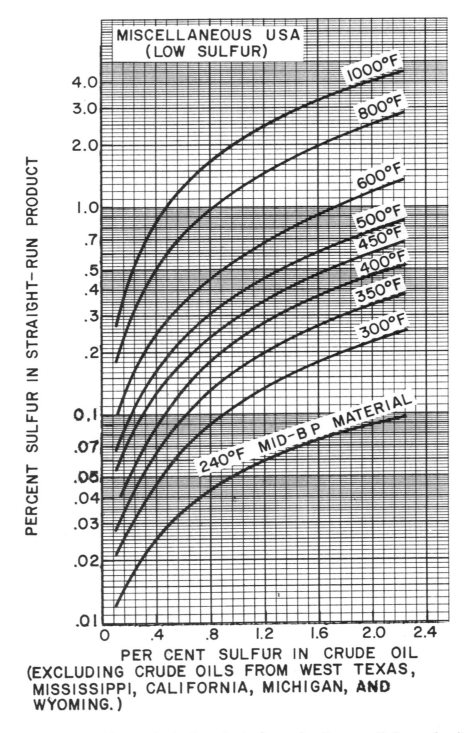

FIG. 4.3a. Sulfur content of products from miscellaneous U.S. crude oils
[Chapter 4, Ref. 2].

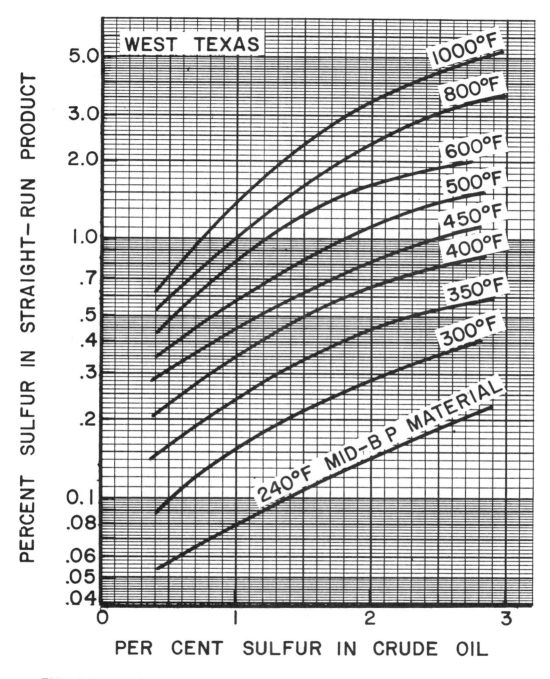

FIG. 4.3b. Sulfur content of products from West Texas crude oils [Chapter 4, Ref. 2].

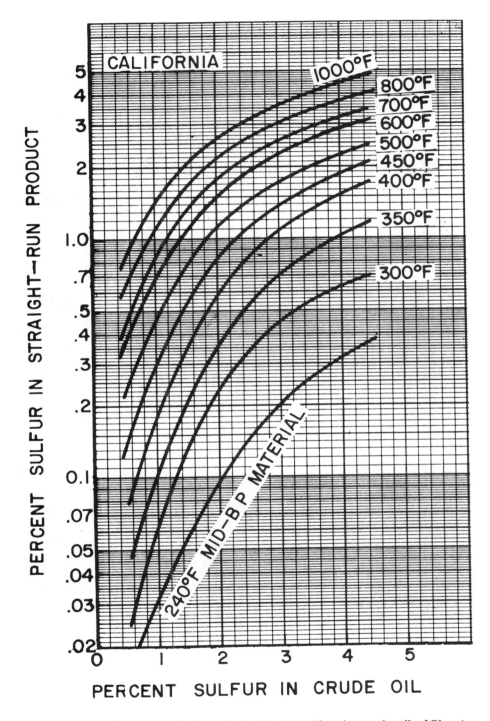

FIG. 4.4a. Sulfur content of products from California crude oils [Chapter 4, Ref. 2].

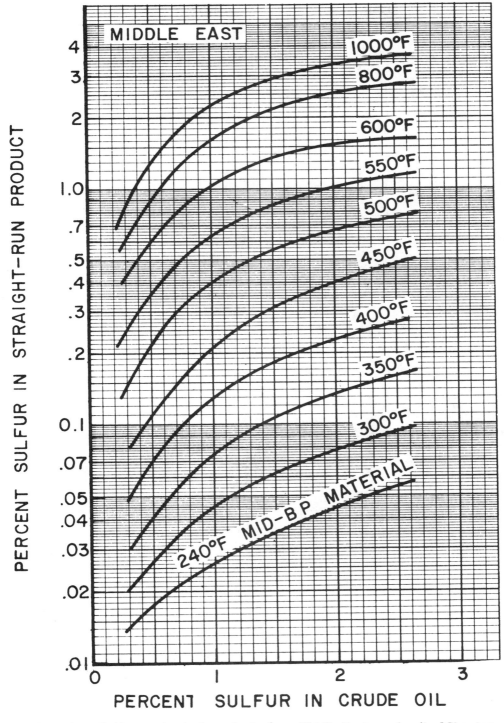

FIG. 4.4b. Sulfur content of products from Middle East crude oils [Chapter 4, Ref. 2].

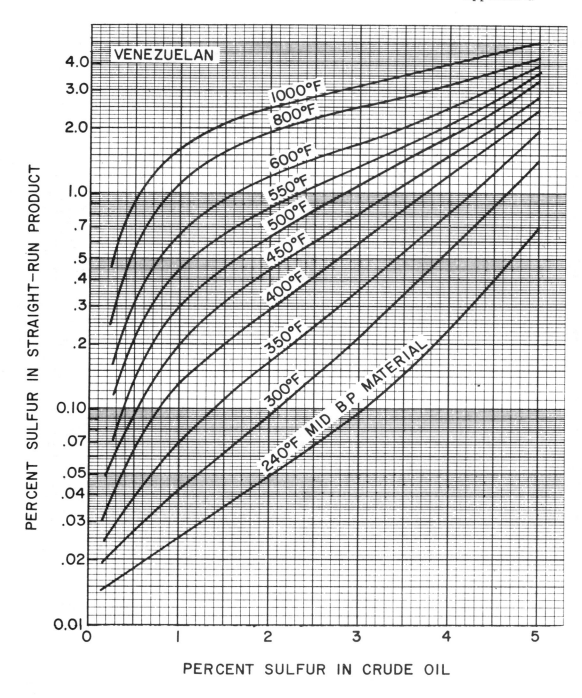

FIG. 4.5. Sulfur content of products from Venezuelan crude oils [Chapter 4, Ref. 2].

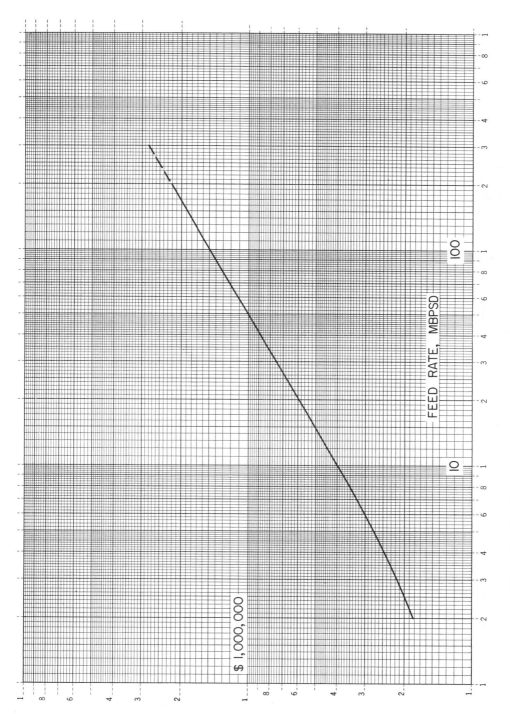

FIG. 4.6. Crude oil desalting units investment costs—1982 U.S. Gulf Coast.

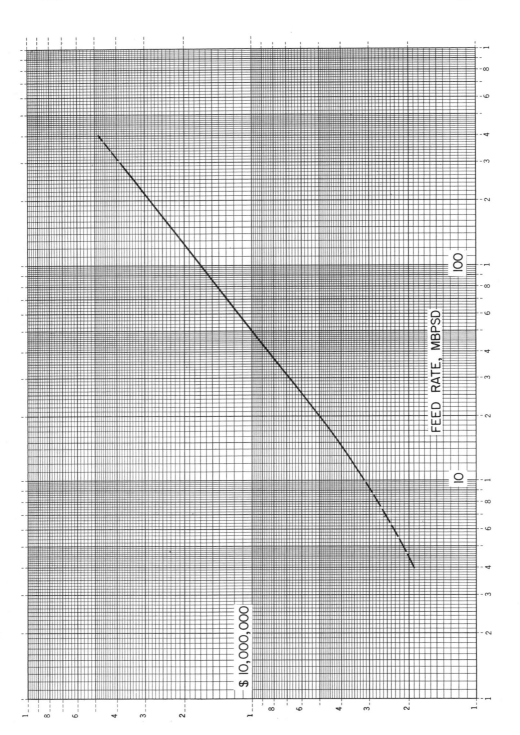

FIG. 4.8. Atmospheric crude distillation units investment cost--1982 U.S. Gulf Coast.

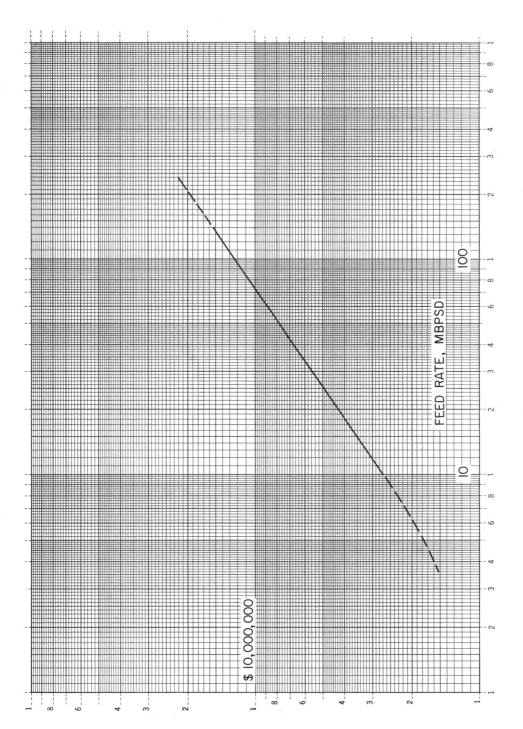

FEED RATE, MBPSD

$ 10,000,000

FIG. 4.10. Vacuum distillation units investment cost—1982 U.S. Gulf Coast.

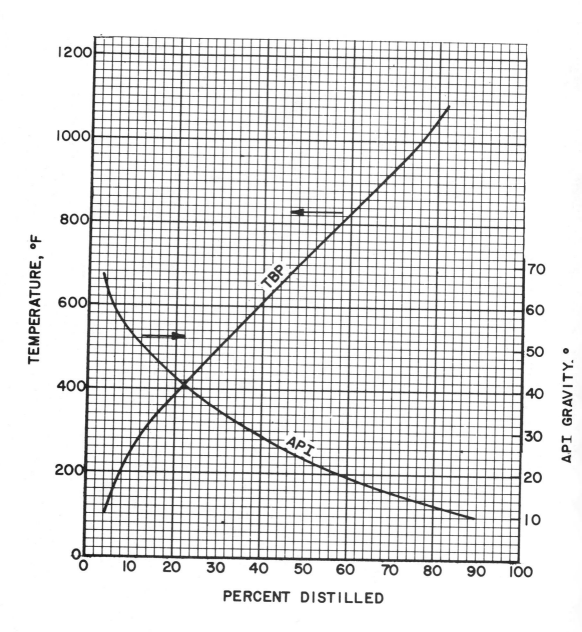

FIG. 4.11. TBP and gravity curves. Crude: North Slope, Alaska; Gravity:
25.7°API; Sulfur: 1.12 wt %.

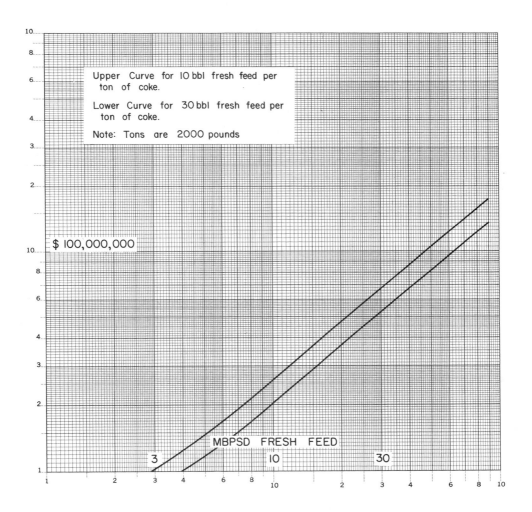

FIG. 5.2. Delayed coking units investment cost--1982 U.S. Gulf Coast.

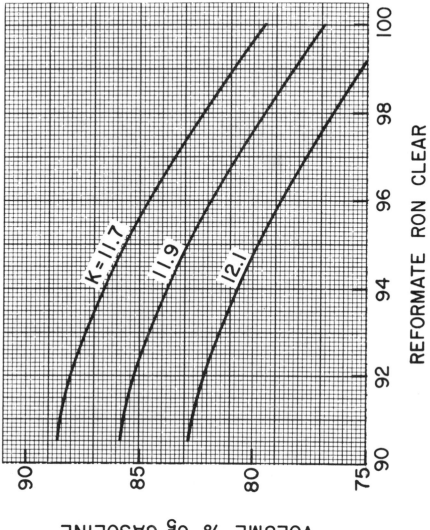

FIG. 6.3. Catalytic reforming yield correlations.

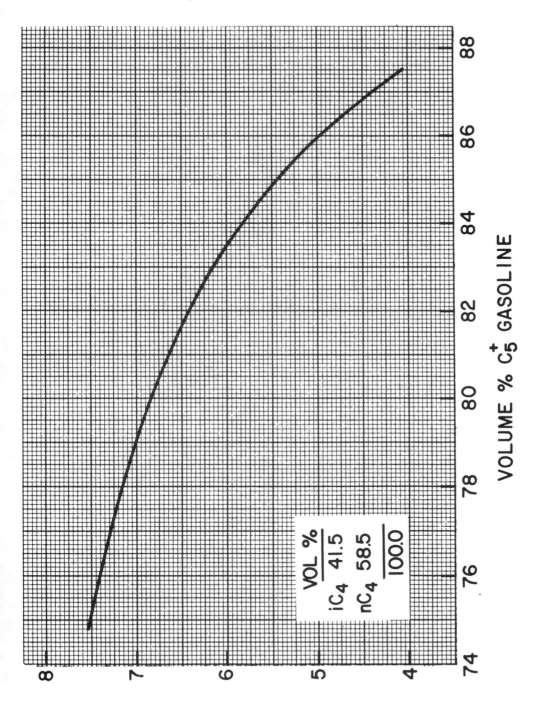

VOLUME % C_5^+ GASOLINE

VOLUME % TOTAL C_4

VOL %	
iC_4	41.5
nC_4	58.5
	100.0

FIG. 6.4. Catalytic reforming yield correlations.

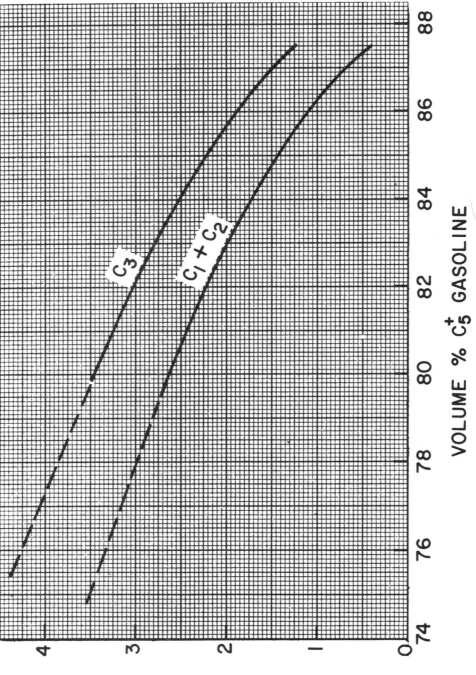

FIG. 6.5. Catalytic reforming yield correlations.

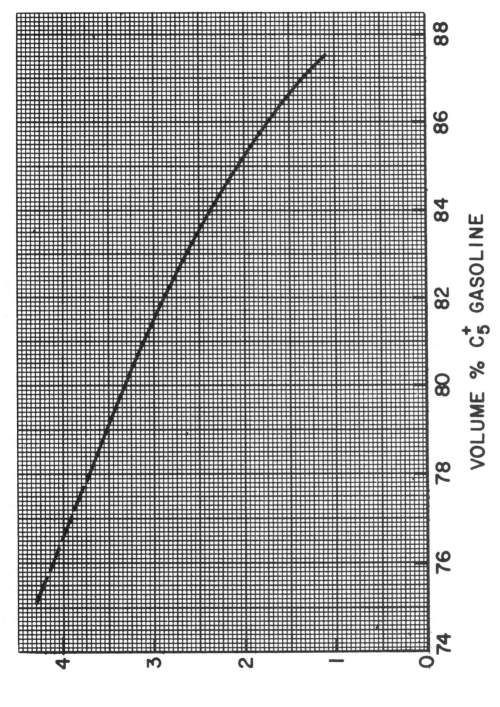

FIG. 6.6. Catalytic reforming yield correlations.

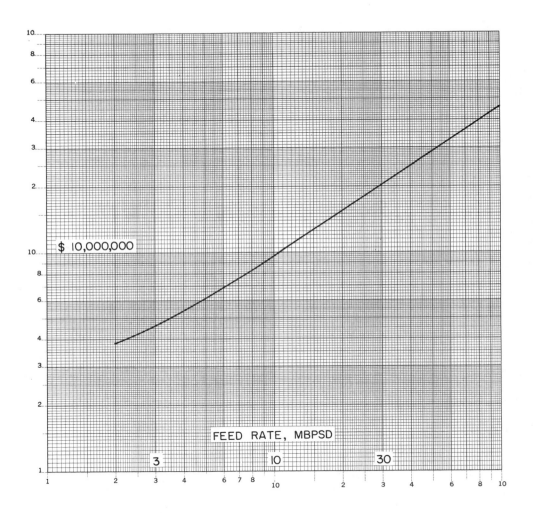

FIG. 6.7. Catalytic reforming units investment cost—1982 U.S. Gulf Coast.

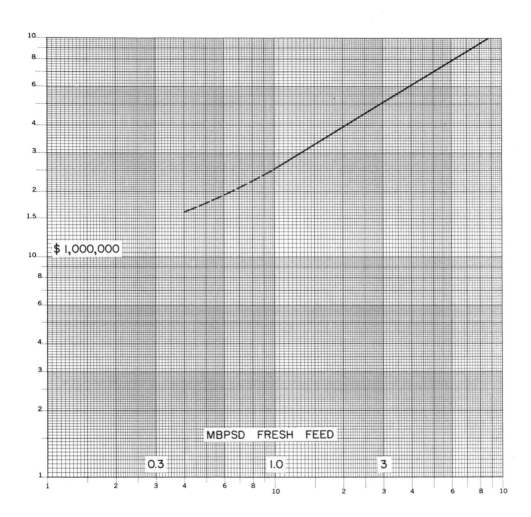

FIG. 6.9. Paraffin isomerization units (platinum catalyst type) investment cost—1982 U.S. Gulf Coast.

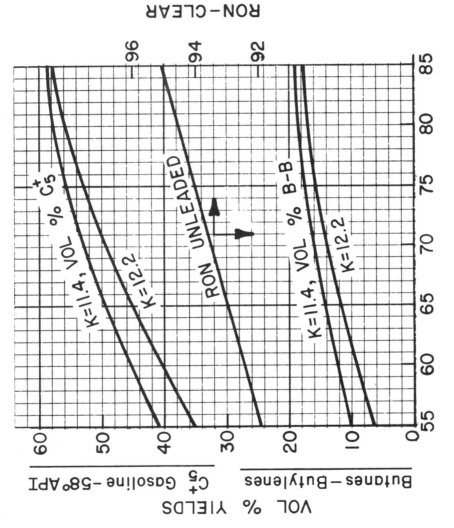

FIG. 7.9. Catalytic cracking yields Silica-alumina catalyst (butanes, butylenes, C_5^+ gasoline).

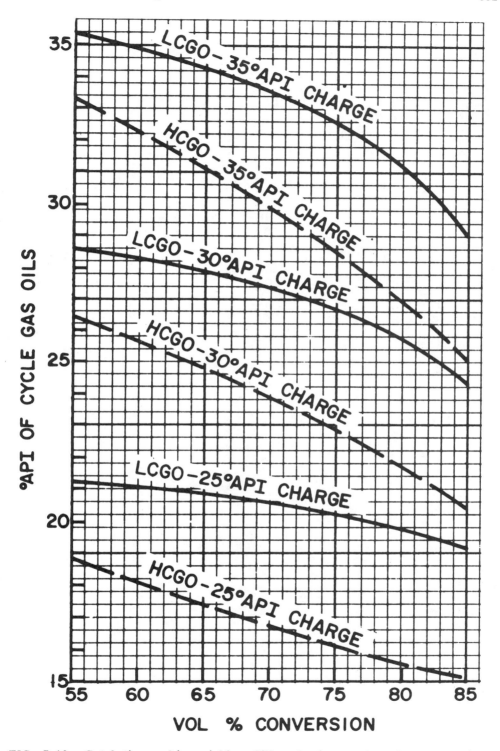

FIG. 7.10. Catalytic cracking yields. Silica-alumina catalyst (cycle gas oils).

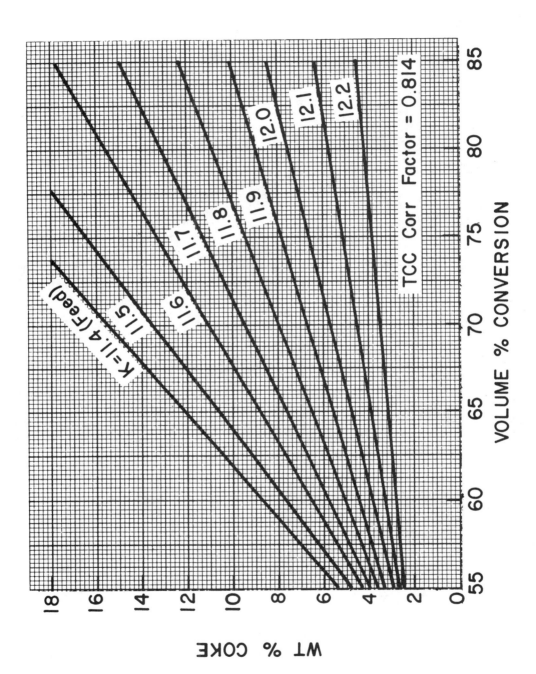

FIG. 7.11. Catalytic cracking yields   Silica-alumina catalyst (coke).

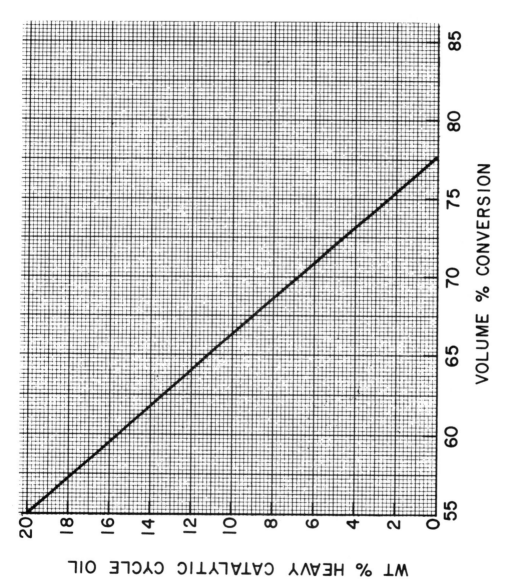

FIG. 7.12. Catalytic cracking yields. Silica-alumina catalyst (heavy catalytic cycle oil).

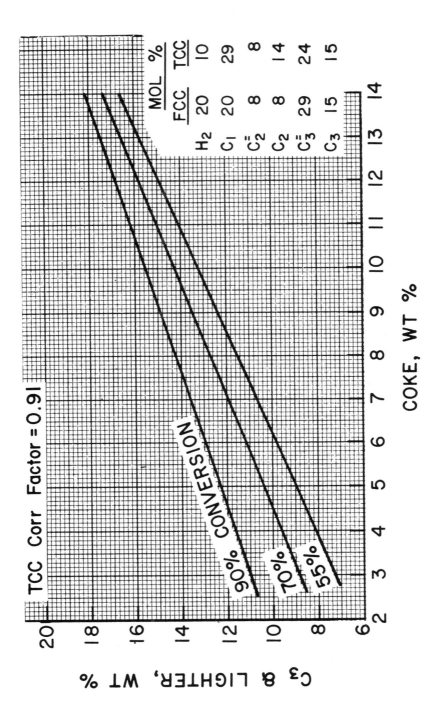

FIG. 7.13. Catalytic cracking yields   Silica-alumina catalyst (C₃ and lighter).

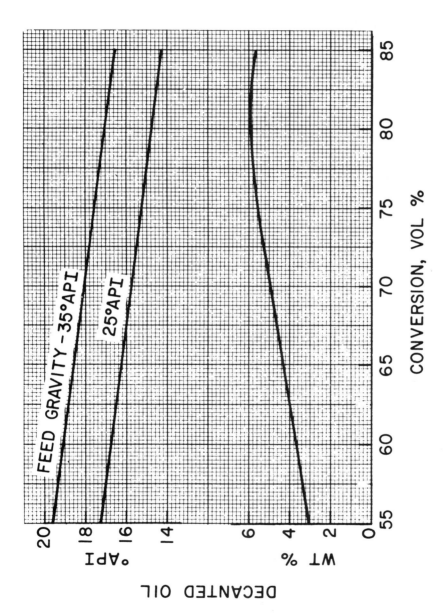

FIG. 7.14.   Catalytic cracking yields.   Silica-alumina catalyst (decanted oil).

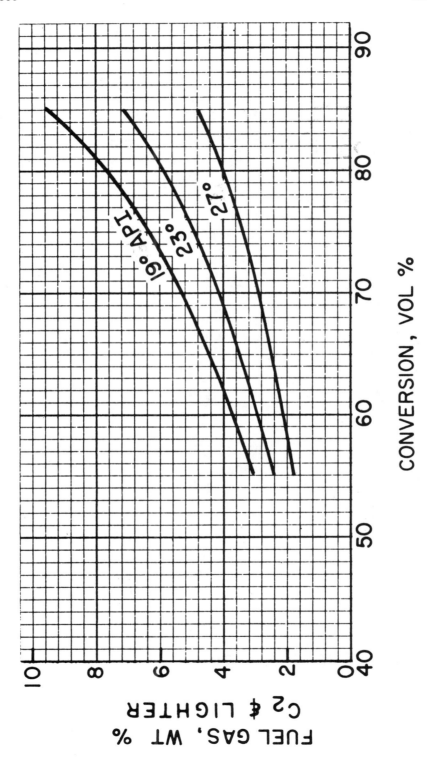

Fig. 7.15.   Catalytic cracking yields   Zeolite catalyst (fuel gas).

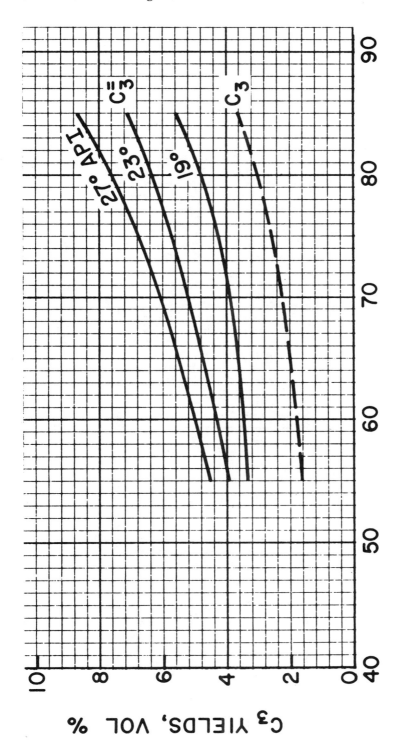

FIG. 7.16. Catalytic cracking yields. Zeolite catalyst (C_3)

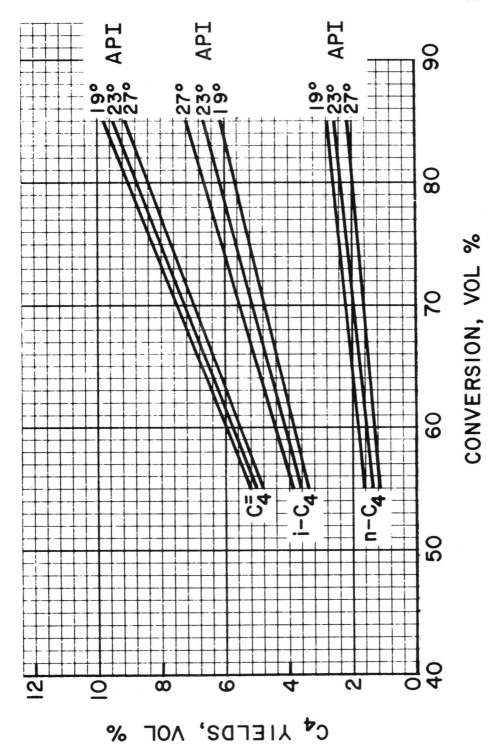

FIG. 7.17. Catalytic cracking yields   Zeolite catalyst (C_4).

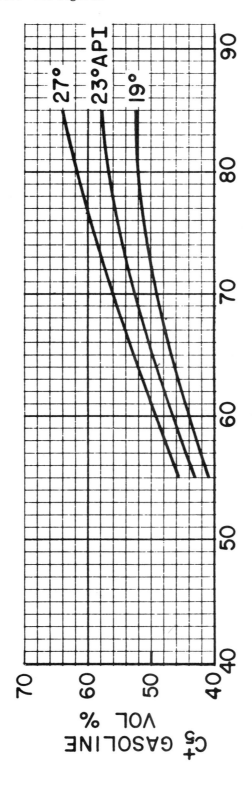

FIG. 7.18. Catalytic cracking yields. Zeolite catalyst (C_5^+ gasoline).

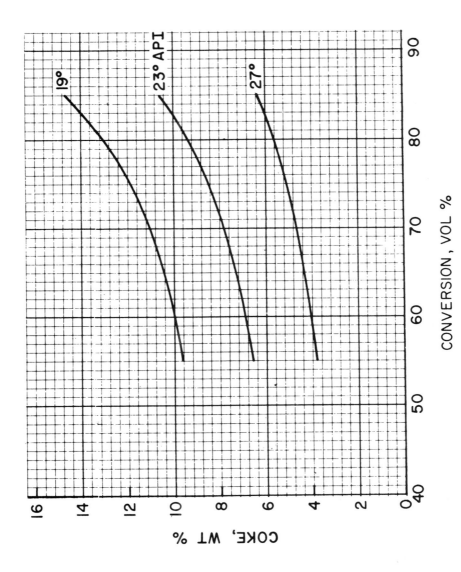

FIG. 7.19. Catalytic cracking yields   Zeolite catalyst (coke).

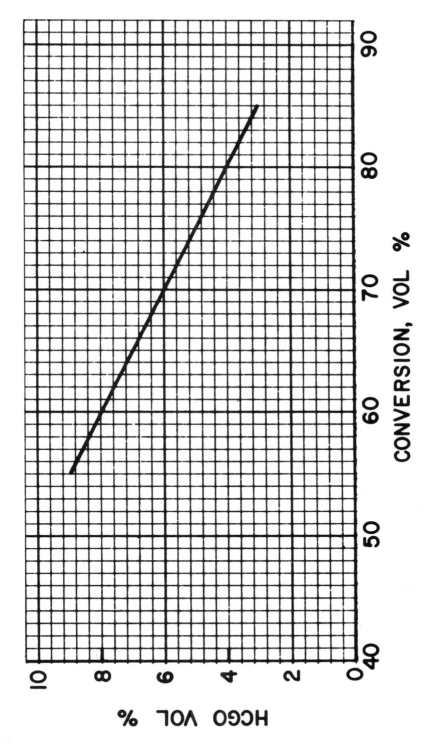

FIG. 7.20. Catalytic cracking yields. Zeolite catalyst (heavy cycle gas oil).

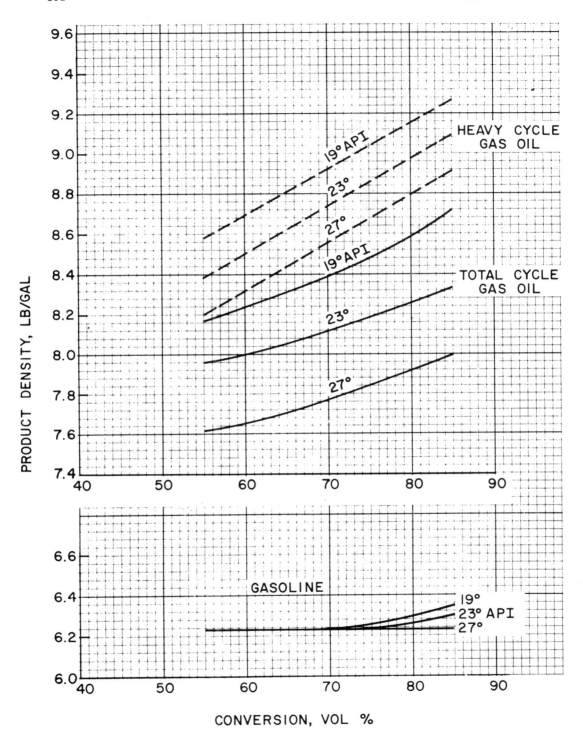

FIG. 7.21. FCC product gravity. Zeolite catalyst.

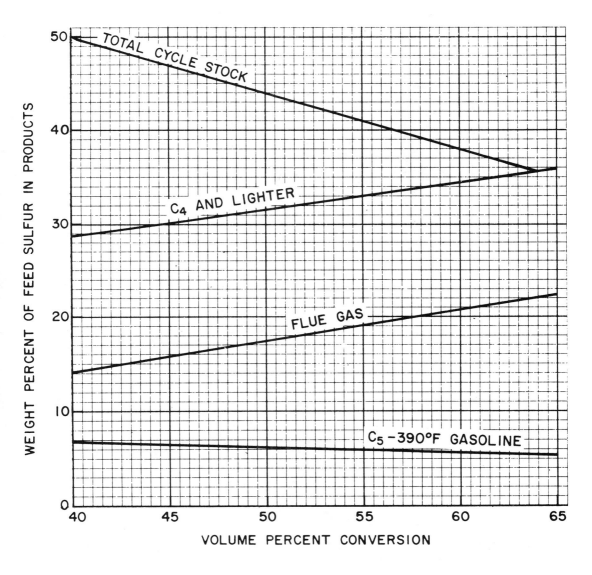

FIG. 7.22. Distribution of sulfur in catalytic cracking products [Chapter 7, Ref. 26].

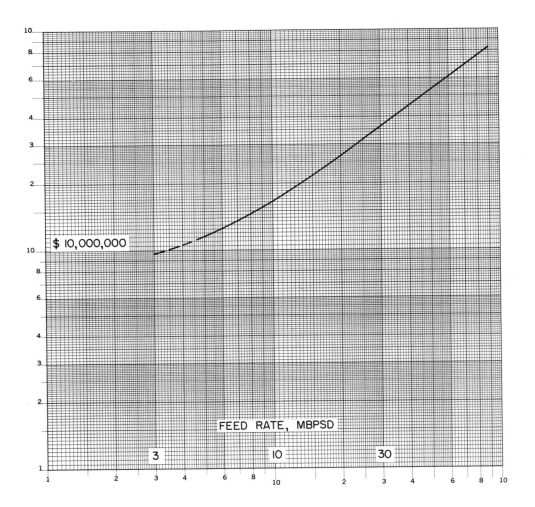

FIG. 7.23. Fluid catalytic cracking units investment cost—1982 U.S. Gulf
Coast.

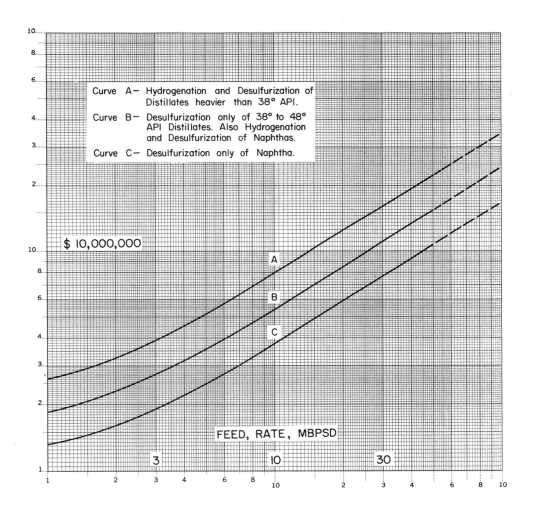

FIG. 8.2. Catalytic desulfurization and hydrogenation units investment
cost—1983 U.S. Gulf Coast.

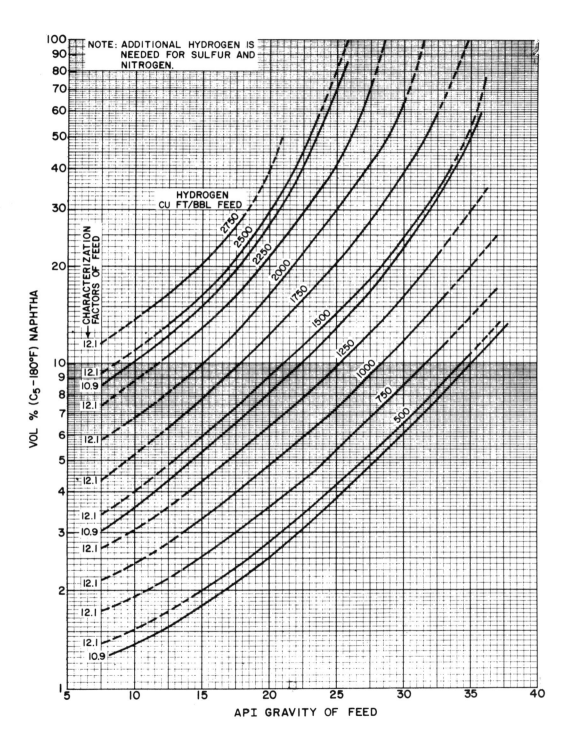

FIG. 9.4. Approximate hydrogen required for hydrocracking.

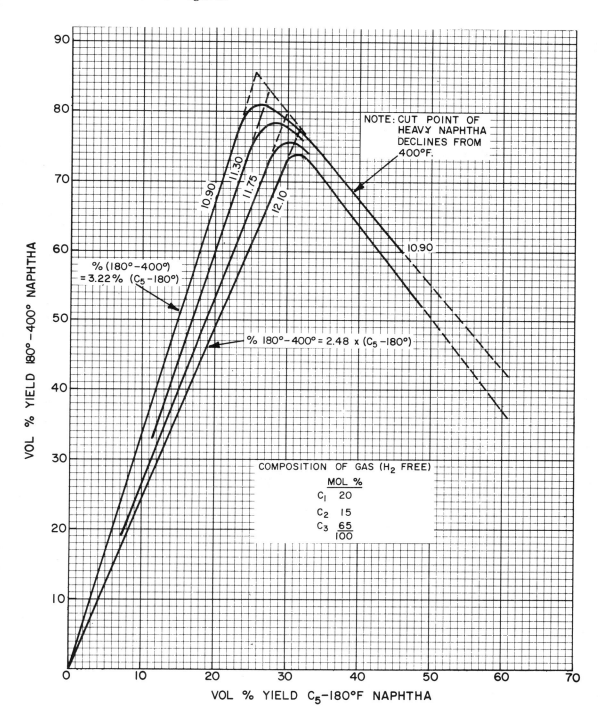

FIG. 9.5. Relationship between the yields of $(C_5-180°F)$ and $(180-400°F)$ hydrocrackates [Chapter 9, Ref. 12].

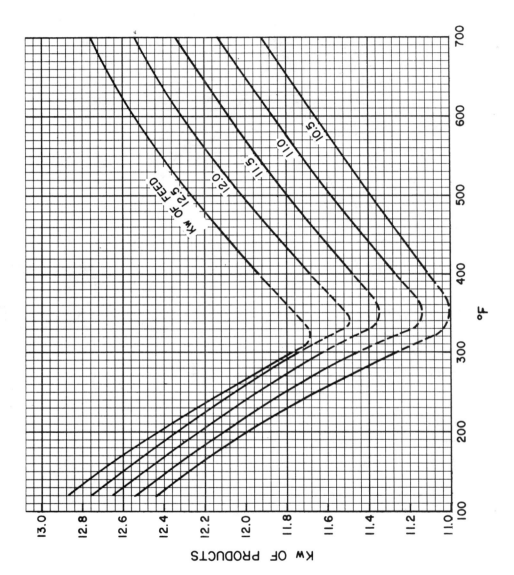

FIG. 9.6.   Characterization factor of hydrocracker products [Chapter 9, Ref. 10].

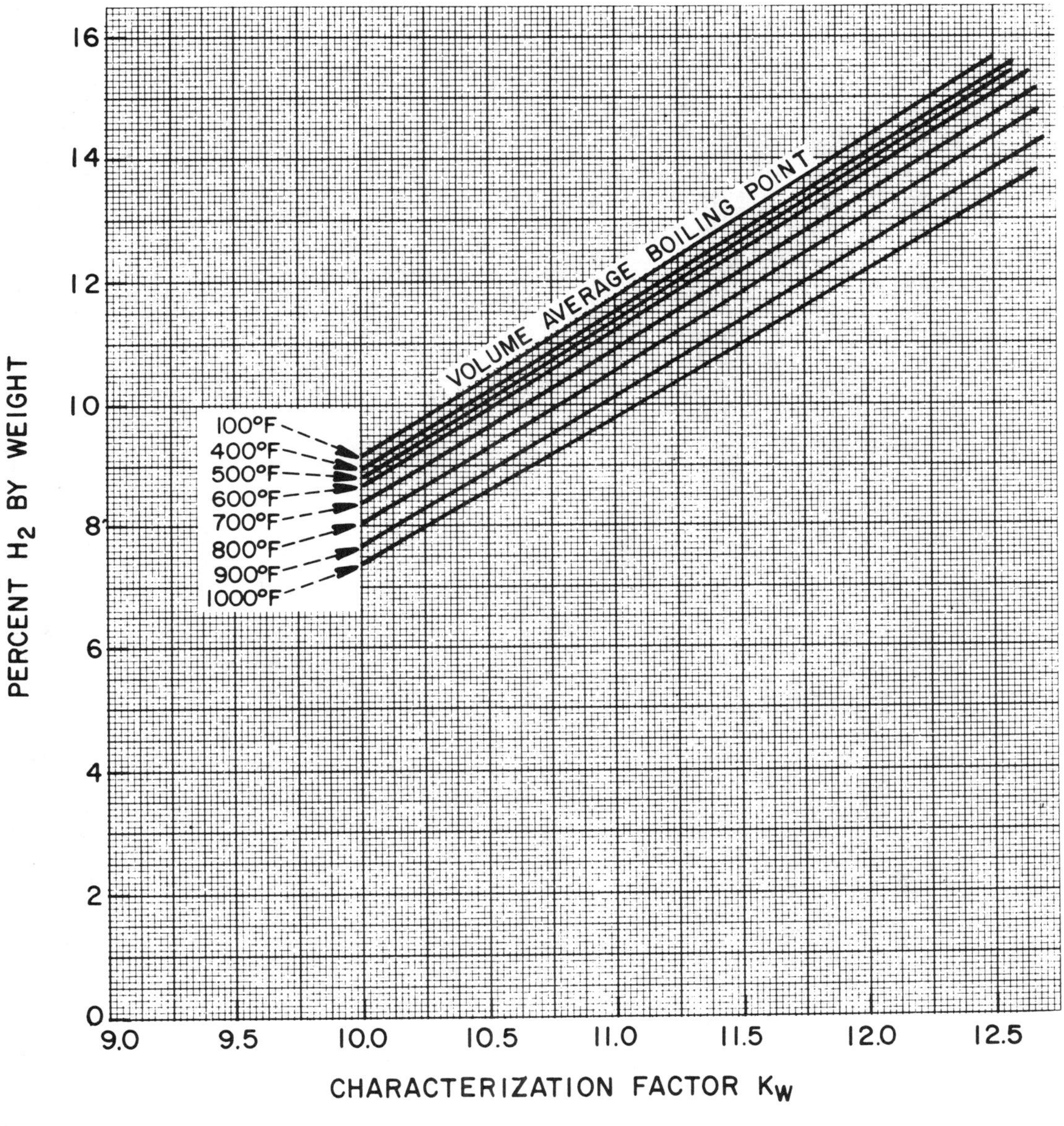

FIG. 9.7. Hydrogen content of hydrocarbons   From "Industrial Chemical
Calculations." D. A. Hogan and K. M. Watson, copyright 1938,
John Wiley and Sons. Reprinted by permission of John Wiley
and Sons, Inc.

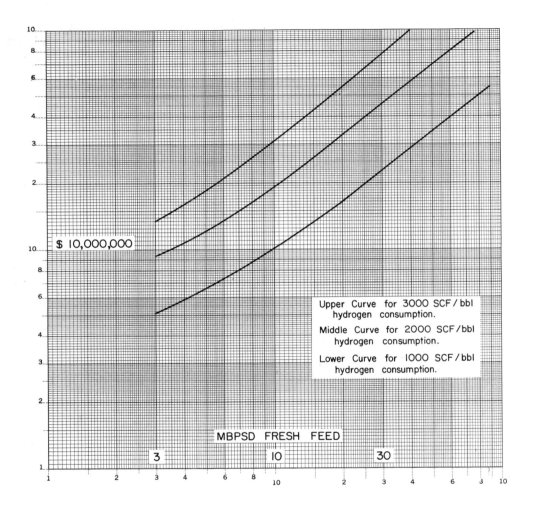

FIG. 9.8. Catalytic hydrocracking units investment cost—1982 U.S. Gulf Coast.

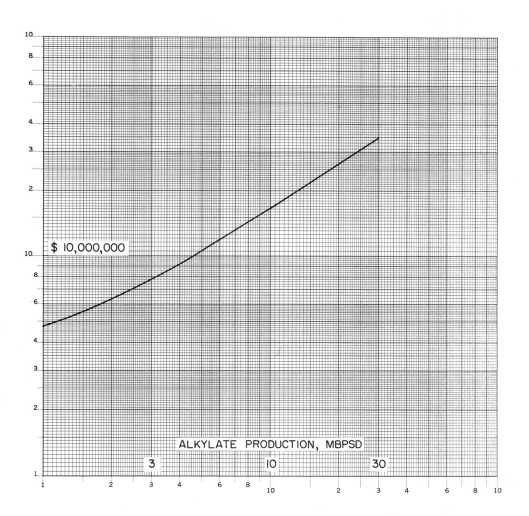

FIG. 10.6. Alkylation units investment cost—1982 U.S. Gulf Coast.

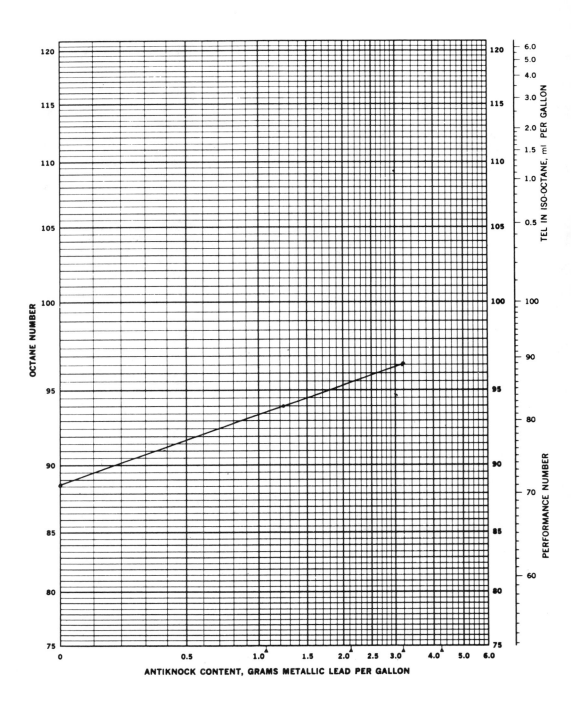

FIG. 11.1.   Tetraethyllead requirement calculation . Chart reproduced with
permission of Ethyl Corporation.

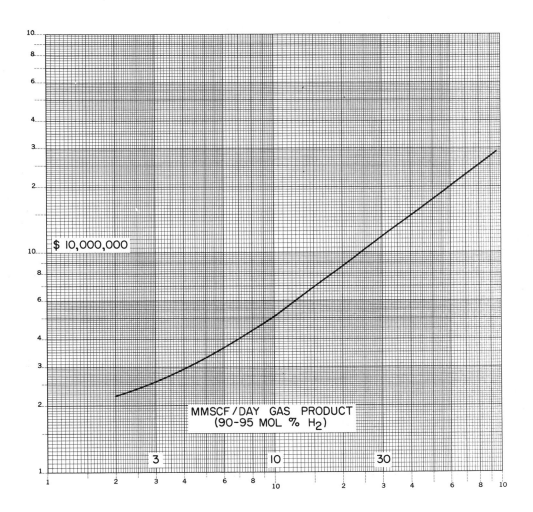

FIG. 12.2. Hydrogen production by steam-methane reforming investment
        cost—1982 U.S. Gulf Coast.

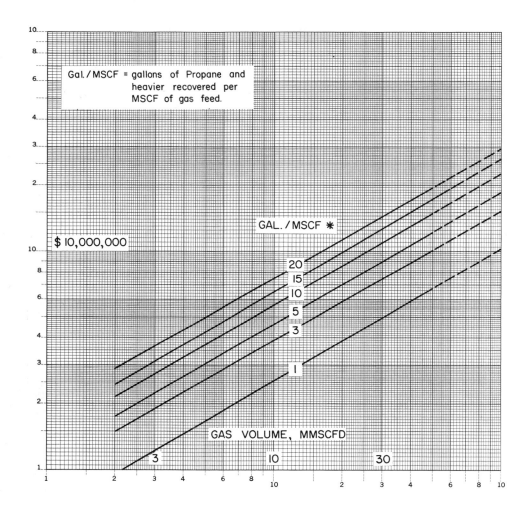

FIG. 12.4.   Refinery gas processing units investment cost—1982 U.S. Gulf
Coast.

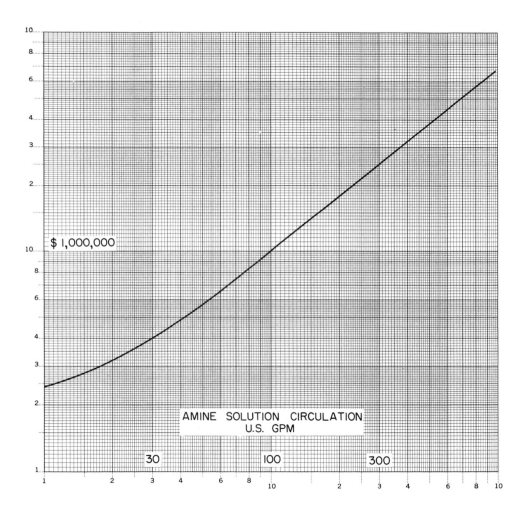

FIG. 12.5. Amine-gas treating units investment cost—1982 U.S. Gulf Coast.

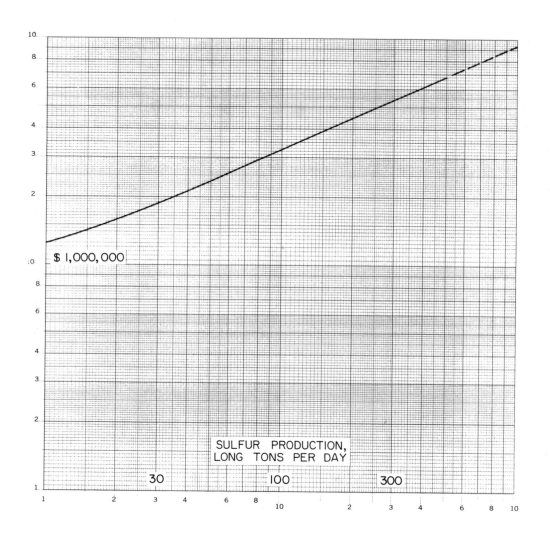

FIG. 12.9. Claus sulfur plant investment cost—1982 U.S. Gulf Coast.

AUTHOR INDEX

Numbers in parentheses are reference numbers and indicate that an author's work is referred to although his name is not cited in the text. Italicized numbers give the page on which the complete reference is cited.

A

Akbar, M., 72(14), *75*
Albright, L. F., 167(12), 169(13), 169(14), 172(14), 172(15), *183*
Allan, D. E., 54(3), 66(3), *74*, 72 (13), *75*
Allen, H. I., 195(4), *202*
Anderson, S., 115(24), *129*
Andrews, J. W., 177(18), *183*
Asseff, P. A., 234(2), *246*
Asselin, G. F., 91(6), *98*, 248(2), *259*

B

Ballard, W. P., (17), *75*
Barton, W. J., 72(13), *75*
Beck, R. R., 112(14), *128*
Bent, R. D., (7), *98*, 124(26), *129*
Bhasin, D. P., 115(25), *129*
Black, J. H., 285(7), *294*

B (continued right column)

Bland, W. F., 5(2), 9(5), 13(6), *14-15*
Blazek, J. J., 104(4), 112(16), *128*
Bohl, G., 115(25), *129*
Bonnifay, P. L., 177(18), *183*
Bour, G., 91(6), *98*
Broughton, D. B., 248(2), *259*
Brownstein, A. M., (9), *259*
Burns, M. D., (25), *232*
Bushnell, J. D., 234(1), 241(1), (7), *246*

C

Cameron, L. C., 211(10), *232*
Castiglioni, B. P., (32), *129*
Celestinos, J. A., 146(8), *158*
Chandlev, W. B., 19(5), *30*
Chase, J. D., 261(6), *280*
Chen, N. Y., 244(6), *246*
Chilton, C. H., 276(12), *280*, 281(1), *294*
Chittenden, D. H., 193(3), *202*

SUBJECT INDEX